Electric Circuits, Systems, and Motors

Timothy A. Bigelow

Electric Circuits, Systems, and Motors

 Springer

Timothy A. Bigelow
Iowa State University
Ames, IA, USA

The solutions will be available at https://www.springer.com/gp/book/9783030313548.

ISBN 978-3-030-31357-9   ISBN 978-3-030-31355-5   (eBook)
https://doi.org/10.1007/978-3-030-31355-5

This Springer imprint is published by the registered company Springer Nature Switzerland AG
The registered company address is: Gewerbestrasse 11, 6330 Cham, Switzerland

Preface

The goal of this book is to provide a basic introduction to circuits and motors for students in electrical engineering as well as other majors that need a basic introduction to circuits and motors. Unlike most other textbooks that highlight only circuit theory, the book goes into detail on many practical aspects of working with circuits including electrical safety and the proper method to measure the relevant circuit parameters using modern measurement systems. Many books provide very little information on how to make safe and valid circuit measurements. This book has a chapter on measurements as well as many other practical hints scattered throughout the other chapters.

The book also has detailed discussions of transformers, balanced three-phase circuits, motors (including stepper motors), and generators as these are critical topics in industrial electronics, robotics, and mechatronics. These concepts may be critical for mechanical and industrial engineering students as well as computer/software engineering students programming robotic systems. When discussing these topics, the book provides sufficient background theory while not expanding beyond what the students actually need to solve circuit problems. For example, many texts dive into a detailed discussion of magnetic circuit analysis prior to discussing transformers even though a precise knowledge of flux flow is not needed to solve transformer circuits. In this text, we introduce just enough magnetic field concepts so that the physics of the transformers is understood while keeping the focus on using transformers in circuits. Likewise, our discussion of motors focuses on their basic analysis and the simple physics of their operation without burying the students in the physics of magnetism and magnetic materials.

In addition to covering the basic circuit concepts, the book also provides the students with the necessary mathematics to correctly analyze the circuit concepts being presented. The chapter on phasor domain circuit analysis begins with a detailed review of complex numbers as many students are weak in this area. Also, many of the example problems show very detailed solution steps to help remind the students of how to solve systems of equations if they have forgotten. This would allow the instructor to focus more on the circuit concepts and refer the students to the book for the detailed steps.

Ames, USA   Timothy A. Bigelow

Contents

Circuit Model Components

1

Electric circuits are an integral part of our modern world. Even the simplest children's toy often has lights and speakers to attract the attention of its intended recipients. Computers surround us and integrate with our automobiles and smartphones. Almost everything we buy is manufactured and packaged by computer-controlled conveyor/robotic systems including much of our food. Embedded sensors in buildings and bridges, autonomous control of production and transportation systems, and advanced manufacturing are the future, and as such, every engineer needs a basic understanding of how circuits operate and how they can be utilized to maximize productivity and innovation in their field. In this chapter, we will review the basic models that help us simplify the flow and control of electricity in circuits. These models will then be used in future chapters to build on our understanding of how circuits can be interpreted and designed.

1.1 Voltage Sources

Electric circuits involve controlling the movement of electrons (i.e., current) under the force induced by electric potential differences (i.e., voltage). Therefore, one of the most important models needed to understand electric circuits is the ideal voltage source. The voltage source has units of volts (V), and the symbol for the ideal voltage source is shown in Fig. 1.1. This voltage source produces a constant voltage, V_s, regardless of the load or current flow out of its terminals. Therefore, shorting out the terminals of an ideal voltage source would result in infinite current which of course is impossible (see Ohm's Law in Eq. (1.1)). Therefore, the ideal voltage source can only be used to model real voltage sources under certain constraints. For example, a battery can normally be modeled as an ideal independent voltage source with a resistor to account for losses (see Sect. 1.3). However, at high currents, the battery itself will overheat and possibly explode. Explosions are not captured by the ideal voltage source model. In fact, many real power supplies have specific circuitry added that will limit the voltage if the current gets too high to protect the source.

In addition to being an ideal voltage source, the model shown in Fig. 1.1 is also independent. This means that it will produce the same voltage regardless of what else is happening in the circuit. In more complicated circuits, however, the voltage sources need to be modeled as depending on currents or voltages elsewhere in the circuit. The most basic example of this would be an amplifier, where the output voltage should be a scale multiple of the input. Dependent voltage sources are used to model voltage sources whose output depends on either voltage or current elsewhere in the circuit. The symbols for dependent voltage sources are shown in Fig. 1.2. The voltage-dependent voltage source

© Springer Nature Switzerland AG 2020
T. A. Bigelow, *Electric Circuits, Systems, and Motors*,
https://doi.org/10.1007/978-3-030-31355-5_1

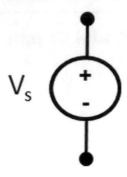

Fig. 1.1 Circuit symbol for ideal independent voltage source

$$V_s = A \cdot v_g \qquad V_s = A \cdot i_g$$

Voltage-Dependent   Current-Dependent
Voltage Source   Voltage Source

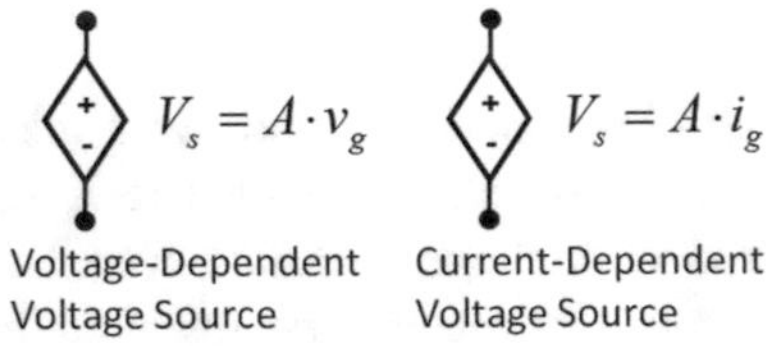

Fig. 1.2 Circuit symbol for dependent voltage sources

produces a voltage that depends on a voltage elsewhere in the circuit while the current-dependent voltage source produces a voltage that depends on a current elsewhere in the circuit.

1.2 Current Sources

While voltage source models may be easier to grasp due to our familiarity with batteries, in order to capture the behavior of different circuit elements and sensors, we need to be able to have a circuit component that produces a constant current regardless of the load. This is the purpose of current source models. Also, just like we saw for voltage sources, current sources can be independent or dependent. The independent current sources produce a constant current regardless of all other currents/voltages in the circuit while the dependent current courses produce a current that depends on another voltage or current in the circuit. The circuit models for the different types of current sources are shown in Fig. 1.3, where I_s is the current being generated in units of amps (A). The arrow for each current source shows the direction of the current. In circuits, the convention has been to use the flow of a fictitious positive charge when defining the current as opposed to the true flow direction of the electrons, which have a negative charge. Therefore, current flows from higher positive potential to lower potential when analyzing circuits using the basic circuit models.

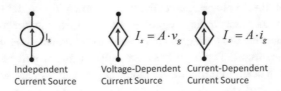

Independent   Voltage-Dependent Current-Dependent
Current Source   Current Source   Current Source

Fig. 1.3 Circuit symbols for current sources

1.3 Resistors

Resistors are used to model losses in the circuit where energy changes from electrical to some other form. Usually, the energy lost goes into heat, but other forms of energy transfer can also be modeled. For example, a resistor is used in a circuit model for a piezoelectric transducer that models the transfer of electrical energy into acoustical energy. The circuit symbol for the resistor is shown in Fig. 1.4, where R is the resistance of the resistor and has units of ohms (Ω).

When designing circuits, losses can be deliberately added to the circuit to control the flow of current by adding resistors to the circuit. A typical resistor is shown in Fig. 1.5. The colored stripes on the resistors are used to indicate the value of the resistor. The most common approach is to use 4 color bands, but some have more bands to specify the resistance value more precisely. When read from left

Fig. 1.4 Circuit symbol for a resistor

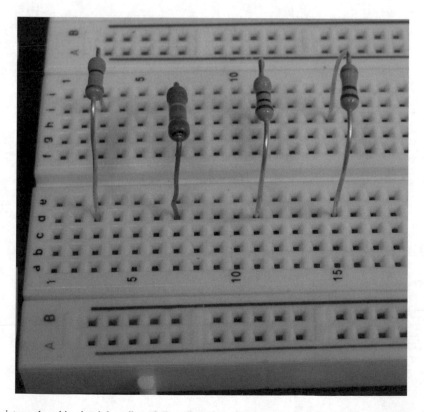

Fig. 1.5 Resistors placed in circuit breadboard. Breadboards and resistors such as these are very common in laboratory courses when first learning about circuits as they allow for rapid circuit construction

to right, the last band gives the tolerance or uncertainty in the resistor value while the first three bands specify the resistance value. The band code for resistors with 4 stripes is given in Table 1.1.

Table 1.1 Band code for resistors with 4 colored stripes

Color	1st Band (1st Digit of Resistance)	2nd Band (2nd Digit of Resistance)	3rd Band (Multiplier)
Black	--	0	1 Ω
Brown	1	1	10 Ω
Red	2	3	100 Ω
Orange	3	4	1 kΩ
Yellow	4	6	10 kΩ
Green	5	5	100 kΩ
Blue	6	6	1 MΩ
Violet	7	7	10 MΩ
Grey	8	8	100 MΩ
White	9	9	1 GΩ
Tolerance Band			
Gold		±5% tolerance	
Silver		±10% tolerance	

Example 1.1 Find the resistance value for each of the resistors below from the resistor's color code

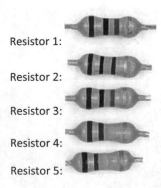

Resistor 1:

Resistor 2:

Resistor 3:

Resistor 4:

Resistor 5:

Solution:

Resistor 1: Colors = Brown Black Orange Gold
                      (1)   (0)   (1 kΩ) (±5%) = 10 kΩ ±5%
Resistor 2: Colors = Brown Black Blue   Silver
                      (1)   (0)   (1 MΩ) (±10%) = 10 MΩ ±10%
Resistor 3: Colors = Brown Black Green   Gold
                      (1)   (0)   (100 kΩ)   (±5%) = 1 MΩ ±5%
Resistor 4: Colors = Red   Red   Yellow Gold
                      (2)   (2)   (10 kΩ) (±5%) = 220 kΩ ±5%
Resistor 5: Colors = Brown Red   Yellow Gold
                      (1)   (2)   (10 kΩ) (±5%) = 120 kΩ ±5%

> **Lab Hint**: Observing the color bands on a resistor can be challenging. Therefore, it is often wise to confirm the resistance values with a multi-meter while building the circuit. The multi-meter will also allow you to select resistors that have the same value from your lab kit when building circuits that require matched resistors for optimal performance.

Example 1.2 What should be the color bands for a 47 Ω ± 5% resistor with 4 bands?

Solution: 47 Ω = (4) (7) × (1 W) (±5%) = Yellow, Violet, Black, Gold

> **Lab Hint**: Most lab resistors can only tolerate about 250 mW. Therefore, you should always be mindful of the amount of power you expect each resistor to absorb. If more power flow is expected, then a resistor with a higher power rating will be needed.

1.3.1 Ohm's Law

Ohm's Law, shown in Eq. (1.1), with the variables illustrated in Fig. 1.6 gives the relationship between voltage and current for a resistor.

$$V = I \cdot R \tag{1.1}$$

In this equation, V is the voltage across the resistor, and I is the current flowing through the resistor. The terminal where the current enters is always at a higher potential (i.e., higher voltage) than the terminal at which the current leaves for the resistor.

Sometimes it is useful to compare the flow of charge to the flow of water when trying to understand circuits. You can think of the voltage source as a water tower. The water tower has a pump that continuously replenishes the water while the height of the water tower provides pressure on the waterline. Without the pump, the water tower would be very similar to a battery. The diameters of the pipes connected to the water tower are the resistance. Higher tower heights (i.e., potential) will increase the pressure on the line. Higher pressures for the same resistance will result in faster water flow rates as will increasing the pipe diameter (i.e., lowering the resistance) while maintaining the pressure.

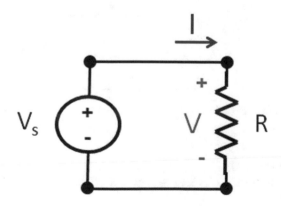

Fig. 1.6 Circuit diagram illustrating Ohm's law

Example 1.3 What is the current, i_s, for the circuit shown in Fig. 1.7?

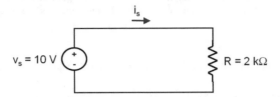

Fig. 1.7 Circuit diagram for Example 1.3

Solution:
Ohm's Law is given by $V = I \cdot R$
When applied to this problem: $v_s = i_s \cdot R \Rightarrow i_s = \frac{v_s}{R} = 5$ mA

Example 1.4 What is the voltage, v_s, for the circuit shown in Fig. 1.8?

Fig. 1.8 Circuit diagram for Example 1.4

Solution:
Ohm's Law is given by $V = I \cdot R$
When applied to this problem: $v_1 = I_s \cdot R \Rightarrow v_1 = 4.5$ V

Example 1.5 What is R for the circuit shown in Fig. 1.9 if the current from the source, i_s, is 1.5 mA?

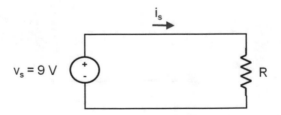

Fig. 1.9 Circuit diagram for Example 1.5

Solution:
Ohm's Law is given by $V = I \cdot R$
When applied to this problem: $v_s = i_s \cdot R \Rightarrow R = \frac{v_s}{i_s} = 6$ kΩ

Example 1.6   (a) Draw the circuit model for the circuit shown in Fig. 1.10. If the voltage at the terminals of the battery is 8.9 V and the current flow out of the battery is 105 mA, what is the total resistance of the resistor and light bulbs?

Fig. 1.10 Photograph of the circuit for Example 1.6

Solution:

(a) Resistors are needed to model the losses in the battery, the losses in the resistor, and the generation of light and heat by the light bulbs. Normally, the resistance of the wires can be neglected and is therefore not included in this example. Therefore, the circuit can be modeled as

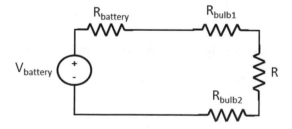

(b) Ohm's Law is given by $V = I \cdot R$

When applied to this problem: Total Resistance $= \frac{8.9\,\text{V}}{105\,\text{mA}} = 84.76\,\Omega$

1.3.2 Resistor Networks

As can be seen in Example 1.6, it is rare for a circuit model to consist of only one resistor. Therefore, it is common to simplify the circuit models by replacing a set of resistors by their equivalent resistance before solving for voltages and/or currents.

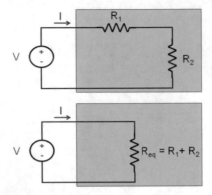

Fig. 1.11 Resistors in series being replaced by their equivalent resistance

Resistors in Series: Fig. 1.11 shows two resistors in series. Circuit elements are in series when the same current MUST pass through both elements. If there is any other path for the current to flow, then the elements are NOT in series. The resistors in series, R_1 and R_2, can be replaced by their equivalent resistance, R_{eq}, given by

$$R_{eq} = R_1 + R_2 \tag{1.2}$$

In order to be equivalent, the voltage, V, and current, I, outside of the blue box must remain the same. The equivalent resistance of any number of resistors, provided they are all in series, can be extrapolated from this basic relationship and is given by

$$R_{eq} = R_1 + R_2 + R_3 + \cdots + R_{N-1} + R_N$$
$$= \sum_{k=1}^{N} R_k \tag{1.3}$$

Resistors in series always have a higher resistance than the resistance of each individual resistor.

$$R_{eq} > \max(R_1, R_2, R_3, \cdots, R_{N-1}, R_N) \tag{1.4}$$

Example 1.7 For each of the following circuits, determine if the resistors are in series. If they are in series, find the equivalent resistance (Fig. 1.12).

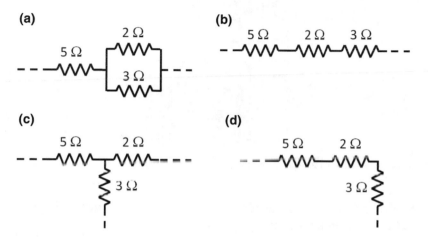

(a)   **(b)**   **(c)**   **(d)**

Fig. 1.12 Circuit diagram for Example 1.7

Solution:

(a) Not in series, the current from the 5 Ω resistor is split between the 2 and 3 Ω resistors.
(b) In series, the current must flow through all three resistors. $R_{eq} = 5\,\Omega + 2\,\Omega + 3\,\Omega = 10\,\Omega$.
(c) Not in series, the current from the 5 Ω resistor is split between the 2 and 3 Ω resistors.
(d) In series, the current must flow through all three resistors. $R_{eq} = 5\,\Omega + 2\,\Omega + 3\,\Omega = 10\,\Omega$.

Resistors in Parallel: Fig. 1.13 shows two resistors in parallel. Circuit elements are in parallel when the voltage across both elements MUST be the same due to how they are wired together. In order to be in parallel, both terminals on one element must be connected to the terminals on the other element. The equivalent resistance for two resistors in parallel is given by

$$R_{eq} = \frac{1}{\frac{1}{R_1} + \frac{1}{R_2}} = \frac{R_1 R_2}{R_1 + R_2} \tag{1.5}$$

When multiple resistors all share the same two terminals, then all of the resistors are in parallel and the equivalent resistance is given by

$$R_{eq} = \frac{1}{\frac{1}{R_1} + \frac{1}{R_2} + \frac{1}{R_3} + \cdots + \frac{1}{R_{N-1}} + \frac{1}{R_N}} = \frac{1}{\sum\limits_{k=1}^{N} \frac{1}{R_k}} \tag{1.6}$$

Resistors in parallel always have a resistance that is smaller than the smallest resistance of the set.

$$R_{eq} < \min(R_1, R_2, R_3, \cdots, R_{N-1}, R_N) \tag{1.7}$$

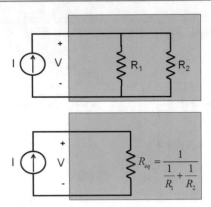

Fig. 1.13 Resistors in parallel being replaced by their equivalent resistance

Example 1.8 For each of the following circuits, determine if the resistors are in parallel. If they are in parallel, find the equivalent resistance (Fig. 1.14).

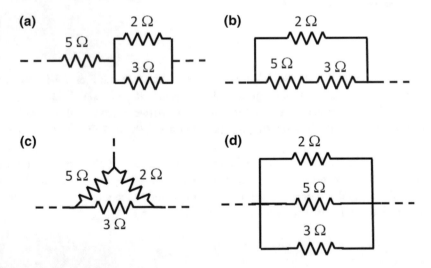

Fig. 1.14 Circuit diagram for Example 1.8

Solution:

(a) Not in parallel. The 2 and 3 Ω resistors are in parallel as they share both terminals, but the 5 Ω resistor is not.
(b) Not in parallel. None of the resistors share both terminals.
(c) Not in parallel. None of the resistors share both terminals.
(d) In parallel. All of the resistors share the same terminal and will have the same voltage drop.

$$R_{eq} = \frac{1}{\frac{1}{2\,\Omega} + \frac{1}{5\,\Omega} + \frac{1}{3\,\Omega}} = 0.968\ \Omega.$$

Example 1.9 Find the equivalent resistance between terminals a and b by repeatedly combining resistors in series and in parallel (Fig. 1.15).

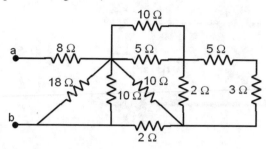

Fig. 1.15 Circuit diagram for Example 1.9

Solution:

For these types of problems, you want to start with the resistors the furthest from the desired terminals and simplify the resistances as you approach the terminals. For the circuit shown, this means working from right to left.

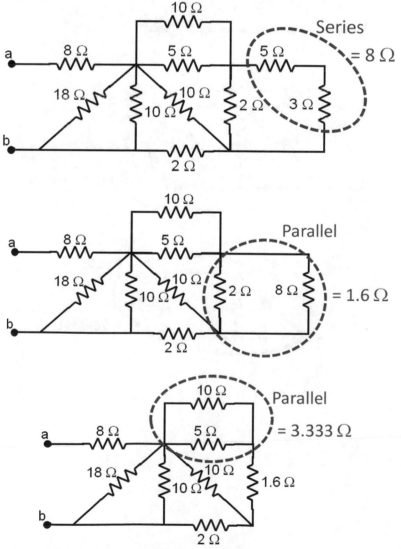

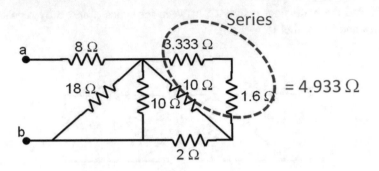

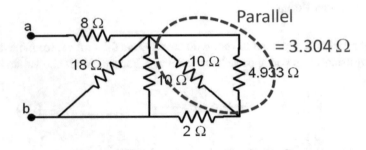

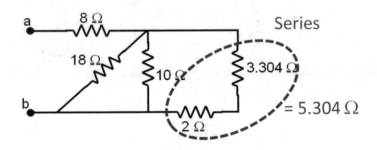

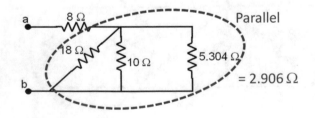

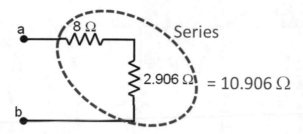

Example 1.10   Find the equivalent resistance between terminals a and b by repeatedly combining resistors in series and in parallel (Fig. 1.16).

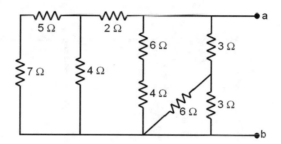

Fig. 1.16   Circuit diagram for Example 1.10

Solution:

Once again, you want to start with the resistors the furthest from the desired terminals and simplify the resistances as you approach the terminals. For the circuit shown, this means working from left to right. This time, however, we will combine all series/parallel resistors at each step.

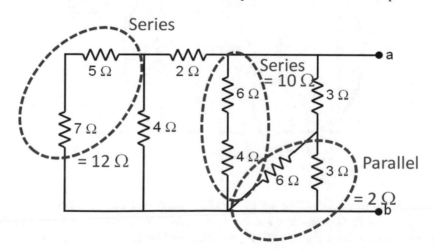

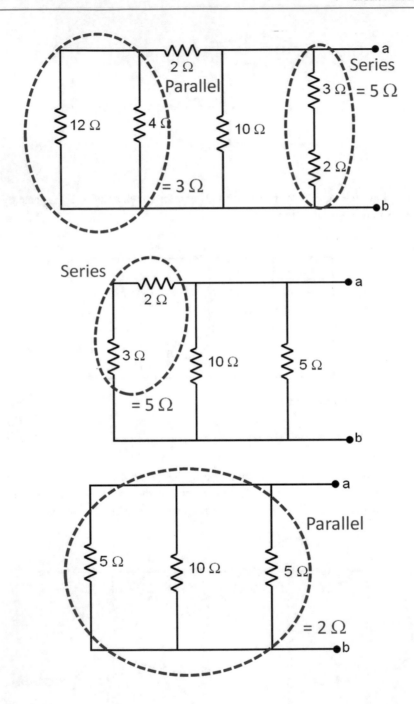

Delta-Wye Transformations: Another set of transformations that can be useful when simplifying resistive networks are the Delta-to-Wye and the Wye-to-Delta transformations. Delta and Wye circuit topologies are common when analyzing 3-phase circuits and bridge circuits such as those commonly used in strain gauges. Both topologies consist of three connection points which are denoted a, b, and c as shown in Fig. 1.17.

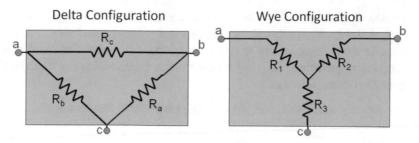

Fig. 1.17 Delta and Wye circuit topologies

The Delta configuration resembles the greek letter Δ while the Wye configuration resembles the letter Y. However, it is the connection of the terminals and not the shape that is important when identifying the topologies in a circuit. Any Delta connected set of resistors can be replaced by a Wye-connected set of resistors without altering any of the voltages/currents in the circuit provided the Wye-connected resistors are given by

$$
\begin{aligned}
R_1 &= \frac{R_b R_c}{R_a + R_b + R_c} \\
R_2 &= \frac{R_a R_c}{R_a + R_b + R_c} \\
R_3 &= \frac{R_a R_b}{R_a + R_b + R_c}
\end{aligned}
\tag{1.8}
$$

Likewise, any Wye-connected resistors can be replaced by Delta connected resistors if the resistances in the Delta configuration are given by

$$
\begin{aligned}
R_a &= \frac{R_1 R_2 + R_1 R_3 + R_2 R_3}{R_1} \\
R_b &= \frac{R_1 R_2 + R_1 R_3 + R_2 R_3}{R_2} \\
R_c &= \frac{R_1 R_2 + R_1 R_3 + R_2 R_3}{R_3}
\end{aligned}
\tag{1.9}
$$

One way to remember these formulas is to draw the Wye configuration over the Delta configuration as shown in Fig. 1.18. Notice that the R_1 resistor is between the R_b and the R_c resistor. Likewise, the R_2 resistor is between the R_a and R_c resistors while the R_3 resistor is between the R_a and R_b resistors. Therefore, to find the resistor value for the Wye configuration, you just multiply the Delta-connected resistors that would be on either side and then divide by the total resistance. Similarly, the R_1 resistor

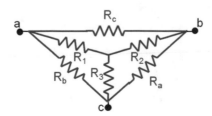

Fig. 1.18 Wye configuration overlaid on Delta configuration to illustrate one way to remember the formulas

connects to the a-node and "points" at the R_a resistor, the R_2 resistor connects to the b-node and "points" at the R_b resistor, and the R_3 resistor connects to the c-node and "points" at the R_c resistor. Hence, the values for the Delta-connected resistors can be found from the Wye-connected resistors by dividing the common numerator in Eq. (1.9) by the resistor connected to the opposite terminal that "points" at the desired delta resistance.

Example 1.11 Identify the Wye connected and Delta connected resistors in the following circuit (Fig. 1.19).

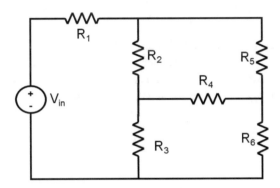

Fig. 1.19 Circuit diagram for Example 1.11

Solution:

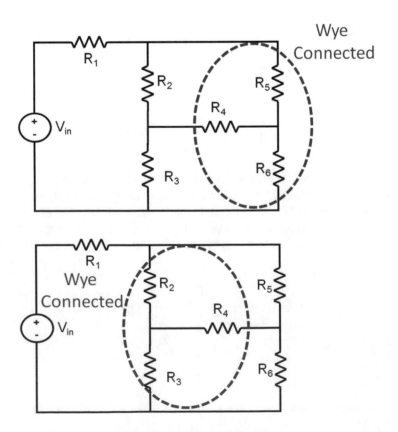

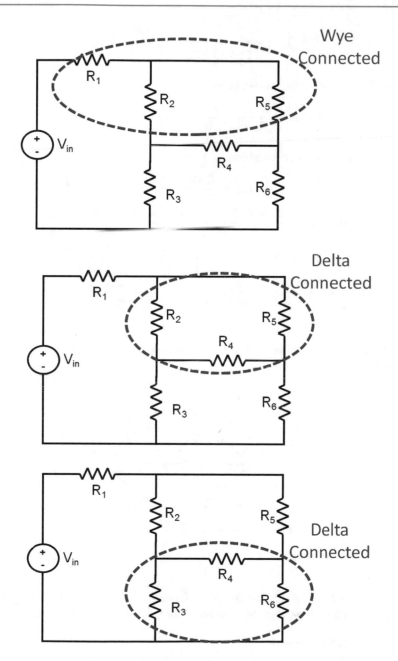

Example 1.12 Prove that the resistance R_1 in the Wye configuration is related to the resistances from the Delta configuration as given by Eq. (1.8).

Solution: In order to be equivalent, the resistance between each terminal must be the same for both circuits. Consider the case when terminal c is not connected. Under these conditions, the resistance between terminals a and b for the Wye configuration is given by $R_1 + R_2$. These resistors add as if they are in series because all of the current must pass through both R_1 and R_2 as there is no path for the current out node c. Likewise, the resistance between the terminals in the Delta configuration is given by $(R_a + R_b)$ in parallel with R_c. Equating these resistance values gives

$$R_1 + R_2 = \frac{1}{\frac{1}{(R_a + R_b)} + \frac{1}{R_c}} = \frac{R_c(R_a + R_b)}{(R_a + R_b) + R_c}$$

Repeating these steps for the other nodes gives

$$R_1 + R_3 = \frac{R_b(R_a + R_c)}{R_a + R_b + R_c} R_2 + R_3 = \frac{R_a(R_b + R_c)}{R_a + R_b + R_c}$$

These three equations can then be simplified as

$$(R_1 + R_3) - (R_2 + R_3) = \frac{R_b(R_a + R_c)}{R_a + R_b + R_c} - \frac{R_a(R_b + R_c)}{R_a + R_b + R_c}$$

$$(R_1 - R_2) = \frac{R_a R_b + R_b R_c - R_a R_b - R_a R_c}{R_a + R_b + R_c} = \frac{R_b R_c - R_a R_c}{R_a + R_b + R_c}$$

$$(R_1 + R_2) + (R_1 - R_2) = \frac{R_c(R_a + R_b)}{R_a + R_b + R_c} + \frac{R_b R_c - R_a R_c}{R_a + R_b + R_c}$$

$$R_1 = \frac{R_a R_c + R_b R_c + R_b R_c - R_a R_c}{2(R_a + R_b + R_c)} = \frac{R_b R_c}{R_a + R_b + R_c}$$

Example 1.13   Find the equivalent resistance between terminals a and b using a Delta-Wye transformation (Fig. 1.20).

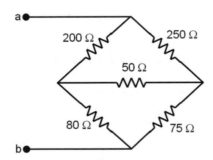

Fig. 1.20 Circuit diagram for Example 1.13

Solution:
Either the top three resistors (200 Ω, 250 Ω, and 50 Ω) or the bottom three resistors (50 Ω, 80 Ω, and 75 Ω) can be replaced by Wye-connected resistors. We will replace the top three.

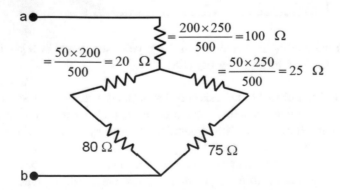

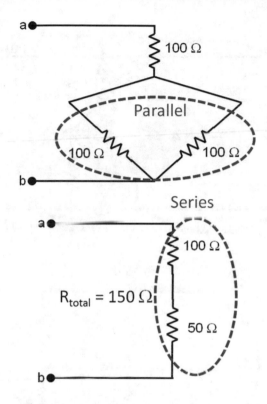

Example 1.14 Find the equivalent resistance between terminals a and b (Fig. 1.21).

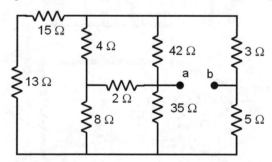

Fig. 1.21 Circuit diagram for Example 1.14

Solution:

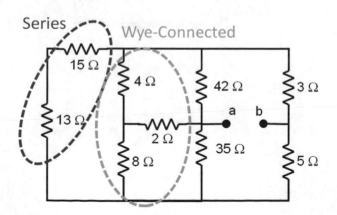

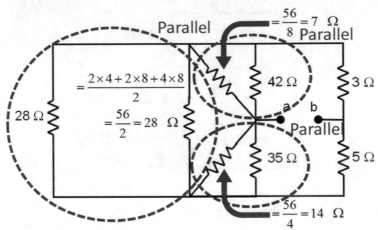

Parallel

$$= \frac{56}{8} = 7 \ \Omega$$ Parallel

$$= \frac{2 \times 4 + 2 \times 8 + 4 \times 8}{2}$$

$$= \frac{56}{2} = 28 \ \Omega$$

28 Ω   42 Ω   3 Ω

a   b   Parallel

35 Ω   5 Ω

$$= \frac{56}{4} = 14 \ \Omega$$

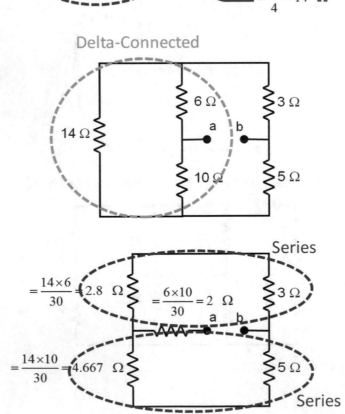

Delta-Connected

14 Ω   6 Ω   3 Ω

a   b

10 Ω   5 Ω

Series

$$= \frac{14 \times 6}{30} = 2.8 \ \Omega$$ $$= \frac{6 \times 10}{30} = 2 \ \Omega$$ 3 Ω

a   b

$$= \frac{14 \times 10}{30} = 4.667 \ \Omega$$ 5 Ω

Series

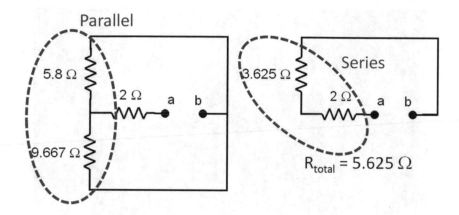

1.4 Capacitors

The circuit symbol for the capacitor is shown in Fig. 1.22. The capacitor is used to model the connection of one part of the circuit to another part of the circuit by electric fields. This is most commonly accomplished by adding a capacitor to the circuit such as those shown in Fig. 1.23. However, even when no capacitors are present, parasitic capacitances due to the electric fields between different conductors in the circuit can impact circuit performance by adding noise and unexpected feedback to the circuit. All that is needed to create a capacitance is two conductors at different voltages separated by an insulator. The relationship between current and voltage for a capacitor is given by

$$i(t) = C \cdot \frac{dv(t)}{dt} \ or \ v(t) = \frac{1}{C} \int_{-\infty}^{t} i(\tau)d\tau \tag{1.10}$$

where $v(t)$ is the time-varying voltage across the capacitor, $i(t)$ is the time-varying current flowing into the capacitor, and C is the capacitance with units of Farads (F). In order to have current flow into a capacitor, the voltage across the capacitor must be varying in time. If the voltage is constant with respect to time, as in DC circuits, capacitors act as open circuits and can be ignored during the circuit analysis. However, even when the voltage does not vary with time, electrical energy can be stored in the electric fields of the capacitor. Given the dependence of the current on the rate of change of the

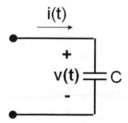

Fig. 1.22 Circuit symbol for a capacitor

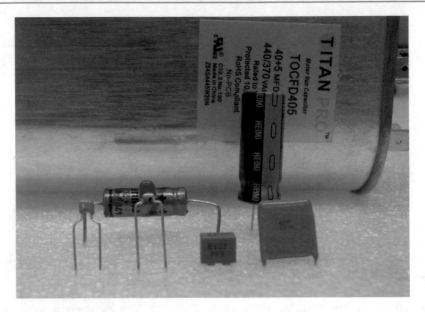

Fig. 1.23 Examples of different capacitors

voltage, the voltage across the capacitors cannot change instantaneously as this would require infinite current.

The physical size of the capacitor depends on both the overall capacitance as well as the maximum voltage that the capacitor can support. One important type of capacitors is electrolytic capacitors such as the cylindrical black/gray and blue capacitors shown in Fig. 1.23. These capacitors have high capacitances for the same voltage/size, but they must always be connected with their positive voltage terminal at a higher voltage than their negative voltage terminal. This is not normally a problem as most time-varying circuits operate about a DC bias point with one terminal of the capacitor always at a higher voltage than the other terminal. However, electrolytic capacitors cannot be used if the time-varying signal will exceed the bias voltages reversing the polarity on the capacitor.

> **Lab Hint**: The value of the capacitance for a capacitor is often printed on the side of the capacitor. In addition, most multi-meters can directly measure the capacitance. Capacitors can be a potential source of injury if not handled properly as they can store energy and maybe a source of electric shock even when the power is off.

Example 1.15   A current of 1 mA flows into a 1 mF capacitor for 1 s as shown in Fig. 1.24, what is the voltage across the capacitor as a function of time?

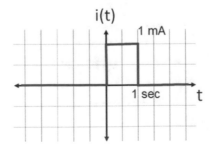

Fig. 1.24 Figure for Example 1.15 showing current flow into the capacitor as a function of time

Solution:

$$v(t) = \frac{1}{C} \int\limits_{-\infty}^{t} i(\tau)d\tau = \frac{1}{1\text{ mF}} \int\limits_{-\infty}^{t} i(\tau)d\tau$$

$t < 0$

$i(\tau) = 0 \Rightarrow v(t) = 0$

$0 \leq t < 1\ s$

$i(\tau) = 1\text{ mA}$

$$v(t) = \frac{1}{1\text{ mF}} \left(\overbrace{\int\limits_{-\infty}^{0} i(\iota)d\tau}^{=0} + \int\limits_{0}^{t} (1\text{ mA})d\tau \right) = \left(1 \frac{V}{s}\right) \cdot t$$

$t \geq 1\ s$

$$v(t) = \frac{1}{1\text{ mF}} \left(\overbrace{\int\limits_{-\infty}^{0} i(\tau)d\tau}^{=0} + \int\limits_{0}^{1} (1\text{ mA})d\tau + \overbrace{\int\limits_{1}^{t} i(\tau)d\tau}^{=0} \right) = 1\ V$$

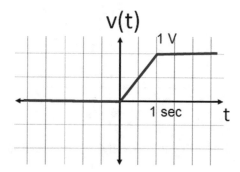

Example 1.16 A voltage pulse given by $v(t) = 2\exp\left(-\left(\frac{t}{200\ \mu s}\right)^2\right)$ V is applied to a 3 μF ca-

pacitor. Plot the current flow into the capacitor and find the maximum current.

Solution:

$$i(t) = C \cdot \frac{dv(t)}{dt} = C \cdot \frac{d}{dt}\left(2\exp\left(-\left(\frac{t}{200\ \mu s}\right)^2\right) V\right)$$

$$= (3 \cdot 10^{-6}) \cdot 2 \cdot \exp\left(-\left(\frac{t}{200\ \mu s}\right)^2\right) \cdot \left(-2\left(\frac{t}{(200\ s)^2}\right)\right)$$

$$= -300t \cdot \exp\left(-\left(\frac{t}{200\ \mu s}\right)^2\right) A$$

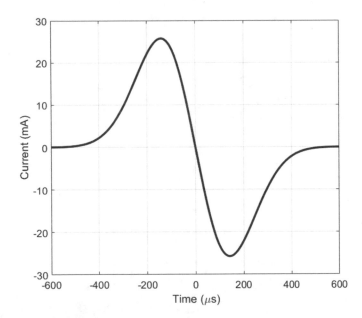

Now find the maximum current by taking the derivative of the current and setting the derivative equal to zero.

$$\frac{d}{dt}i(t) = -300 \cdot \exp\left(-\left(\frac{t_{max}}{200\ \mu s}\right)^2\right) - 300t_{max} \cdot \exp\left(-\left(\frac{t_{max}}{200\ \mu s}\right)^2\right)\left(-2t_{max}\left(\frac{1}{200\ \mu s}\right)^2\right) = 0$$

$$1 + t_{max} \cdot \left(-2t_{max}\left(\frac{1}{200\ \mu s}\right)^2\right) = 0 \Rightarrow 2t_{max}^2 = (200\ \mu s)^2 \Rightarrow t_{max} = \pm\frac{200\ \mu s}{\sqrt{2}}$$

$$i(t_{\max}) = \mp 300 \left(\frac{200~\mu s}{\sqrt{2}}\right) \cdot \exp\left(-\left(\frac{1}{\sqrt{2}}\right)^2\right) = \mp 25.73~\text{mA}$$

1.4.1 Capacitors in Series and in Parallel

Circuit models with capacitors can also be simplified when multiple capacitors are in series or in parallel. Remember, circuit elements are in series when the same current MUST pass through both elements, and circuit elements are in parallel when both terminals are connected so that the voltage across both elements are the same. Figure 1.25 shows three capacitors in series and in parallel. The equivalent capacitance for capacitors in series is given by

$$C_{eq} = \frac{1}{\frac{1}{C_1} + \frac{1}{C_2} + \frac{1}{C_3} + \cdots + \frac{1}{C_{N-1}} + \frac{1}{C_N}} = \frac{1}{\sum\limits_{k=1}^{N} \frac{1}{C_k}} \tag{1.11}$$

while the equivalent capacitance for capacitors in parallel is given by

$$C_{eq} = C_1 + C_2 + C_3 + \cdots + C_{N-1} + C_N = \sum_{k=1}^{N} C_k \tag{1.12}$$

Notice that capacitors in series decrease the overall capacitance while capacitors in parallel increase the overall capacitance. This is the opposite of the equivalents found for resistors. However, the reversed relationship makes sense if we compare Ohm's law ($V = IR$) to the expression for the voltage for a capacitor ($v(t) = \frac{1}{C} \int\limits_{-\infty}^{t} i(\tau)d\tau$). For the resistor, for the same current values, the voltage drop increases as the resistance increases. However, for the capacitor, the voltage drop decreases as the capacitance increases for the same current values over time. Hence, the capacitance and resistance are inversely related relative to each other when solving for the current and/or voltage.

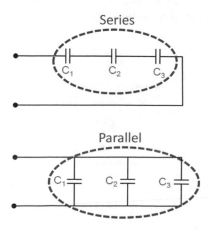

Fig. 1.25 Capacitors in series and in parallel

Example 1.17 You have four capacitors with capacitances of 1 mF, 2 mF, 3 mF, and 4 mF.

(a) What is the total capacitance if the capacitors are connected in series?
(b) What is the total capacitance if the capacitors are connected in parallel?

Solution:

(a) Series connection

$$C_{eq} = \frac{1}{\sum\limits_{k=1}^{N} \frac{1}{C_k}} = \frac{1}{\frac{1}{1\,\text{mF}} + \frac{1}{2\,\text{mF}} + \frac{1}{3\,\text{mF}} + \frac{1}{4\,\text{mF}}} = \frac{1\,\text{mF}}{1 + 0.5 + 0.333333 + 0.25}$$

$$= \frac{1\,\text{mF}}{2.08333} = 0.48\,\text{mF}$$

(b) Parallel connection

$$C_{eq} = \sum\limits_{k=1}^{N} C_k = 1\,\text{mF} + 2\,\text{mF} + 3\,\text{mF} + 4\,\text{mF} = 10\,\text{mF}$$

1.5 Inductors

In addition to storing energy in electric fields as in capacitors, energy can also be stored in magnetic fields. An inductor models the impact of energy storage in magnetic fields on circuit voltages and currents. All connecting wires in a circuit have some parasitic inductance, but the values are normally small. If more inductance is needed, wires can be round around a magnetic core to create a specific inductance value. Most circuit designs try to avoid adding inductors to a circuit as they tend to require a lot of space relative to other components. Therefore, inductors are usually found when modeling other components, such as the coils in a motor and are less common as stand-alone circuit components. The circuit model for an inductor is shown in Fig. 1.26.

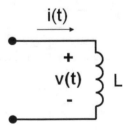

Fig. 1.26 Circuit symbol for an inductor

The relationship between current and voltage for an inductor is given by

$$i(t) = \tfrac{1}{L} \int\limits_{-\infty}^{t} v(\tau)d\tau \quad or \quad v(t) = L \cdot \tfrac{di(t)}{dt} \tag{1.13}$$

where, once again, $v(t)$ is the time-varying voltage across the inductor, $i(t)$ is the time-varying current flowing into the inductor, and L is the inductance with units of Henrys (H). In order to have a voltage drop across an inductor, the current must be changing in time. A constant current, such as in DC circuits, results in no voltage drop, and the inductor can be replaced in the circuit model as a short circuit or simple wire. The dependence of voltage on the time rate of change of current also means that the current cannot change instantaneously in an inductor as this would require infinite voltage.

Example 1.18 A current flowing through a 2 mH inductor increases linearly to 1 mA in 0.5 ms before returning to 0 mA 1 ms later as shown in Fig. 1.27. What is the voltage across the inductor as a function of time?

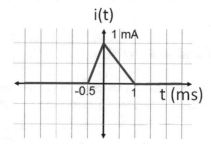

Fig. 1.27 Figure for Example 1.18 showing current flow into the inductor as a function of time

Solution:

$$v(t) = L \cdot \frac{di(t)}{dt}$$

The current is piecewise continuous, so we need to split the solution into four regions.

Region 1 : $t \le -0.5$ ms

$i(t) = 0 \Rightarrow v(t) = 0$

Region 2 : -0.5 ms $< t \le 0$ ms

$Slope = \frac{1 \text{ mA}}{0.5 \text{ ms}}$

$Intercept = 1$ mA

$i(t) = \overbrace{(2 \text{ A/s})t + 1 \text{ mA}}$

$\Rightarrow v(t) = 2 \text{ } mH \frac{d}{dt}((2 \text{ A/s})t + 1 \text{ mA}) = 4 \text{ mV}$

Region 3 : 0 ms $< t \le 1$ ms

$Slope = \frac{-1 \text{ mA}}{1 \text{ ms}}$

$Intercept = 1$ mA

$i(t) = \overbrace{(-1 \text{ A/s})t + 1 \text{ mA}}$

$\Rightarrow v(t) = 2 \text{ } mH \frac{d}{dt}((-1 \text{ A/s})t + 1 \text{ mA}) = -2 \text{ mV}$

Region 4 : $t > 1$ ms

$$i(t) = 0 \Rightarrow v(t) = 0$$

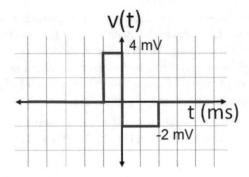

1.5.1   Inductors in Series and in Parallel

Just like resistors and capacitors, inductors in series or in parallel can be simplified by finding an equivalent inductance. The equivalent inductance for inductors in series is given by

$$L_{eq} = L_1 + L_2 + L_3 + \cdots + L_{N-1} + L_N = \sum_{k=1}^{N} L_k \tag{1.14}$$

while the equivalent inductance for inductors in parallel is given by

$$L_{eq} = \frac{1}{\frac{1}{L_1} + \frac{1}{L_2} + \frac{1}{L_3} + \cdots + \frac{1}{L_{N-1}} + \frac{1}{L_N}} = \frac{1}{\sum_{k=1}^{N} \frac{1}{L_k}} \tag{1.15}$$

The formulas for inductors in series/parallel mirror those for resistors in series/parallel. This is because for the same current expression, the voltage increases as the inductance increases just like voltage increase with resistance increase that occurs for the resistor.

1.6   Problems

Problem 1.1: What would be the color code values for a resistor with a resistance of 470 $\Omega \pm 5\%$?

Problem 1.2: If a resistor has the color bands of red, white, orange, and silver what is the expected minimum, maximum, and average resistance value?

Problem 1.3: If 5 V is applied across a 1 kΩ resistor, what is the current flow in the resistor?

Problem 1.4: If 3 mA is flowing in 7 kΩ resistor, what is the voltage drop across the resistor?

Problem 1.5: If a current if 4 mA flows when 15 V is applied to an unknown load, what is the resistance of the load?

Problem 1.6: Draw the equivalent circuit for a battery and a light bulb connected to a battery. Use a unique resistor for each circuit element while assuming that all of the wires are lossless.

Problem 1.7: A flashlight operates at a voltage of 3 V while drawing a current of 0.3 A. What is the effective resistance of the flashlight circuit?

Problem 1.8: Label the portions of the following resistor combinations that are in series or in parallel.

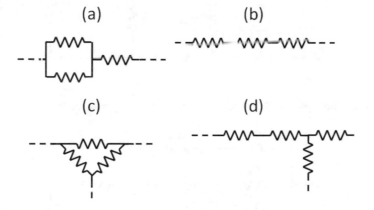

Problem 1.9: Label the portions of the following resistor combinations that are in series or in parallel.

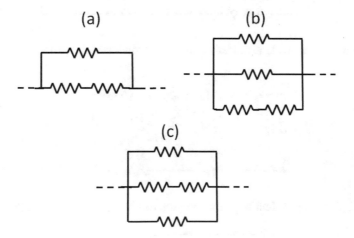

Problem 1.10: What is the equivalent resistance of a 100 Ω, 300 Ω, and 500 Ω resistor in series?

Problem 1.11: What is the equivalent resistance of a 100 Ω, 300 Ω, and 500 Ω resistor in parallel?

Problem 1.12: Find the equivalent resistance for the following resistor network.

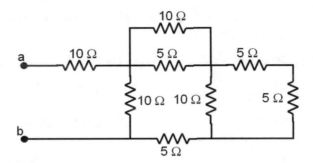

Problem 1.13: Find the equivalent resistance for the following resistor network.

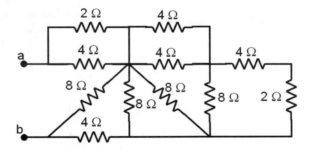

Problem 1.14: Find the equivalent resistance for the following resistor network.

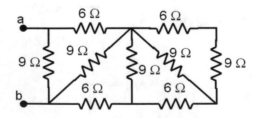

Problem 1.15: Find the equivalent resistance for the following resistor network.

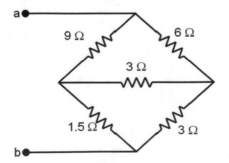

Problem 1.16: Find the equivalent resistance seen by the voltage source for the following circuit.

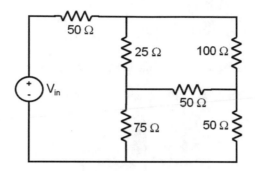

Problem 1.17: Find the equivalent resistance for the following resistor network.

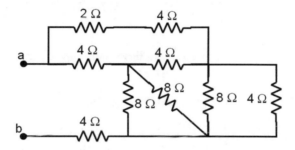

Problem 1.18: The current flowing in a 20 μF capacitor is given by

$$i(t) = \begin{cases} 0 & t \le 0 \text{ s} \\ 2\sin(2\pi t) \text{ mA} & 0 < t \le 0.5 \text{ s} \\ 0 & t > 0.5 \text{ s} \end{cases}$$

What is the voltage across the capacitor as a function of time?

Problem 1.19: The current flowing in a 100 μF capacitor is given by the function shown below. What is the voltage across the capacitor as a function of time?

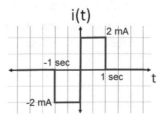

Problem 1.20: The current flowing in a 5 mF capacitor is given by

$$i(t) = \begin{cases} 0 & t < 0 \text{ s} \\ 4 \exp(-t) \text{ mA} & t \geq 0 \text{ s} \end{cases}$$

What is the voltage across the capacitor as a function of time and the maximum voltage over all time?

Problem 1.21: The voltage across a 4 mF capacitor is shown below.

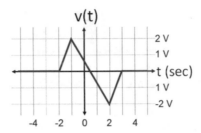

What is the current flowing into the capacitor as a function of time as well as the maximum current over all time?

Problem 1.22: The voltage across a 2 mF capacitor is given by

$$v(t) = \begin{cases} 5 \sin(10\pi t) \ V & 0 < t \leq 0.2 \text{ s} \\ 0 & \text{else} \end{cases}$$

What is the current flowing into the capacitor as a function of time?

Problem 1.23: You have four capacitors with capacitances of 3 mF, 7 mF, 9 mF, and 11 mF.

(a) What is the total capacitance if the capacitors are connected in series?
(b) What is the total capacitance if the capacitors are connected in parallel?

Problem 1.24: What happens to the capacitance when two equal capacitors are placed in parallel?

Problem 1.25: What happens to the inductance when two equal inductors are placed in parallel?

Problem 1.26: You have four inductors with inductances of 3 mH, 7 mH, 9 mH, and 11 mH.

(a) What is the total inductance if the inductors are connected in series?
(b) What is the total inductance if the inductors are connected in parallel?

Problem 1.27: The current flowing into a 3 mH inductor capacitor is given by

$$i(t) = \begin{cases} (5t^2 - 5) \text{ mA} & -1 \leq t \leq 1 \text{ s} \\ 0 & \text{else} \end{cases}$$

What is the voltage across the inductor as a function of time?

Problem 1.28: The voltage across an 8 mH inductor capacitor is given by $v(t) = 4\exp(-200|t|)$ V. What is the current flowing into the inductor as a function of time?

Electrical Safety

2

Most of the time electrical circuits present very little danger to the student or practicing professional. Occasionally, an integrated circuit component can poke you in the finger when being placed into a breadboard or the circuit components can become hot when not wired properly, but the risk of electric shock is normally low. However, there are times when higher voltages are used. In addition, a breakdown or bypassing of skin resistance can lead to dangerous currents. Therefore, it is important to be mindful of these risks when working with electric circuits. Relying on the misinformation presented in movies and television can be very dangerous.

2.1 Dangers of Electric Current

The most important concept to know related to electrical safety is that it is the electric current that is dangerous and not the voltage. High voltages present a danger because they can drive high currents. However, it is possible to safely touch high voltages without danger if there is no power/current behind them. Remember, real voltage sources are limited by the amount of current they can provide. For example, rubbing your feet on the carpet to buildup static electricity can lead to a voltage of over 20,000 V while a Van de Graaff generator can achieve even higher voltages. However, the number of charged particles available is very small mitigating the danger. The second important concept to know is that it is the current travelling through the heart that is the most dangerous as it can produce ventricular fibrillation. If the current path does not include the heart, much higher currents can flow with minimal injury.

The possible biological effects as a function of electric current at 60 Hz for a 1 to 3 s exposure are shown in Fig. 2.1. At the low end, there is the threshold of perception that varies based on individual and current path that spans from about 0.5 to 10 mA. This is the minimum current that an individual can detect. It is normally felt as a tingling sensation as the current excites the never endings in the skin.

As the current increases, it will soon surpass the "let-go" current which spans from about 6 mA to 100 mA for most people. This is the maximum current that can be experienced while still allowing the person to voluntarily stop touching the source of the current. Therefore, the "let-go" current is a measure of electrical safety. If a person cannot voluntarily withdraw from the current, injury and death can occur. However, even if the "let-go" current is not exceeded, pain and muscle fatigue can occur as well as secondary injury from reflex withdrawals. The "let-go" current varies from person to person and also depends on frequency. For example, the lowest "let-go" currents occur for

© Springer Nature Switzerland AG 2020
T. A. Bigelow, *Electric Circuits, Systems, and Motors*,
https://doi.org/10.1007/978-3-030-31355-5_2

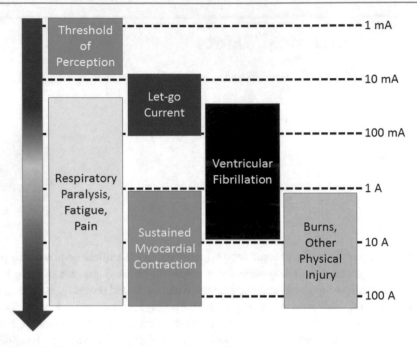

Fig. 2.1 Physiological effects of electricity as a function of electric current

frequencies between 10 and 1,000 Hz. Higher currents are needed to prevent "let-go" outside of this frequency range. Also, the "let-go" current tends to be higher in men than in women.

Beginning around 20 mA, the current flow will result in respiratory paralysis due to an involuntary contraction of the respiratory tissue. If the current is below the "let-go" current, the danger is normally minimal as the victim can interrupt the current flow. However, if the current continues to flow, asphyxiation and death can occur. The most common cause of death from current flow, however, is ventricular fibrillation which occurs for currents in the 75 mA to 6 A range. These currents produce rapid and disorganized ventricular contractions and depolarization waves randomly sweep over the ventricles of the heart. These contractions prevent the heart from pumping blood and may continue even after the current has been removed. Typically, a defibrillator is needed to depolarize all of the heart cells simultaneously to restore normal heartbeats. Ventricular fibrillation is the major cause of death for victims of electric shock. Ironically, currents greater than about 1 A will produce a sustained myocardial contraction which is the basis for a defibrillator. Therefore, the "treatment" for one electric shock is another electric shock to restore heart function. However, the exposure must be relatively brief to avoid tissue damage and allow for the heart and breathing to resume. The highest current levels can also result in skin burns as well as sever muscle contractions that may pull the muscle off of the bone. In addition, the brain and nervous tissue lose all functions.

2.2 Importance of Skin Resistance

When it comes to protection from electric shock, our skin is our best line of defense. Normal dry skin has a resistance of 15 kΩ to 1 MΩ for 1 cm^2 of skin. Once past the skin barrier, however, the internal resistance between the limbs is on the order of 300–500 Ω. Therefore, anything that reduces the skin resistance such as moisture or wounds dramatically increases the risk of injury from electric shock.

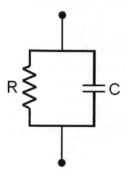

Fig. 2.2 Simple circuit model for skin

This is why the risk of electric shock is much greater when swimming or bathing. However, the sudden application of a voltage can greatly reduce the effective resistance of the skin. A simple circuit model for the skin is shown in Fig. 2.2. The capacitance results from the low-conductivity of the epidermis separating the conductivity of the subcutaneous layers and the applied voltage source. As was shown in Chap. 1, the current in a capacitor is governed by the time-rate of change of the applied voltage according to the equation $i(t) = C \cdot \frac{dv(t)}{dt}$. Therefore, a sudden application of voltage can result in relatively high current flows. These high current flows can damage the skin, lowering the skin resistance, and allowing for higher currents to continue to flow resulting in injury or death.

Example 2.1   If a section of skin has a resistance of 50 kΩ, what is the current flow when 500 V is applied? Is this current flow dangerous?

Solution: Using Ohm's Law we have

$$V = I \cdot R \Rightarrow I = \frac{V}{R} = \frac{500 \text{ V}}{50 \text{ k}\Omega} = 10 \text{ mA}$$

A current of 10 mA is above the threshold for perception, and it may be above the "let-go" current for some individuals. If the individual cannot "let-go," the sustained current flow could result in damage to the skin that would allow larger and more dangerous currents to flow.

Example 2.2   If a section of skin has infinite resistance and a capacitance of 30 nF, what is the current flow when the voltage across the skin changes by 500 V in 10 μs? Is this current flow dangerous?

Solution: Using the current/voltage relationship for a capacitor we have:

$$i(t) = C \cdot \frac{dv(t)}{dt} = C \frac{\Delta V}{\Delta t} = \left(30 \times 10^{-9} \text{ F}\right) \frac{500 \text{ V}}{10 \times 10^{-6} \text{ s}} = 1.5 \text{ A}$$

A current of 1.5 A is greater than the "let-go" current so the individual will not be able to disconnect the current flow. In addition, burns will quickly damage the skin causing even higher currents to flow. Ventricular fibrillation and respiratory paralysis will also occur. Therefore, serious harm will likely occur.

Movie Mistake: Movies and TV shows frequently get their science wrong when it comes to electric shock. For example, in an old murder drama (CSI, Season 2, Episode 3, "Overload") the lead "scientist" concludes that high iron content in the blood reduced the blood resistance so that there were no skin burns. This made the electrocution look like a suicide. However, sodium and other ions (NOT iron) are the most important components when setting the conductivity of the plasma which is the most conductive part of the blood. Furthermore, changes in blood chemistry will not change the resistance of the skin which is where the skin burns would occur.

2.3   Importance of Grounding

One of the most common causes of electric shock is poor grounding. Figure 2.3 shows an image of a typical wall outlet in the United States that can provide proper grounding. The outlet consists of three slots connected to three separate wires. The two narrow slots are for the "hot" (black wire) and neutral (white wire) while the rounded slot is for the ground wire. For many outlets, the neutral slot is slightly larger than the "hot" slot. The "hot" wire has an RMS voltage of 120 V relative to both the neutral and ground wires. While the neutral and ground wires are at approximately the same potential, the ground wire is usually connected to the outside chassis of the device. When everything is working properly, the "hot" wire provides the power while the neutral wire provides a return path for the current with no current flow along the grounding wire. If there is an electric short, however, the ground wire provides a path for the current so that charge does not build-up on the chassis. Without this grounding wire, anyone that touched the outside of the device could receive an electric shock should a short occur. Therefore, you should never use any adapters or extension cords that eliminate the ground connection.

Fig. 2.3 Diagram of typical wall outlet in the United States

2.4   Electrical Safety Guidelines

In order to protect yourself from electric shock, the most important action that you can take is to keep your mind on your work. You need to be aware of your surroundings and the work that needs to be accomplished. You should make sure that none of your power cords have any frayed insulation. Whenever possible, you should turn off the power when working on your circuit. Many companies require a lockout-tagout procedure when working with high voltage equipment so that the power does not get accidently turned back on while the circuits are being repaired. This policy typically requires the worker to:

(1) Inform co-workers that the power will be shutting off.
(2) Identify the breakers that need to be turned off to cut the power.
(3) Turn off the power and place a lock so that the power cannot be turned back on. The lock is normally tagged with the workers identification as well.
(4) Confirm that the power is off to the equipment needing work.

In a lab setting at lower voltages (~ 120 V_{rms} or less), the lockout-tagout procedure may not be practical. However, it is still important to observe some basic safe practices. Being mindful of your work and working environment should still be your top priority. Whenever possible you should disconnect the power before working on the circuit. Most of the time, this would mean unplugging the power supply from the wall so that it cannot be accidently switched on. This also ensures that all of the voltages in the circuit are low. After unplugging, you may need to wait some time for the capacitors to discharge. The time required will depend on the circuit connections, so you should be aware of how the capacitors should normally discharge. Also, you should assume the capacitors are charged until you have been able to confirm that the charge has dissipated.

There are times in the lab when the power needs to be on in order to solve a circuit problem or explore a concept. When the voltages in the circuit are less than about 30 V_{rms}, the danger of electric shock is minimal unless there is a reduction in skin resistance. This is why most lab classes use voltages on the order of 10 V or less. For these low voltages, there is a slight risk from other hazards. For example, poor wiring could cause an integrated circuit to get very hot yielding minor burns when touched or a capacitor could burst if exposed to voltages higher than its rated values. However, the risk of any serious injury is very low.

If the power needs to be on for higher voltages in the 30 to 120 V_{rms} range, however, the risk of electric shock is much more significant and other precautions should be taken. First, you should remove all watches, rings, and necklaces as these can provide an efficient conduction path. Sweat often builds under these accessories which are often metal. The moisture in the sweat reduces the skin resistance increasing the risk of electric shock. Second, you should not apply power until you have double-checked all of the connections. Specifically, you should check for any short circuits. You should also use a multi-meter to confirm that the resistance between the largest voltages and ground are in the expected range as low resistances can be a sign of a dangerous short circuit. This can be more challenging with motors as they typically have a relatively low resistance. If you are working in a group, you should also inform all of your group members before turning the power on. Otherwise, they may reach for the circuit as you apply the power potentially subjecting them to an electric shock. Once the power is on, you should use only ONE probe to check the voltages in the circuit and you should not touch the metal portion of the probe. The other probe should be connected to ground prior to applying any power to the circuit and it should remain secured to the ground during the testing. If the ground probe becomes disconnected during the course of the measurement, then you should place the probe that you were using to check the voltages down, and use only one hand to reconnect the ground probe. You should NEVER operate with a probe in each hand when the power is on as this will provide a path through your heart should you accidently touch any exposed wires.

> **Lab Hint:** When working with higher voltages, it is very easy to abandon safe practices when working with a difficult circuit. Therefore, you should keep one hand in your pocket when checking the voltages in the circuit. This will reduce the likelihood of accidently touching the

circuit with your other hand. You should also wear rubber-soled shoes without open toes to prevent an inadvertent current path through your feet. Long pants that cover your legs are also advisable if you like to rest your feet on your lab bench.

2.5   Step Potential Versus Touch Potential

One of the aspects of electric shock that is often ignored in movies and television is that you don't need to touch a high-voltage source to be shocked. Just standing close can be dangerous. Consider the energized but grounded object shown in Fig. 2.4. This could be a high-voltage power line that has fallen on the ground. One person in the figure is touching the high-voltage object, and they will receive an electric shock due to the potential difference between the charged object and the ground. This potential difference is called the touch potential. However, the second person who is just standing near the object will also receive a shock due to the potential difference between their feet. This second potential difference is called the step potential and is typically ignored in movies and television shows.

For example, Fig. 2.5 shows the voltage as a function of distance as one moves away from a 10 kV conductor that has fallen on clay soil. Notice that the voltage falls very quickly with distance. However, if you are standing about 1 foot away with a separation of your feet of about 1 foot, then there would be a 2 kV difference in potential between your legs. If you move back to 3 feet with the same foot separation, then the potential difference drops to about 250 V. The step potential makes approaching high-power lines very dangerous and presents a significant challenge when aiding those that have been injured by high-voltage lines.

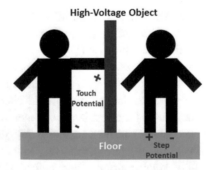

Fig. 2.4 Diagram illustrating the difference between step potential and touch potential

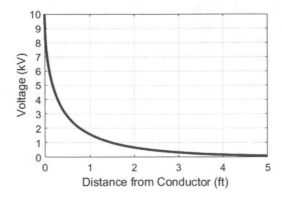

Fig. 2.5 Example change in voltage with distance as one moves away from a conductor that has fallen

2.6 Emergency Treatment for Victims of Electric Shock

So far, we have discussed various methods to prevent electric shock from occurring. However, electric shocks can still occur. If you observe an electric shock with high-voltage lines, then you need to stay away from the injured person until the power to the lines has been shutoff due to the dangers of step potential. Also, if you feel a tingling sensation in your legs, then you are likely still too close to the high-voltage line and you should move further back from the danger until the power is off. When moving away, you should take small careful steps as a long stride will result in a greater step potential difference. DO NOT try to hop on one foot, as some have suggested, as falling could place you at an even greater risk.

If high-voltage lines are not involved, the risks from step-potential will be much less. Therefore, your first priority will be to separate the victim from the current source if possible. In many cases, this means unplugging the electrical device or shutting off the circuit breaker. If you can't turn off the power to the circuit, then you should try to move the current away from the injured person by using a non-conducting object. Once the person is away from the current source, then you can begin first aid treatment if you are appropriately trained. Specifically, CPR should be given if needed as well as the treatment for any bleeding or burns. You should also treat the person for medical shock as you wait for emergency personnel. You should not move the individual unless absolutely necessary as they may have other injuries such as fractures.

2.7 Problems

Problem 2.1: If a section of skin has a resistance of 50 kΩ, what is the current flow when 300 V is applied? Is this current flow dangerous?

Problem 2.2: If a section of skin has a resistance of 10 kΩ, what is the current flow when 240 V is applied? Is this current flow dangerous?

Problem 2.3: If a section of skin has a resistance of 1.2 kΩ, what is the current flow when 120 V is applied? Is this current flow dangerous?

Problem 2.4: If a section of skin has infinite resistance and a capacitance of 25 nF, what is the current flow when the voltage across the skin changes by 120 V in 5 μs? Is this current flow dangerous?

Problem 2.5: Which part of the outlet is the electric ground?

Problem 2.6: What is the basic lockout-tagout procedure?

Problem 2.7: Why should you remove watches and rings when working with powered electric circuits?

Problem 2.8: Why might you want to place one hand in your pocket when working with electric circuits?

Problem 2.9: What is step potential and why is it potentially dangerous?

Problem 2.10: The voltage along the ground as a function of distance from a downed power line is given by $v(x) = 5000 \exp(-0.2x)$ V where x is in feet. What is step potential between your legs separated by 1 foot if your closest leg is 4 feet from the power line?

DC Circuit Analysis

3

The analysis of electric circuits is essential for both the design and understanding of electrical devices. As such, numerous different approaches have been developed to calculate the voltages, currents, and powers associated with each circuit model element when analyzing the circuit. These different approaches can be used to analyze circuits under a wide range of excitation conditions. However, when first learning circuits, it is best to limit our attention to DC circuits and thus avoid the complexity that comes with more complicated excitation conditions. DC refers to direct current, and DC circuits have voltages and currents that do not vary with time. Most battery-operated circuits operate at DC, so DC circuits do have some practical applications in addition to their use in illustrating basic circuit analysis concepts. As has been done by other authors, we will use uppercase letters (i.c., I and V) to denote currents and voltages that do not change with time and lower case letters (i.e., i and v) to denote values that have a time-varying component.

3.1 Circuit Analysis Terminology

Before discussing the different circuit analysis techniques, we need to define some terminology.

3.1.1 Node

A node in a circuit is a "point" at which 2 or more circuit elements are connected. When three or more circuit elements join, it is called an essential node. In circuit models, it is sometimes denoted as a black dot, but in other instances only the crossing of the lines from the different circuit elements are shown. Examples of possible connections for a node in a circuit are shown in Fig. 3.1. From a practical standpoint, however, a node is never a single point. Instead, all of the conducting wires between the circuit elements are part of the same node and can be thought of as a single "point" when performing the circuit analysis. Hence, the large ground plane that is common on many printed circuit boards is all one node and would be modeled as a single point prior to analyzing the circuit.

When determining the currents and voltages in a circuit, the most important node is the ground node. This node serves as a reference point for voltages at all other nodes in a circuit. Unless one of the nodes is already labeled as the ground node, you may assume that the node with the most elements directly connected to it is the ground node. Normally, the ground node is also connected to the metal chassis of the device, but there are times in circuit analysis when it is useful to distinguish

© Springer Nature Switzerland AG 2020
T. A. Bigelow, *Electric Circuits, Systems, and Motors*,
https://doi.org/10.1007/978-3-030-31355-5_3

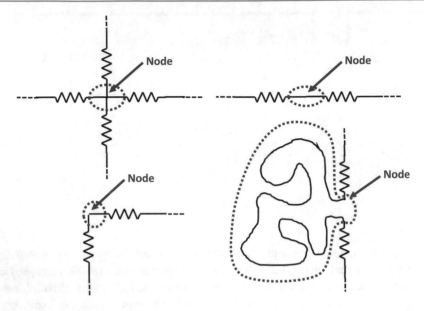

Fig. 3.1 Illustrations of a node in a circuit

Fig. 3.2 Circuit symbols for the ground or reference node

between the chassis node and the ground node in the circuit. Also, for battery-operated devices, the selection of the ground node is less critical as there is no connection to a common grounding point (i.e., earth ground) through the chassis. Distinguishing between earth ground, circuit ground, and chassis ground is important when reducing electronic noise in a circuit but is beyond the scope of this introductory text. The symbol for the ground node is shown in Fig. 3.2.

> **Lab Hint**: When working with a lab power supply, one of the output terminals will be labeled as ground. This ground should always be connected to the reference/ground node in your circuit. Also, when working with an oscilloscope, the ground node for the oscilloscope should also be connected to this same reference/ground node when making voltage measurements.

Example 3.1 Select a ground node and label the other nodes in the following circuit. Also, circle the wires corresponding to each node (Fig. 3.3).

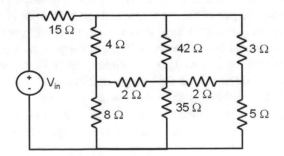

Fig. 3.3 Circuit diagram for Examples 3.1 and 3.4

Solution:

This circuit has 6 nodes. Of these, 3 of the nodes have 4 circuit elements connected. Either of these nodes would be a good choice for a reference/ground node. However, the best choice is the bottom node as this node would be at the lowest potential based on the location of the voltage source. The other nodes are labeled nodes 1 through 5.

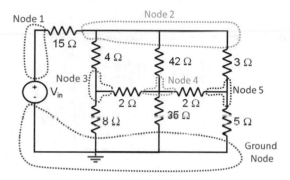

3.1.2 Branch

A branch is one or more circuit elements that are connected between two nodes. All of the circuit elements must be connected in series so that the same current travels through all of the elements between the nodes. Examples of branches are shown in Fig. 3.4. When the branch connects two essential nodes, it is called an essential branch.

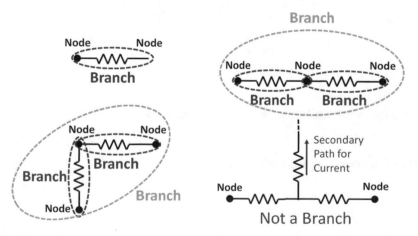

Fig. 3.4 Examples of different circuit branches as well as an example of three circuit elements that do not form a single branch because there is a secondary path for the current

3.1.3 Paths, Loops, and Meshes

A path is a collection of connected branches that do not pass through a connecting node more than once when traced. If the starting and ending node are the same, then the path forms a loop. A mesh is a loop that does not contain within it another loop. The direction of tracing the mesh, loop, or path does not matter.

Example 3.2 Determine all of the meshes and loops for the following circuit (Fig. 3.5).

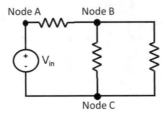

Fig. 3.5 Circuit diagram for Example 3.2

Solution:
There are two meshes in total. A–B–C–A and B–C–B. There is also one additional loop around the outside perimeter of the circuit.

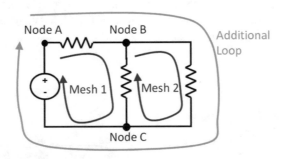

Example 3.3 How many meshes are in the circuit shown in Fig. 3.3?

Solution:
The circuit has 5 meshes as illustrated below.

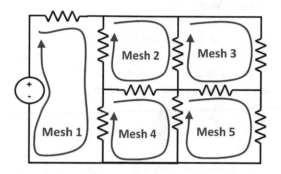

Example 3.4 Identify each of the following traces in Fig. 3.6 as a mesh, loop, or path.

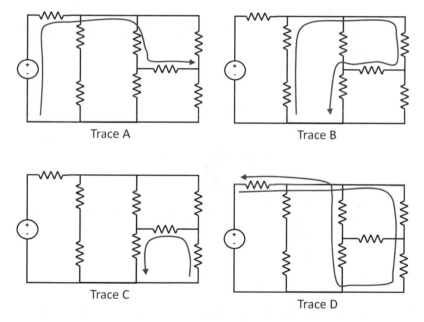

Fig. 3.6 Circuit diagrams for Example 3.4

Solution:

Trace A is a path. It does not pass through a connecting node more than once, but the first and last node are different. Trace B is a loop. Once again, no connecting node is crossed more than once, and the first and last node are the same. Trace B contains 2 meshes so it cannot be a mesh. Trace C is a mesh as it is a loop with no other meshes inside of it. Trace D is not a mesh, loop, or path as it crosses the top node twice; once at the beginning and once at the end of the trace.

3.2   Voltage/Current Dividers

The simplest circuits to analyze are voltage and current dividers. A voltage divider consists of a voltage source in series with a set of resistors with only one mesh/loop as shown in Fig. 3.7. In order to find the current flowing in this circuit, replace the set of resistors in series with its equivalent resistance. Now we have just a voltage source in series with a resistance, so the current I_s is given by

$$I_S = \frac{V_s}{R_{eq}} = \frac{V_s}{\sum_{n=1}^{N} R_n} \tag{3.1}$$

Once the current in the loop is known, the voltage drop across any resistor is just given by the current multiplied by the resistance value.

$$V_k = \frac{V_s R_k}{\sum_{n=1}^{N} R_n} \tag{3.2}$$

Therefore, the total voltage is distributed among the resistors based on the fraction of the resistors value relative to the total resistance in the circuit.

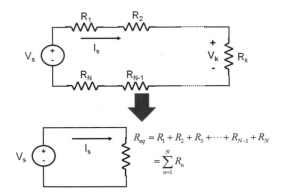

Fig. 3.7 Circuit diagram for the voltage divider

Example 3.5 Find the voltage drop across each resistor for the circuit shown (Fig. 3.8).

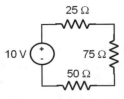

Fig. 3.8 Circuit diagram for Example 3.5

Solution:

$$V_k = \frac{V_s R_k}{\sum_{n=1}^{N} R_n} \Rightarrow \begin{array}{l} V_{25\Omega} = \dfrac{10 \cdot 25}{25 + 75 + 50} = 1.6667 \text{ V} \\[3mm] V_{50\Omega} = \dfrac{10 \cdot 50}{25 + 75 + 50} = 3.3333 \text{ V} \\[3mm] V_{75\Omega} = \dfrac{10 \cdot 75}{25 + 75 + 50} = 5 \text{ V} \end{array}$$

Likewise, a current divider consists of a current source in parallel with a set of resistors with only two nodes as shown in Fig. 3.9. The voltage drop between the two nodes can be found by replacing the resistors in parallel with their equivalent resistance and applying Ohm's Law.

$$V_S = \frac{I_S}{\sum_{n=1}^{N} \frac{1}{R_n}} \tag{3.3}$$

The current flowing in each resistor can then be found by dividing the voltage by the resistance in each branch.

$$I_k = \frac{1}{R_k} \left(\frac{I_S}{\sum_{k=1}^{N} \frac{1}{R_n}} \right) \tag{3.4}$$

Therefore, the current in each branch is inversely proportional to the resistance of the branch with a greater fraction of the current flowing in the branches with the smallest resistances.

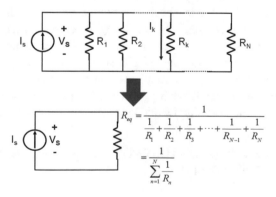

Fig. 3.9 Circuit diagram for current divider

Example 3.6 Find the current through each resistor for the circuit shown (Fig. 3.10).

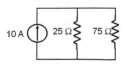

Fig. 3.10 Circuit diagram for Example 3.6

Solution:

$$I_k = \frac{1}{R_k}\left(\frac{I_S}{\sum_{k=1}^{N}\frac{1}{R_n}}\right)$$

$$\Rightarrow \quad \begin{aligned} I_{25\Omega} &= \tfrac{1}{25}\left(\frac{I_S}{\frac{1}{25}+\frac{1}{75}}\right) = \tfrac{1}{25}\left(\frac{75\cdot25\cdot I_S}{75+25}\right) = \left(\frac{75 I_S}{75+25}\right) = 7.5 \text{ A} \\ I_{75\Omega} &= \tfrac{1}{75}\left(\frac{I_S}{\frac{1}{25}+\frac{1}{75}}\right) = \tfrac{1}{75}\left(\frac{75\cdot25\cdot I_S}{75+25}\right) = \left(\frac{25 I_S}{75+25}\right) = 2.5 \text{ A.} \end{aligned}$$

3.3 Node-Voltage Method of Circuit Analysis

The most important circuit analysis technique is the node-voltage method. While some circuits can be simpler to analyze with some of the other techniques, the node-voltage method is directly correlated with how circuit analysis is normally done in the lab. When analyzing circuits in the lab, the first step is to attach the multi-meter or oscilloscope ground to the circuit ground. The voltages are then measured relative to this reference node by placing the other lead at various nodes in the circuit. Calculating the voltages at different nodes of the circuit relative to a reference node is the goal of the node-voltage method. Node-voltage methods also form the basis for most circuit modeling packages as well.

The first step in the node-voltage method is to select a reference node. If no ground node is identified, one of the nodes with the most branches should be selected. After labeling the reference node, the voltages at the other nodes should be labeled. Commonly either numbers (i.e., V_1, V_2, V_3, ...) or letters (i.e., V_a, V_b, V_c, ...) are used to identify the voltages at each node. Once each node is identified, Kirchhoff's Current Law is used to write a single equation for all of the nodes except the reference node. Therefore, if there are N nodes including the ground node, there will be N − 1 equations to solve. Kirchhoff's Current Law states that the algebraic sum of all currents entering any node in a circuit is zero and is a form of charge conservation. When summing the currents, currents entering the node can be assigned a positive sign while currents leaving the node can be assumed negative OR currents entering can be negative with currents leaving being positive. Either convention is fine as long as you are consistent. When I analyze circuits, I typically have currents leaving be positive, and I then add up all of the currents assuming they are all leaving. When writing the equations, the currents should be written in terms of the node voltages using Ohm's Law. The following examples illustrate the steps in more detail.

Example 3.7 Use the Node-Voltage method to find the voltages at each node in the circuit as well as the current flowing in the 20 Ω resistor (Fig. 3.11).

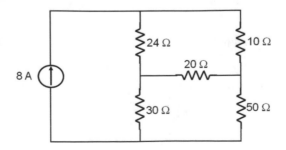

Fig. 3.11 Circuit diagram for Example 3.7

Solution:

The first step is to identify a ground node and label the other nodes as shown.

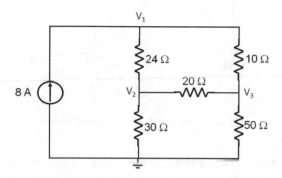

Now, let's write an equation for the currents entering/leaving node 1.

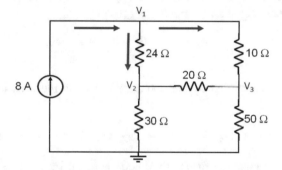

This node has 8 A entering from the current source as well as a current flowing from V_1 to V_2 and a current flowing from V_1 to V_3. According to Ohm's law, the current in a resistor is given by the voltage drop across the resistor divided by the resistance. For the current flowing from V_1 to V_2, the voltage drop is given by $V_1 - V_2$. Likewise, the voltage drop for the current flowing from V_1 to V_3 is $V_1 - V_3$. Therefore, the equation for the currents associated with node 1 is given by

$$8 \text{ A} = \frac{V_1 - V_2}{24} + \frac{V_1 - V_3}{10}$$

If we multiply both sides of this equation by 120, we can simplify the equation as

$$960 = 5(V_1 - V_2) + 12(V_1 - V_3) \Rightarrow 17V_1 - 5V_2 - 12V_3 = 960$$

Now, let's write an equations for the currents entering/leaving node 2. For this node, we will write the equation assuming all of the currents are leaving as shown below.

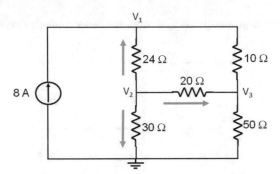

Notice, that by this convention, the current between nodes 1 and 2 is now assumed to flow in the opposite direction. This is acceptable as we are finding the equation for each node independently. For the current flowing from V_2 to V_1, the voltage drop is given by $V_2 - V_1$. Likewise, the voltage drop from V_2 to ground is given by $V_2 - 0$, and the voltage drop from V_2 to V_3 is given by $V_2 - V_3$. Therefore, the node-voltage equation for node 2 is given by

$$0 = \frac{V_2 - V_1}{24} + \frac{V_2 - 0}{30} + \frac{V_2 - V_3}{20}$$

This equation can also be simplified by multiplying through by the least common denominator of 120.

$$0 = 5(V_2 - V_1) + 4V_2 + 6(V_2 - V_3) \Rightarrow -5V_1 + 15V_2 - 6V_3 = 0$$

Similarly, the node-voltage equation for the third node is given by

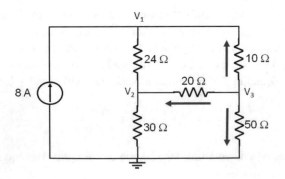

$$0 = \frac{V_3 - V_2}{20} + \frac{V_3 - V_1}{10} + \frac{V_3}{50}$$

The least common denominator for this equation is 100, so if we multiply by 100 we get

$$0 = 5(V_3 - V_2) + 10(V_3 - V_1) + 2V_3 \Rightarrow -10V_1 - 5V_2 + 17V_3 = 0$$

Therefore, the three equations with three unknowns that need to be solved for this circuit are

$$17V_1 - 5V_2 - 12V_3 = 960$$
$$-5V_1 + 15V_2 - 6V_3 = 0$$
$$-10V_1 - 5V_2 + 17V_3 = 0$$

This can be written in matrix form as

$$\begin{pmatrix} 17 & -5 & -12 \\ -5 & 15 & -6 \\ -10 & -5 & 17 \end{pmatrix} \begin{pmatrix} V_1 \\ V_2 \\ V_3 \end{pmatrix} = \begin{pmatrix} 960 \\ 0 \\ 0 \end{pmatrix}$$

Multiplying both sides of the equation by the inverse of the 3 × 3 matrix gives

$$\begin{pmatrix} V_1 \\ V_2 \\ V_3 \end{pmatrix} = \begin{pmatrix} 0.225 & 0.145 & 0.210 \\ 0.145 & 0.169 & 0.162 \\ 0.175 & 0.135 & 0.230 \end{pmatrix} \begin{pmatrix} 960 \\ 0 \\ 0 \end{pmatrix} = \begin{pmatrix} 216\ V \\ 139.2\ V \\ 168\ V \end{pmatrix}$$

Now that the voltages are known, we can find the current flow in the 20 Ω resistor. Since V_3 is greater than V_2, the current will flow from node 3 to node 2 and is given by

$$I_{20\Omega} = \frac{V_3 - V_2}{20\ \Omega} = \frac{168\ V - 139.2\ V}{20\ \Omega} = 1.44\ A$$

Example 3.8 Use the Node-Voltage method to find the voltages at each node in the circuit as well as the current flowing in the 1 Ω resistor (Fig. 3.12).

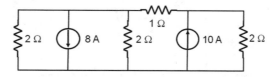

Fig. 3.12 Circuit diagram for Example 3.8

Solution:
Label the bottom node as the reference node as it has the most branches. Next, sketch arrows corresponding to the currents associated with each of the other two nodes. The currents associated with node 1 are shown in blue while the currents associated with node 2 are shown in red.

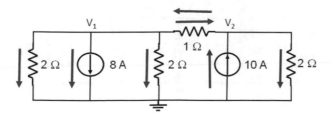

Therefore, the node-voltage equation for the currents associated with node 1 is given by

$$0 = \frac{V_1 - 0}{2} + 8 + \frac{V_1 - 0}{2} + \frac{V_1 - V_2}{1} \Rightarrow 2V_1 - V_2 = -8$$

where we have assumed all of the currents are leaving. Likewise, the node-voltage equation for the second node is given by

$$10 = \frac{V_2 - 0}{2} + \frac{V_2 - V_1}{1} \Rightarrow -2V_1 + 3V_2 = 20$$

These two equations can now be solved.

$$2V_1 - V_2 = -8$$
$$-2V_1 + 3V_2 + \; = 20$$
$$\left.\begin{array}{r} 2V_1 - V_2 = -8 \\ -2V_1 + 3V_2 = 20 \end{array}\right\} + \; \Rightarrow 2V_2 = 12 \Rightarrow V_2 = 6 \text{ V}$$
$$2V_1 - 6 = -8 \Rightarrow V_1 = -1 \text{ V}$$

Since V_2 is greater than V_1, the current will flow from V_2 into V_1 and is given by

$$\frac{V_2 - V_1}{1\Omega} = \frac{6V - -1V}{1\;\Omega} = 7 \text{ A}$$

Example 3.9 Use the Node-Voltage method to find the voltages at each node in the circuit as well as the labeled currents I_A and I_B (Fig. 3.13).

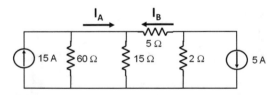

Fig. 3.13 Circuit diagram for Example 3.9

Solution:
The first step is to solve for the node-voltages, so we will focus on the nodes and not on the currents I_A and I_B. The bottom node has the most branches so it will be the reference node. The currents associated with node 1 are shown in blue while the currents associated with node 2 are shown in red.

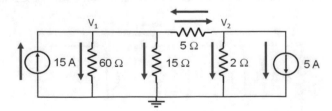

The node-voltage equation for the first node is given by

$$15 = \frac{V_1 - 0}{60} + \frac{V_1 - 0}{15} + \frac{V_1 - V_2}{5} \Rightarrow V_1 + 4V_1 + 12V_1 - 12V_2 = 900 \Rightarrow 17V_1 - 12V_2 = 900$$

while the node voltage equation for the second node is given by

$$0 = 5 + \frac{V_2 - 0}{2} + \frac{V_2 - V_1}{5} \Rightarrow 5V_2 + 2V_2 - 2V_1 = -50 \Rightarrow -2V_1 + 7V_2 = -50$$

Note: We could have kept the current through the 5 Ω resistor flowing in the same direction for both node-voltage equations and still have gotten the same equations as long as we were consistent with our current directions. This is tempting for this problem as I_B is already labeled. However, when obtaining the node-voltage equations, it becomes very tedious to attempt to label all of the node currents despite what you may have learned in your physics classes. Therefore, to discourage this practice, I did not explicitly include I_B in my node voltage equations in any form.

Once obtained, the node-voltage equations can be solved to give the node voltages.

$$\left.\begin{matrix} 17V_1 - 12V_2 = 900 \\ -2V_1 + 7V_2 = -50 \end{matrix}\right\} \Rightarrow \begin{pmatrix} 17 & -12 \\ -2 & 7 \end{pmatrix}\begin{pmatrix} V_1 \\ V_2 \end{pmatrix} = \begin{pmatrix} 900 \\ -50 \end{pmatrix} \Rightarrow \begin{pmatrix} V_1 \\ V_2 \end{pmatrix}$$

$$= \begin{pmatrix} 0.0737 & 0.1263 \\ 0.0211 & 0.1789 \end{pmatrix}\begin{pmatrix} 900 \\ -50 \end{pmatrix} = \begin{pmatrix} 60 \text{ V} \\ 10 \text{ V} \end{pmatrix}$$

Now that we know the node-voltages, we can solve for the currents. Current I_A is flowing along a wire that is part of node 1. Therefore, we cannot just find a voltage drop across a resistor to find I_A. Instead, we need to use Kirchhoff's Current Law again. Specifically, the current from the 15 A current source MUST be split into the current I_A and the current in the 60 Ω resistor. The current flowing from node 1 to ground through the 60 Ω resistor is given by $V_1/60$ Ω. Therefore, the Kirchhoff's current law equation needed to find I_A is given by

$$15 \text{ A} = \frac{V_1}{60 \ \Omega} + I_A \Rightarrow I_A = 15 \text{ A} - \frac{60 \text{ V}}{60 \ \Omega} = 14 \text{ A}$$

Current, I_B can be sound similar to the currents found in our previous examples; however, the current direction for I_B has already been given. Therefore, even though V_1 is greater than V_2, we need to use V_2 minus V_1 when solving for I_B to be consistent with the given current direction.

$$I_B = \frac{V_2 - V_1}{5 \ \Omega} = \frac{10 \text{ V} - 60 \text{ V}}{5 \ \Omega} = -10 \text{ A}$$

3.3.1 Node-Voltage Method for Circuits with Voltage Sources: The Super-Node

Up to this point, all of the circuits solved by the node-voltage method only had current sources. However, we also need to solve circuits with voltage sources. Voltage sources present a greater challenge because we cannot know the current flowing through a voltage source until we have finished solving the circuit. For current sources, the current is given, and for resistors, the current can be found in terms of the node voltages and the resistance. For voltage sources, however, we only know the voltage difference between the two nodes on either side of the voltage source. To solve this dilemma, we will combine the voltage source and its two terminal nodes into a single super-node and sum the currents associated with this super-node. We do not know how much current is flowing through the voltage source, but we do know that the current flowing in the source must both enter and exit from the nodes on either side of the voltage source. When writing the equation, we also include the voltage drop of the voltage source as is indicated by the following examples.

Example 3.10 Use the Node-Voltage method to find the voltages at each node in the circuit (Fig. 3.14).

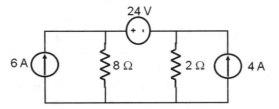

Fig. 3.14 Circuit diagram for Example 3.10

Solution:
The bottom node is the best choice for the reference node as it has the most branches. Due to the voltage source, there is only one node, a super node, in this circuit. If the voltage on the right side of the voltage source is V_1, then the voltage on the left side of the voltage source is $V_1 + 24$ V as shown in the diagram below.

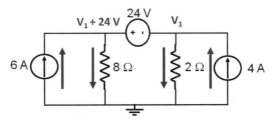

The node-voltage equation is then given by

$$0 = \overbrace{\frac{V_1}{2} - 4}^{\substack{\text{Right of}\\\text{Supernode}}} + \overbrace{\frac{V_1 + 24}{8} - 6}^{\substack{\text{Left of}\\\text{Supernode}}} \Rightarrow 6 + 4 = \frac{V_1}{2} + \frac{V_1 + 24}{8} \Rightarrow 4V_1 + V_1 + 24 = 80 \Rightarrow V_1 = 11.2 \text{ V}$$

Example 3.11 Use the Node-Voltage method to find the voltages, V_1, V_2, V_3, and V_4 at the indicated nodes in the circuit (Fig. 3.15).

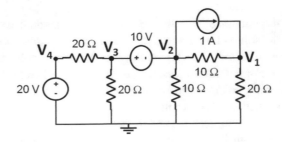

Fig. 3.15 Circuit diagram for Example 3.11

Solution:

This circuit is simpler than it first appears because of the voltage source between V_4 and ground. This means that the voltage at V_4 relative to ground is already known and is given by $V_4 = 20$ V. There is NO NEED to write an equation in terms of the branch currents to find this voltage. One of the most common mistakes students make is to try and write equations for currents at nodes with voltage sources connected to ground. This leads to an erroneous node-voltage equation that is always wrong. The super node due to the voltage source between V_2 and V_3 also reduces the number of equations needed as $V_3 = V_2 + 10$ V. Therefore, only two node-voltage equations are needed to solve this circuit.

Equation for Node V_1:

$$1 \text{ A} = \frac{V_1 - V_2}{10} + \frac{V_1}{20} \Rightarrow 3V_1 - 2V_2 = 20$$

Equation for Node V_2:

$$0 = \frac{V_2 + 10 - V_4}{20} + \frac{V_2 + 10 - 0}{20} + \frac{V_2}{10} + \frac{V_2 - V_1}{10} + 1 \text{ A}$$

$$-1 \text{ A} = \frac{V_2 - 10}{20} + \frac{V_2 + 10}{20} + \frac{V_2}{10} + \frac{V_2 - V_1}{10} \Rightarrow -2V_1 + 6V_2 = -20 \Rightarrow -V_1 + 3V_2 = -10 \Rightarrow -3V_1 + 9V_2 = -30$$

$$\Rightarrow 7V_2 = -10 \Rightarrow V_2 = -1.429 \text{ V}$$
$$\Rightarrow V_3 = V_2 + 10 \text{ V} = 8.571 \text{ V}$$
$$V_1 = 10 - 3V_2 = 5.714 \text{ V}$$

3.3.2 Node-Voltage Method for Circuits with Dependent Sources

Dependent sources are when a voltage or current in a circuit depend on a voltage or current elsewhere in the circuit as was explained in Chap. 1. Dependent sources are the basic model for amplifiers. When a circuit has dependent sources, the steps for finding the node-voltage equations remain the same. A reference node is first identified, and the voltages at the remaining nodes are labeled. The currents associated with each node are then summed to find the node-voltage equations. However, each dependent source introduces an additional equation that also must be solved when solving the system of equations. Once again, this is easiest to illustrate with an example.

Example 3.12 Use the Node-Voltage method to find the voltages at all of the nodes in the circuit shown in Fig. 3.16 as well as the current through the dependent source.

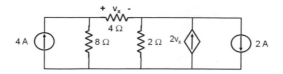

Fig. 3.16 Circuit diagram for Example 3.12

Solution:

The node with the most branches is the node at the bottom, so this will be our reference node. We have two nodes other than the reference node as well as one dependent source, so three equations will need to be generated and solved.

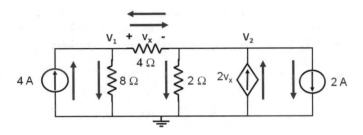

Node 1 Equation:

$$4 \text{ A} = \frac{V_1 - V_2}{4} + \frac{V_1}{8} \Rightarrow 3V_1 - 2V_2 = 32$$

Node 2 Equation:

$$2v_x = \frac{V_2 - V_1}{4} + \frac{V_2}{2} + 2 \text{ A} \Rightarrow 3V_2 - V_1 - 8v_x = -8$$

Dependent Source Equation:

$$v_x = V_1 - V_2$$

Now, substitute the dependent source equation into the Node 2 equation.

$$-9V_1 + 11V_2 = -8$$

Now, combine with the node 1 equation to solve the node voltages.

$$3 \times (3V_1 - 2V_2 = 32) + (-9V_1 + 11V_2 = -8) \Rightarrow 5V_2 = 88 \Rightarrow V_2 = 17.6 \text{ V}$$
$$V_1 = \frac{32 + 2V_2}{3} = 22.4 \text{ V}$$

Now that we know the node voltages, the current in the dependent source is given by

$$2v_x = 2(V_1 - V_2) = 2 \cdot 4.8 = 9.6 \text{ A}$$

Example 3.13 Use the Node-Voltage method to find the voltages at all of the nodes in the circuit as well as the current through the dependent source (Fig. 3.17).

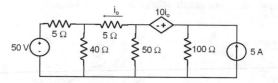

Fig. 3.17 Circuit diagram for Example 3.13

Solution:

The bottom node is once again the best choice for the ground node as it has the most branches connected. Also, the independent 50 V voltage source gives us the voltage at one of the nodes while the dependent voltage source acts as a super node. Therefore, we will have 2 node-voltage equations in addition to the dependent source equation.

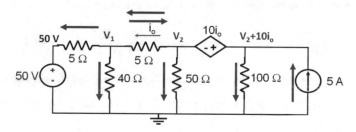

The equations are given by

$$0 = \frac{V_1 - 50}{5} + \frac{V_1 - V_2}{5} + \frac{V_1}{40} \Rightarrow 17V_1 - 8V_2 = 400$$

$$5 \text{ A} = \frac{V_2 - V_1}{5} + \frac{V_2}{50} + \frac{V_2 + 10i_o}{100} \Rightarrow -20V_1 + 23V_2 + 10i_o = 500$$

$$i_o = \frac{V_2 - V_1}{5} \Rightarrow V_1 - V_2 + 5i_o = 0$$

We can solve these equations using matrices

$$\left.\begin{array}{r} 17V_1 - 8V_2 = 400 \\ -20V_1 + 23V_2 + 10i_o = 500 \\ V_1 - V_2 + 5i_o = 0 \end{array}\right\} \Rightarrow \begin{pmatrix} 17 & -8 & 0 \\ -20 & 23 & 10 \\ 1 & -1 & 5 \end{pmatrix} \begin{pmatrix} V_1 \\ V_2 \\ i_o \end{pmatrix} = \begin{pmatrix} 400 \\ 500 \\ 0 \end{pmatrix}$$

$$\Rightarrow \begin{pmatrix} V_1 \\ V_2 \\ i_o \end{pmatrix} = \begin{pmatrix} 0.1004 & 0.0321 & -0.0643 \\ 0.0884 & 0.0683 & -0.1365 \\ -0.0024 & 0.0072 & 0.1855 \end{pmatrix} \begin{pmatrix} 400 \\ 500 \\ 0 \end{pmatrix} = \begin{pmatrix} 56.2249 \text{ V} \\ 69.4779 \text{ V} \\ 2.6506 \text{ A} \end{pmatrix}$$

Now, we need to use Kirchhoff's Current Law to find the current flowing in the dependent source. Since no direction is specified in the problem statement, we will find the current flowing out of the positive voltage terminal of the dependent source as shown below.

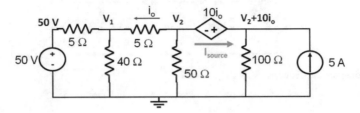

If we sum up the currents at the node to the left of the dependent source, we have

$$0 = i_o + \frac{V_2}{50} + I_{source} \Rightarrow I_{source} = -i_o - \frac{V_2}{50} = -4.04 \text{ A}$$

If instead we use the node to the right of the dependent source, we have

$$\frac{V_2 + 10i_o}{100} = I_{source} + 5 \text{ A} \Rightarrow I_{source} = \frac{V_2 + 10i_o}{100} - 5 = -4.04 \text{ A}$$

Therefore, Kirchhoff's Current Law can be used at either node to find the dependent source current.

Example 3.14 Solve for V_o in the following circuit (Fig. 3.18) using the node-voltage method.

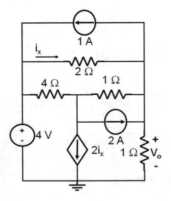

Fig. 3.18 Circuit diagram for Example 3.14

Solution:

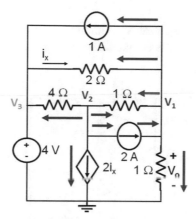

This circuit has two nodes needing node-voltage equations as the voltage at V_3 is already known (i.e., $V_3 = 4$ V). In addition, there is one dependent source equation. These equations are given by

$$i_x = \frac{4 - V_1}{2} \Rightarrow V_1 + 2i_x = 4$$

$$1\,\text{A} + \frac{V_1 - 4}{2} + \frac{V_1 - V_2}{1} + \frac{V_1}{1} = 2\,\text{A} \Rightarrow 5V_1 - 2V_2 = 6$$

$$\frac{V_2 - V_1}{1} + 2\,\text{A} + 2i_x + \frac{V_2 - 4}{4} = 0 \Rightarrow -4V_1 + 5V_2 + 8i_x = -4$$

$$\begin{pmatrix} 1 & 0 & 2 \\ 5 & -2 & 0 \\ -4 & 5 & 8 \end{pmatrix} \begin{pmatrix} V_1 \\ V_2 \\ i_x \end{pmatrix} = \begin{pmatrix} 4 \\ 5 \\ -4 \end{pmatrix} \Rightarrow \begin{pmatrix} V_1 \\ V_2 \\ i_x \end{pmatrix} = \begin{pmatrix} -0.8889 & 0.5556 & 0.2222 \\ -2.2222 & 0.8889 & 0.5556 \\ 0.9444 & -0.2778 & -0.1111 \end{pmatrix} \begin{pmatrix} 4 \\ 6 \\ -4 \end{pmatrix} = \begin{pmatrix} -1.1111\ \text{V} \\ -5.7778\ \text{V} \\ 2.5556\ \text{A} \end{pmatrix}$$

3.4   Mesh-Current Method of Circuit Analysis

The second most common circuit analysis technique is the mesh-current method. This method is based on Kirchhoff's Voltage Law that the sum of all voltages around any mesh or loop in a circuit must be zero. This means that if we start and end at the same point, the total voltage change must be zero. The voltage drops as we traverse the loops are given by Ohm's law and expressed in terms of mesh currents. These loop/mesh currents are normally different from the branch currents as each branch can have more than one mesh current passing through it.

When solving a circuit using mesh-current, the first step is to determine the number of meshes in the circuit and assign a unique mesh current to each mesh. The direction of the mesh current MUST also be specified as the assumed current direction will govern the polarity of the voltage change as we traverse the mesh. Next, you traverse the mesh in the direction of the mesh current and apply Ohm's law to determine the voltage drop across each circuit element. Since a circuit element may have more than one mesh current flowing through it, the total current is first determined by summing the mesh currents prior to calculating the voltage drop. Since Kirchoff's Voltage Law states that the sum of the voltages around the mesh must add up to zero, traversing the entire mesh and setting the sum of the voltages to zero will provide an equation for each mesh in the circuit. Once all of the meshes are traversed, the system of equations can be solved to find the mesh currents. These steps are illustrated by the following examples.

Example 3.15 Use the mesh-current method to find the currents flowing out of the positive voltage terminal of each voltage source for the circuit shown in Fig. 3.19.

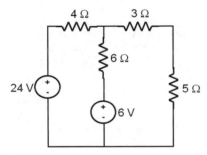

Fig. 3.19 Circuit diagram for Example 3.15

Solution:

This circuit has two meshes. Therefore, we will need two mesh-current equations to solve the circuit. We will assign two mesh currents with the assumed direction of the current flow as shown. The real current flow may be in the opposite direction. If this occurs, then our solved mesh-currents will be negative.

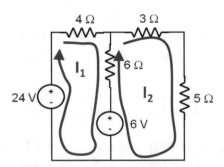

As we traverse the I_1 mesh starting in the upper left hand corner, the first element we encounter is the 4 Ω resistor. The voltage drop across this resistor according to Ohm's law is $(4\ \Omega) \cdot I_1$. Continuing around the loop, the next element is the 6 Ω resistor. However, this element has both mesh current I_1 and mesh current I_2 flowing through it. Therefore, the total current is $I_1 - I_2$ giving a voltage drop of $(6\ \Omega) \cdot (I_1 - I_2)$. Next, we encounter the 6 V source. Since we are entering the positive terminal of the voltage source as we traverse the loop, the voltage drop is positive and is given by 6 V. Lastly, we cross the 24 V source. This time we are entering the negative terminal of the voltage source indicating the voltage is actually increasing, so the voltage drop is given by –24 V. Summing up all of these voltage drops gives us the mesh-current equation

$$4I_1 + 6(I_1 - I_2) + 6 - 24 = 0 \Rightarrow 10I_1 - 6I_2 = 18 \Rightarrow 5I_1 - 3I_2 = 9$$

For the I_2 mesh, we have a 3 Ω resistor, a 5 Ω resistor, a 6 V source and a 6 Ω resistor as we traverse the loop. The voltage drop across each of these is given by $(3\ \Omega) \cdot I_2$, $(5\ \Omega) \cdot I_2$, –6 V, and $(6\ \Omega) \cdot (I_2 - I_1)$, respectively. The currents in the 6 Ω resistor sum as $(I_2 - I_1)$ this time as we are finding the voltage drops from the perspective of the I_2 mesh current. Therefore, the mesh current equation is given by

$$3I_2 + 5I_2 - 6 + 6(I_2 - I_1) = 0 \Rightarrow -6I_1 + 14I_2 = 6 \Rightarrow -3I_1 + 7I_2 = 3$$

We can now solve these two equations with their two unknowns.

$$\left.\begin{array}{c} (5I_1 - 3I_2 = 9) \times 3 \\ (-3I_1 + 7I_2 = 3) \times 5 \end{array}\right\} \Rightarrow -9I_2 + 35I_2 = 27 + 15 \Rightarrow 26I_2 = 42 \Rightarrow I_2 = 1.6154 \text{ A}$$

$$I_1 = \frac{9 + 3I_2}{5} = 2.7692 \text{ A}$$

Now that we know the mesh currents, we can find the currents flowing out of each of the voltage sources. For the 24 V voltage source, the current flowing out of the positive voltage terminal is $I_1 = 2.7692$ A. For the 6 V voltage source, the current out of the positive voltage terminal is given by $(I_2 - I_1) = -1.1538$ V.

3.4.1 Mesh-Current Method for Circuits with Current Sources

Current sources will reduce the number of equations that need to be solved when performing mesh-current analysis, but the generation of the equations prior to solving can be more difficult. We don't know the voltage drop across a current source until the circuit has been solved, so we cannot include the voltage drop across a current source in our equations. However, we do not need to write a mesh-current equation if the mesh-current can be directly determined from the current source which is the case if only one mesh current is passing through each current source. However, we will need to be creative in drawing our current loops so that every branch has at least one current, AND only one current passes through each current source as will be illustrated in the following examples.

Example 3.16 Use the mesh-current method to find the currents flowing out of the positive voltage terminal of the voltage source (Fig. 3.20).

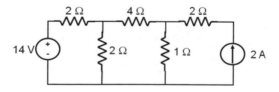

Fig. 3.20 Circuit diagram for Example 3.16

Solution:

The circuit has three meshes. Therefore, we need three mesh-current equations. We also want only one mesh-current through the branch with the current source. Therefore, we will define the following mesh currents.

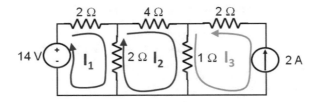

Since I_3 is the only mesh current through the 2 A source and I_3 is in the same direction as the 2 A source, $I_3 = 2$ A. The remaining two equations can now be generated by summing the voltages around the remaining two loops.

I_1 Loop Equation:

$$2I_1 + 2(I_1 - I_2) - 14 = 0 \Rightarrow 4I_1 - 2I_2 = 14 \Rightarrow 2I_1 - I_2 = 7$$

I_2 Loop Equation:

$$4I_2 + 1(I_2 + I_3) + 2(I_2 - I_1) = 0 \Rightarrow -2I_1 + 7I_2 = -I_3 \Rightarrow 2I_1 - 7I_2 = 2$$

I_2 and I_3 add in this equation as they are flowing in the same direction as they pass through the 1 Ω resistor. Solving these equations gives

$$(2I_1 - I_2 = 7) - (2I_1 - 7I_2 = 2) \Rightarrow 6I_2 = 5 \Rightarrow I_2 = 0.83333 \text{ A}$$

$$I_1 = \frac{7 + I_2}{2} = 3.9167 \text{ A}$$

The current out of the positive voltage terminal is I_1.

Example 3.17 Use the mesh-current method to find the mesh currents and then solve for V_o, the voltage across the 4 Ω resistor for Fig. 3.21.

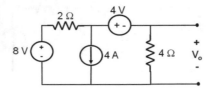

Fig. 3.21 Circuit diagram for Example 3.17

Solution:

This circuit has two meshes and therefore two mesh-current equations are needed to solve the circuit. If we draw mesh current I_1 as shown, then in order to still have a loop current through every branch and NOT have two loop currents through the current source, then the second loop current will need to skirt around the outside loop as shown below.

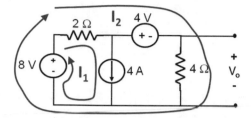

The first mesh-current equation is then given by $I_1 = 4$ A while the second equation is given by

$$2(I_2 + I_1) + 4 + 4I_2 - 8 = 0 \Rightarrow 2I_1 + 6I_2 = 4 \Rightarrow I_2 = \frac{2 - I_1}{3} = -0.6667 \text{ A}$$

The voltage, V_o, is given by

$$V_o = 4I_2 = -2.6667 \text{ A}$$

Example 3.18 Use the Mesh-Current method to find the mesh currents for Fig. 3.22.

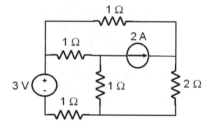

Fig. 3.22 Circuit diagram for Example 3.18

Solution:
This circuit has three meshes, so three equations are needed. If we define, I_1 and I_2 as shown, then the third loop-current will need to pass through the 1 Ω resistor without passing through the current source. Therefore, there are two possible options for I_3. Both options are shown below.
Option 1 Solution:

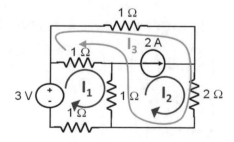

I_1 Loop Equation:

$$1(I_1 - I_3) + 1(I_1 - I_2 - I_3) + I_1 - 3 = 0 \Rightarrow 3I_1 - I_2 - 2I_3 = 3$$

I_2 Loop Equation:

$$I_2 = 2 \text{ A}$$

I_3 Loop Equation:

$$I_3 + 2(I_3 + I_2) + 1(I_3 + I_2 - I_1) + 1(I_3 - I_1) = 0 \Rightarrow -2I_1 + 3I_2 + 5I_3 = 0$$

$$\begin{pmatrix} 3 & -1 & -2 \\ 0 & 1 & 0 \\ -2 & 3 & 5 \end{pmatrix} \begin{pmatrix} I_1 \\ I_2 \\ I_3 \end{pmatrix} = \begin{pmatrix} 3 \\ 2 \\ 0 \end{pmatrix} \Rightarrow \begin{pmatrix} I_1 \\ I_2 \\ I_3 \end{pmatrix} = \begin{pmatrix} 0.4545 & -0.0909 & 0.1818 \\ 0 & 1 & 0 \\ 0.1818 & -0.6364 & 0.2727 \end{pmatrix} \begin{pmatrix} 3 \\ 2 \\ 0 \end{pmatrix}$$

$$= \begin{pmatrix} 1.1818 \text{ A} \\ 2 \text{ A} \\ -0.7273 \text{ A} \end{pmatrix}$$

Option 2 Solution:

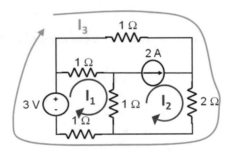

I_1 Loop Equation:

$$1I_1 + 1(I_1 - I_2) + (I_1 + I_3) - 3 = 0 \Rightarrow 3I_1 - I_2 + I_3 = 3$$

I_2 Loop Equation:

$$I_2 = 2 \text{ A}$$

I_3 Loop Equation:

$$1I_3 + 2(I_3 + I_2) + 1(I_3 + I_1) - 3 = 0 \Rightarrow I_1 + 2I_2 + 4I_3 = 3$$

$$\begin{pmatrix} 3 & -1 & 1 \\ 0 & 1 & 0 \\ 1 & 2 & 4 \end{pmatrix} \begin{pmatrix} I_1 \\ I_2 \\ I_3 \end{pmatrix} = \begin{pmatrix} 3 \\ 2 \\ 3 \end{pmatrix} \Rightarrow \begin{pmatrix} I_1 \\ I_2 \\ I_3 \end{pmatrix} = \begin{pmatrix} 0.3636 & 0.5455 & -0.0909 \\ 0 & 1 & 0 \\ -0.0909 & -0.6364 & 0.2727 \end{pmatrix} \begin{pmatrix} 3 \\ 2 \\ 3 \end{pmatrix}$$

$$= \begin{pmatrix} 1.9091 \text{ A} \\ 2 \text{ A} \\ -0.7273 \text{ A} \end{pmatrix}$$

At first, it may seem as if the two different approaches do not agree as the mesh current I_1 is different. However, the mesh currents are only mathematical tools to solve the circuit. To do a real comparison, the actual currents flowing in the branches needs to be calculated. For example, if we calculate the branch current flowing out of the 3 V voltage source in option 1, we have that the branch current is $I_1 = 1.1818$ A. For option 2, however, the branch current is given by $I_1 + I_3 = 1.1818$ A. Therefore, both solutions agree.

3.4.2 Mesh-Current Method for Circuits with Dependent Sources

Dependent sources function the same way for both the node-voltage and the mesh-current method. Namely, the dependent source just introduces an additional equation into the system of equations that would need to be solved concurrently as is illustrated in the following examples.

Example 3.19 Use the mesh-current method to find the mesh currents in Fig. 3.23.

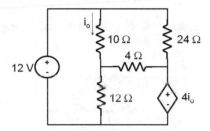

Fig. 3.23 Circuit diagram for Example 3.19

Solution:
The circuit has three meshes and one dependent source, so four equations will be needed to solve the circuit. The circuit has no current sources, so the mesh-currents can just be defined as

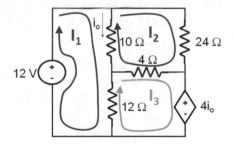

Dependent Source Equation:

$$i_o = (I_1 - I_2)$$

I_1 Loop Equation:

$$10(I_1 - I_2) + 12(I_1 - I_3) - 12 = 0 \Rightarrow 22I_1 - 10I_2 - 12I_3 = 12 \Rightarrow 11I_1 - 5I_2 - 6I_3 = 6$$

I_2 Loop Equation:

$$24I_2 + 4(I_2 - I_3) + 10(I_2 - I_1) = 0 \Rightarrow -10I_1 + 38I_2 - 4I_3 = 0 \Rightarrow -5I_1 + 19I_2 - 2I_3 = 0$$

I_3 Loop Equation:

$$4(I_3 - I_2) + 4i_o + 12(I_3 - I_1) = 0 \Rightarrow -12I_1 - 4I_2 + 16I_3 + 4i_o = 0$$
$$\Rightarrow -8I_1 - 8I_2 + 16I_3 = 0 \Rightarrow I_1 + I_2 - 2I_3 = 0$$

$$\begin{pmatrix} 11 & -5 & -6 \\ -5 & 19 & -2 \\ 1 & 1 & -2 \end{pmatrix} \begin{pmatrix} I_1 \\ I_2 \\ I_3 \end{pmatrix} = \begin{pmatrix} 6 \\ 0 \\ 0 \end{pmatrix} \Rightarrow \begin{pmatrix} I_1 \\ I_2 \\ I_3 \end{pmatrix} = \begin{pmatrix} 0.1875 & 0.0833 & -0.6458 \\ 0.0625 & 0.0833 & -0.2708 \\ 0.1250 & 0.0833 & -0.9583 \end{pmatrix} \begin{pmatrix} 6 \\ 0 \\ 0 \end{pmatrix}$$

$$= \begin{pmatrix} 1.125 & A \\ 0.375 & A \\ 0.75 & A \end{pmatrix}$$

Example 3.20 Use the mesh-current method to find the mesh currents in Fig. 3.24.

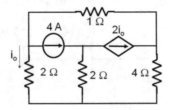

Fig. 3.24 Circuit diagram for Example 3.20

Solution:

The circuit has three meshes and one dependent source, so four equations are needed to solve the circuit. If we assign currents so that only one loop current goes through each current source, the only possible configuration is given by

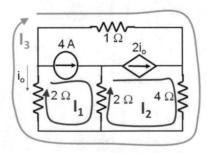

Dependent Source Equation:

$$i_o = -(I_1 + I_3)$$

I_1 Loop Equation:

$$I_1 = 4\ A$$

I_2 Loop Equation:

$$I_2 = 2i_o = -2(I_1 + I_3) = -8 - 2I_3$$

I_3 Loop Equation:

$$1(I_3) + 4(I_3 + I_2) + 2(I_3 + I_1) = 0 \Rightarrow 7I_3 + 4I_2 + 2I_1 = 0$$
$$7I_3 + 4(-8 - 2I_3) + 2 \cdot 4 = 0$$
$$-I_3 + -24 = 0 \Rightarrow I_3 = -24 \text{ A}.$$

3.5 Circuit Simplification and Source Transformations

Frequently when analyzing circuits, knowing the currents and voltages everywhere is not as critical as knowing a subset of currents and voltages at specific locations in the circuit. Therefore, one attractive approach is to simplify the circuit so that it has fewer nodes and/or branches while keeping the critical nodes/branches unaltered. Reducing the number of nodes/branches will reduce the number of node-voltage or mesh-current equations that need to be solved. Source transformations are one tool that is used to simplify many circuits when coupled with finding equivalent resistances. The goal is to replace independent voltage sources with equivalent independent current sources and independent current sources with equivalent independent voltage sources. In addition, voltage sources in series are combined into equivalent single voltage sources and current sources in parallel are combined into single equivalent current sources. Along the way, resistors in series and/or parallel are also simplified and Delta/Y transformations are performed as needed.

3.5.1 Combining Voltage Sources in Series and Current Sources in Parallel

When two voltage sources are in series, as shown in Fig. 3.25, we can use Kirchhoff's Voltage Law to find the equivalent voltage source V_o. Summing the voltages around the loop of the top circuit gives us $-V_2 - V_1 + V_o = 0$. Therefore, the voltage V_o must be given by

$$V_o = V_1 + V_2 \tag{3.5}$$

Likewise, if we have two current sources in parallel, as shown in Fig. 3.26, we can use Kirchhoff's current law to find the equivalent current source I_o. Summing the currents at the top node gives

$$I_o = I_1 + I_2 \tag{3.6}$$

if the two circuits inside the blue boxes are to be equivalent.

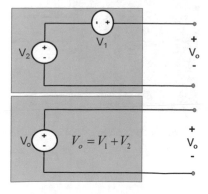

Fig. 3.25 Combining voltage sources in series

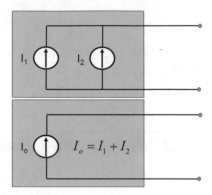

Fig. 3.26 Combining current sources in parallel

3.5.2 Transforming Independent Current/Voltage Sources

In addition to combining voltage and current sources, a voltage source with a resistor in series is equivalent to a current source with a resistor in parallel if one is only interested in the currents/voltages away from the source. In order to be equivalent, the current and voltage at the terminals must be the same as shown in Fig. 3.27. If nothing is connected to the output terminals, then there will be no current flow in the top circuit. Without current flow, there can be no voltage drop across the resistor and the voltage across the terminals must be given by $V_o = V_S$. Likewise, for the bottom circuit, all of the current flow must be through the resistor because there is no path for the current through the output terminals. Therefore by Ohm's Law, the voltage across the terminals must be given by $V_o = I_S R_S$. Equating the two voltages means that

$$I_S = \frac{V_S}{R_S} \tag{3.7}$$

in order for the two circuits inside of the blue boxes to be equivalent. We can now use these source transformations to simplify circuits.

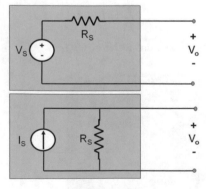

Fig. 3.27 Independent voltage/current source equivalents

Example 3.21 Find the voltage V_o in Fig. 3.28 by applying source transformations to simplify the circuit.

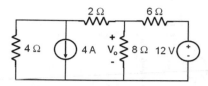

Fig. 3.28 Circuit diagram for Example 3.21

Solution:

We can first replace the current source with a voltage source,

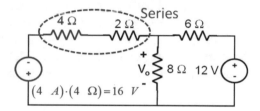

After combining the resistors in series, we can then replace the two voltage sources with current sources.

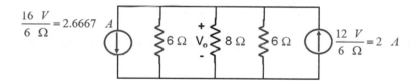

Notice that the five remaining circuit elements are all in parallel as they all share both nodes. Therefore, the current sources can be combined into a single new current source. Also, we can combine the two 6 Ω resistors into a new equivalent resistance. However, we DO NOT want to include the 8 Ω resistor as we need to find the voltage V_o. If we replace the 8 Ω resistance with some equivalent, the voltage we care about will be buried in the equivalence and we will not be able to solve for it. The new equivalent circuit is given by

We can now write a very simple node-voltage equation to solve for V_o assuming the bottom node is our ground node.

$$\frac{V_o}{8} + \frac{V_o}{3} + 0.6667 = 0 \Rightarrow V_o(0.45833) = -0.6667 \Rightarrow V_o = -1.4545 \text{ V}.$$

Example 3.22 Find the current flowing out of the 10 V source, I_o, in Fig. 3.29 by applying source transformations to simplify the circuit.

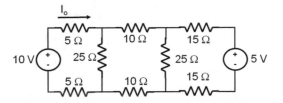

Fig. 3.29   Circuit diagram for Example 3.22

Solution:

Since we are finding the current associated with the 10 V source, we cannot include this source in any of our source transformations. Instead, we will focus on the 5 V source. The branch with the 5 V source consists of two 15 Ω resistors in series. However, if we are ONLY interested in the voltages outside of the branch, then the order of the elements in series along the branch does not matter. The total voltage drop will remain the same. Therefore, we can redraw the circuit as

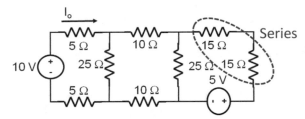

The 15 Ω resistors can now be combined into a 30 Ω resistor. We can also transform the 5 V voltage source into a current source.

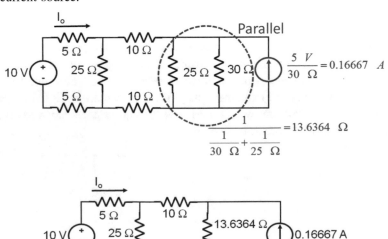

Now apply another source transformation on the current source.

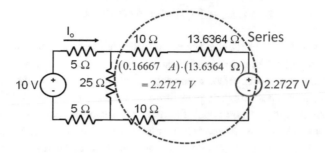

Combining the three resistors in series gives

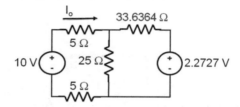

Applying another source transformation gives

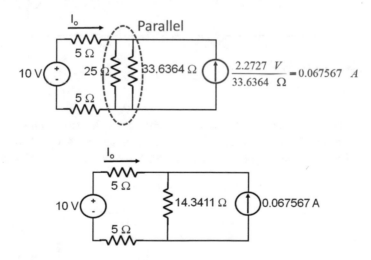

Applying one more source transformation gives

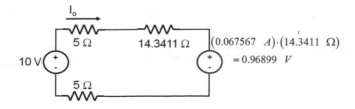

We can now write a single mesh current equation to solve for I_o.

$$5I_o + 14.3411I_o + 0.96899 + 5I_o - 10 = 0$$
$$24.3411I_o = 9.031 \Rightarrow I_o = 0.371 \text{ A}$$

When solving circuits, it is rare to reduce a circuit to a single mesh or a pair of nodes prior to solving the circuit. Usually, repetitive source transformations take longer than just solving a set of node-voltage or mesh-current equations. Also, when dependent sources are present, the circuit cannot be simplified down to a single mesh or node pair. Source transformations DO NOT apply to dependent sources, and the current/voltage upon which the dependent source depends CANNOT be buried by source transformations. Therefore, source transformations are usually only used to remove some of the loops/meshes prior to solving the circuit as is illustrated in the next example.

Example 3.23 Find the current I_A by applying source transformations to simplify the circuit in Fig. 3.30 prior to using the node-voltage method to solve the circuit.

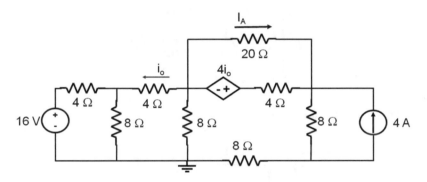

Fig. 3.30 Circuit diagram for Example 3.23

Solution:
This circuit has 4 nodes/super-nodes and 4 meshes where the mesh-current would not be known as well as a dependent source. Therefore, 5 equations would need to be found and solved if we did not simplify the circuit first. However, applying a source transformation to the 16 V source and the 4 A source gives us

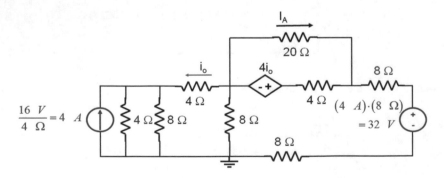

Combining the resistors in series and in parallel gives

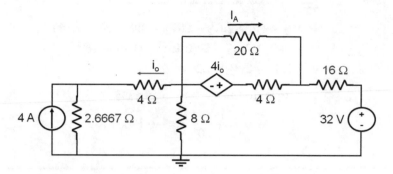

Doing one more source transformation gives

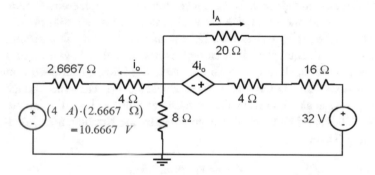

This circuit can be solved with only two node-voltage equations and the dependent source equation.

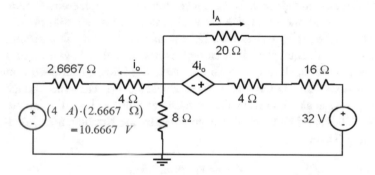

$$i_o = \frac{V_1 - 10.6667}{6.6667} = 0.15V_1 - 1.6$$

$$\frac{V_1 - 10.6667}{6.6667} + \frac{V_1}{8} + \frac{V_1 - V_2}{20} + \frac{V_1 + 4i_o - V_2}{4} = 0$$

$$\Rightarrow 6V_1 + 5V_1 + 2V_1 - 2V_2 + 10V_1 + 40i_o - 10V_2 = 64$$

$$\Rightarrow 23V_1 + 40(0.15V_1 - 1.6) - 12V_2 = 64 \Rightarrow 29V_1 - 12V_2 = 128$$

$$\frac{V_2 - V_1}{20} + \frac{V_2 - (V_1 + 4i_o)}{4} + \frac{V_2 - 32}{16} = 0$$
$$\Rightarrow 4V_2 - 4V_1 + 20V_2 - 20V_1 - 80i_o + 5V_2 = 160$$
$$\Rightarrow -24V_1 + 29V_2 - 80(0.15V_1 - 1.6) = 160$$
$$\Rightarrow -36V_1 + 29V_2 = 32$$

$$V_1 = 10.0147 \ V \quad V_2 = 13.5355 \ V$$
$$I_A = \frac{V_1 - V_2}{20} = -0.17604 \ A$$

3.6   Thevenin and Norton Equivalent Circuits

The most useful circuit simplification is the conversion of a complex circuit into its Thevenin or Norton equivalent. This allows for a quick calculation of currents and voltages when varying loads are connected. The Thevenin Equivalent for a DC circuit consists of a voltage source in series with a resistor while the Norton Equivalent is a current source in parallel with a resistor as shown in Fig. 3.31. If the output terminals are left open (i.e., no current flow), then there would be no voltage drop across the resistor and the terminal voltage would be the Thevenin voltage. Therefore, the Thevenin voltage, V_{Th}, is also known as the open-circuit voltage, V_{OC}. For the Norton Equivalent, open terminals would result in a terminal voltage given by the Norton current, I_N, multiplied by the Thevenin resistance. Therefore,

$$V_{OC} = V_{Th} = I_N \cdot R_{Th} \tag{3.8}$$

Likewise, if the terminals are shorted so that all of the current would flow out of the terminals, then the current leaving the terminals would be the Norton current. Therefore, the Norton current, I_N, is also known as the short-circuit current, I_{SC}. For the Thevenin equivalent model, the Thevenin voltage divided by the Thevenin resistance would also give the short circuit. Therefore,

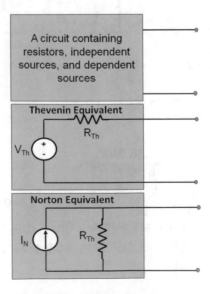

Fig. 3.31 Thevenin and Norton Equivalent circuit models

$$I_{SC} = I_N = \frac{V_{Th}}{R_{Th}} \tag{3.9}$$

Hence, the open-circuit voltage and short-circuit current can be used to find either the Thevenin Equivalent or the Norton Equivalent circuit, and a simple source transformation can be done to translate between the two models.

Example 3.24 Find Thevenin and Norton equivalent for terminals a and b for the following circuit (Fig. 3.32) by first finding the open-circuit voltage and the short-circuit current.

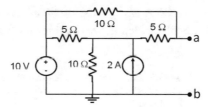

Fig. 3.32 Circuit diagram for Example 3.24

Solution:

First, find the open-circuit voltage using the node-voltage method.

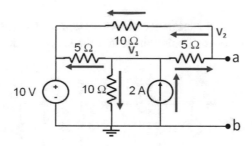

With this labeling of the nodes, the open-circuit voltage is given by V_2. The node-voltage equations for the circuit can now be written as

$$\frac{V_1 - 10}{5} + \frac{V_1}{10} + \frac{V_1 - V_2}{5} = 2 \Rightarrow 5V_1 - 2V_2 = 40 \Rightarrow V_1 = \frac{40 + 2V_2}{5}$$

$$\frac{V_2 - V_1}{5} + \frac{V_2 - 10}{10} = 0 \Rightarrow -2V_1 + 3V_2 = 10 \Rightarrow$$

$$-2\left(\frac{40 + 2V_2}{5}\right) + 3V_2 = 10 \Rightarrow 11V_2 = 130 \Rightarrow V_2 = V_{OC} = 11.8182 \text{ V}$$

Now, find the short-circuit current. To do this, we need to redraw the circuit with the short in place.

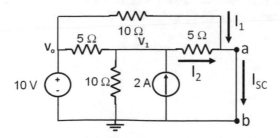

From Kirchhoff's current law, we know that $I_{SC} = I_1 + I_2$. We can also find I_1 directly from Ohm's Law because the voltage V_o is 10 V while point a is connected to the ground node at 0 V. Therefore, $I_1 = \frac{10 \text{ V}}{10 \text{ } \Omega} = 1$ A. To find the current I_2, however, we need to find V_1. V_1 is NOT the same as the open-circuit case because the circuit wiring has changed with the addition of the short circuit. The node-voltage equation for V_1 is now given by

$$\frac{V_1 - 10}{5} + \frac{V_1}{10} + \frac{V_1}{5} = 2 \Rightarrow 5V_1 = 40 \Rightarrow V_1 = 8 \text{ V}$$

Therefore, I_2 is given by $I_2 = \frac{8 \text{ V}}{5 \text{ } \Omega} = 1.6$ A giving a short-circuit current of $I_{SC} = 2.6$ A and a Thevenin resistance of $R_{Th} = \frac{V_{OC}}{I_{SC}} = 4.5455 \text{ } \Omega$. Hence, we can draw the Thevenin and Norton equivalents as

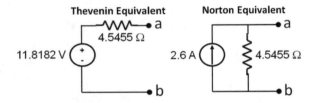

Example 3.25 Find Thevenin and Norton equivalent for terminals a and b for the following circuit (Fig. 3.33) by first finding the open-circuit voltage and the short-circuit current.

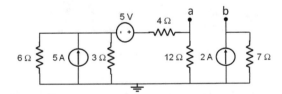

Fig. 3.33 Circuit diagram for Example 3.25

Solution:

First, find the open-circuit voltage using the node-voltage method. Notice that in this case, the node connected to terminal b is separate from the rest of the circuit and is given by the 2 A current flowing in the 7 Ω resistor

$$2 \text{ A} = \frac{V_b}{7} \Rightarrow V_b = 14 \text{ V}$$

Also, if we know the voltage of the super-node containing the 5 V source, then the voltage at terminal a can just be found by the voltage divider equation. Therefore, we only need to write one node-voltage equation.

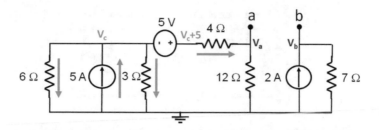

$$5\ A = \frac{V_c}{6} + \frac{V_c}{3} + \frac{V_c + 5}{16} \Rightarrow 240 = 8V_c + 16V_c + 3V_c + 15 \Rightarrow 27V_c = 225 \Rightarrow V_c = 8.3333\ \text{V}$$

Applying the voltage divider equation for terminal a then gives

$$V_a = (8.3333\ \text{V} + 5\ \text{V})\frac{12}{4 + 12} = 10\ \text{V}$$

Therefore, the open-circuit voltage is given by

$$V_{OC} = V_a - V_b = -4\ \text{V}$$

Now, find the short-circuit current as labeled below. We will once again use the node-voltage method to find the voltages and then find the current from the voltages.

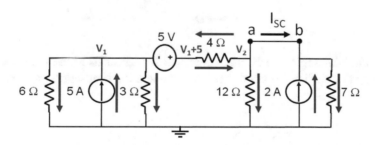

$$5\ \text{A} = \frac{V_1}{6} + \frac{V_1}{3} + \frac{V_1 + 5 - V_2}{4} \Rightarrow 9V_1 - 3V_2 = 45 \Rightarrow 3V_1 - V_2 = 15 \Rightarrow V_2 = 3V_1 - 15$$

$$2\ \text{A} = \frac{V_2}{7} + \frac{V_2}{12} + \frac{V_2 - V_1 - 5}{4} \Rightarrow -21V_1 + 40V_2 = 273$$

$$\Rightarrow -21V_1 + 40(3V_1 - 15) = 273 \Rightarrow 99V_1 = 873 \Rightarrow V_1 = 8.8182\ \text{V}$$

$$V_2 = 3V_1 - 15 = 11.4545\ \text{V}$$

Summing the currents at terminal b gives

$$I_{SC} + 2\ \text{A} = \frac{V_2}{7\ \Omega} \Rightarrow I_{SC} = -0.3636\ \text{A}$$

$$R_{Th} = \frac{V_{OC}}{I_{SC}} = 11\ \Omega$$

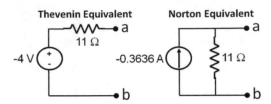

However, the negative sign in the source terms just means that the polarity of the voltage and direction of the current were just opposite from what was assumed. We can always change the assumed polarity/direction to give

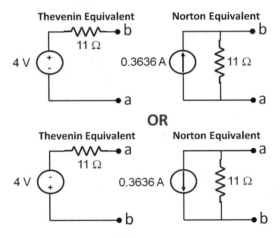

Example 3.26 Find Thevenin and Norton equivalent for terminals a and b for the following circuit (Fig. 3.34) by first finding the open-circuit voltage and the short-circuit current.

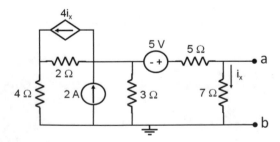

Fig. 3.34 Circuit diagram for Example 3.26

Solution:
Once again, find the open-circuit voltage using the node-voltage method. Since the voltage drop across the 7 Ω resistor can be found once i_x is known, two node-voltage equations and one dependent source equation are needed to solve the circuit.

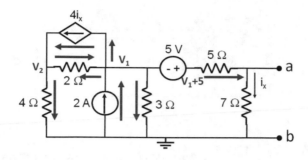

$$\frac{V_1 - V_2}{2} + 4i_x + \frac{V_1}{3} + \frac{V_1+5}{12} - 2 \wedge \Rightarrow 6V_1 - 6V_2 + 48i_x + 4V_1 \mid V_1 + 5 = 24$$

$$\Rightarrow 11V_1 - 6V_2 + 48i_x = 19$$

$$\frac{V_2 - V_1}{2} - 4i_x + \frac{V_2}{4} = 0 \Rightarrow -2V_1 + 3V_2 - 16i_x = 0$$

$$i_x = \frac{V_1+5}{12} \Rightarrow V_1 - 12i_x = -5$$

Now, combine these three equations into a matrix.

$$\begin{pmatrix} 11 & -6 & 48 \\ -2 & 3 & -16 \\ 1 & 0 & -12 \end{pmatrix} \begin{pmatrix} V_1 \\ V_2 \\ i_x \end{pmatrix} = \begin{pmatrix} 19 \\ 0 \\ -5 \end{pmatrix} \Rightarrow \begin{pmatrix} V_1 \\ V_2 \\ i_x \end{pmatrix} = \begin{pmatrix} 0.12 & 0.24 & 0.16 \\ 0.1333 & 0.6 & -0.2667 \\ 0.01 & 0.02 & -0.07 \end{pmatrix} \begin{pmatrix} 19 \\ 0 \\ -5 \end{pmatrix}$$

$$= \begin{pmatrix} 1.48 \text{ V} \\ 3.8667 \text{ V} \\ 0.54 \text{ A} \end{pmatrix}$$

The open-circuit voltage is then given by

$$V_{OC} = V_a - V_b = V_a - 0 = (7 \ \Omega)i_x = 3.78 \text{ V}$$

We can now find the short-circuit current.

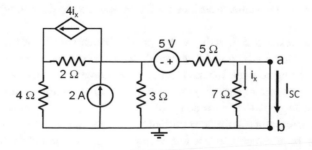

However, with terminals a and b shorted, no current will flow through the 7 Ω resistor ($i_x = 0$). Therefore, there will be no current flow through the dependent source. This allows us to replace the 7 Ω resistor and the dependent source by open circuits. This allows us to redraw the circuit in a simpler form prior to solving the node-voltage equations.

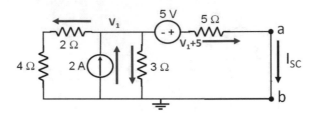

$$\frac{V_1}{6} + \frac{V_1}{3} + \frac{V_1 + 5}{5} = 2 \text{ A} \Rightarrow 21V_1 = 30 \Rightarrow V_1 = 1.4286 \text{ V}$$

The short-circuit current is then given by

$$I_{SC} = \frac{V_1 + 5}{5 \ \Omega} \Rightarrow I_{SC} = 1.2857 \text{ A}$$

$$R_{Th} = \frac{V_{OC}}{I_{SC}} = 2.94 \ \Omega$$

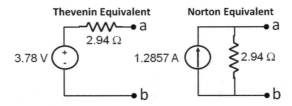

As can be seen from Example 3.26, there are times when finding either the short-circuit current or open-circuit voltage is considerably easier than finding the other value. In Example 3.26, the short-circuit current calculation only required solving one node-voltage equation while finding the open-circuit voltage required solving two node-voltage equations and a dependent source equation. There is, however, an alternate approach that allows for the Thevenin resistance to be solved directly. Therefore, one could find either the open-circuit voltage or the short-circuit current (not both), and then use the alternative method to get the Thevenin resistance. This approach can be simpler than finding both the open-circuit voltage and the short-circuit current. Also, there are times when all we really need to know is the Thevenin resistance such as when we are designing for maximum power transfer to a load.

To find the Thevenin resistance directly, we first need to deactivate all of the independent sources. The dependent sources (diamonds) remain unaltered, but all of the independent sources (circles) are removed. When deactivating sources, independent voltage sources are replaced by short circuits while independent current sources are replaced by open circuits. A short circuit has no voltage drop across it so when a voltage source goes to 0 V, it acts like a wire. Similarly, an open circuit has no current flow. Therefore, a current source with zero current would act like an open circuit. Once the independent sources have been deactivated, a test voltage, V_{Test}, is applied to the terminals and the resulting current, I_{Test}, is calculated. The Thevenin resistance is then given by

$$R_{Th} = \frac{V_{Test}}{I_{Test}} \tag{3.10}$$

It is also possible to apply a test current, I_{Test}, and calculate the resulting voltage across the terminals, V_{Test}, but usually a test voltage will result in fewer equations when using the node-voltage method to solve the circuit.

Example 3.27 Directly find the Thevenin resistance from Example 3.24 by deactivating the independent sources.

Solution:
Replacing the 10 V source with a short circuit and the 2 A source with an open circuit gives

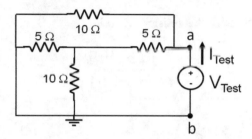

As is often the case with independent sources deactivated, the easiest way to solve this circuit is to simplify the resistor network as all we need to know is I_{Test}.

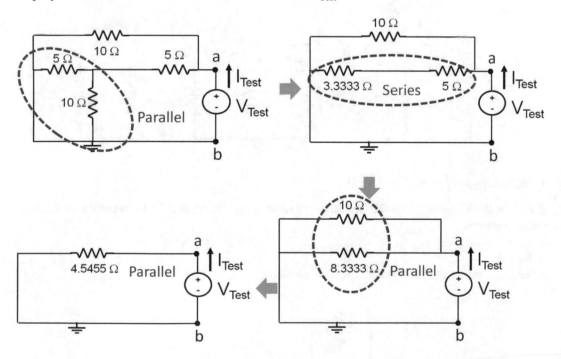

Therefore, $I_{Test} = \frac{V_{Test}}{4.5455\,\Omega} \Rightarrow R_{Th} = 4.5455\ \Omega$

Example 3.28 Directly find the Thevenin resistance from Example 3.25 by deactivating the independent sources.

Solution:
After deactivating the independent sources, we have

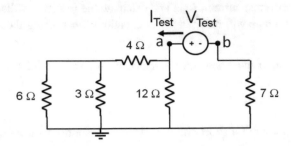

where once again we can simplify this resistor network.

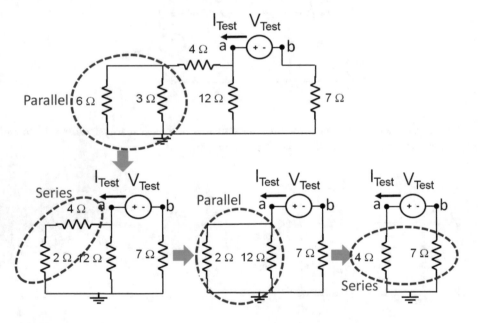

Therefore, $I_{\text{Test}} = \frac{V_{\text{Test}}}{11\,\Omega} \Rightarrow R_{\text{Th}} = 11\ \Omega$

Example 3.29 Directly find the Thevenin resistance from Example 3.26 by deactivating the independent sources.

Solution:
After deactivating the independent sources, we have

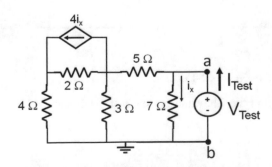

This circuit cannot be solved by simplifying the resistor network due to the presence of the dependent source. Therefore, we will solve the circuit using the node-voltage method. However, once the dependent voltage source has been applied, the current i_x is known as V_{Test} is applied directly across the 7 Ω resistor. Hence, $i_x = \frac{V_{\text{Test}}}{7\,\Omega}$. Also, since V_{Test} can be anything, we can select $V_{\text{Test}} = 7$ V $\Rightarrow$ $i_x = 1$ A to simplify the equations slightly.

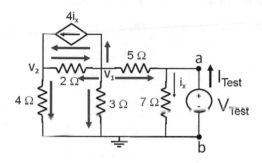

$$\frac{V_1 - V_2}{2} + 4i_x + \frac{V_1}{3} + \frac{V_1 - V_{\text{Test}}}{5} = 0 \Rightarrow$$

$$31V_1 - 15V_2 + 120i_x - 6V_{\text{Test}} = 0 \Rightarrow 31V_1 - 15V_2 = -78$$

$$\frac{V_2 - V_1}{2} - 4i_x + \frac{V_2}{4} = 0 \Rightarrow 3V_2 - 2V_1 - 16i_x = 0 \Rightarrow -2V_1 + 3V_2 = 16$$

$$\begin{pmatrix} 31 & -15 \\ -2 & 3 \end{pmatrix} \begin{pmatrix} V_1 \\ V_2 \end{pmatrix} = \begin{pmatrix} -78 \\ 16 \end{pmatrix} \Rightarrow \begin{pmatrix} V_1 \\ V_2 \end{pmatrix} = \begin{pmatrix} 0.0476 & 0.2381 \\ 0.0317 & 0.4921 \end{pmatrix} \begin{pmatrix} -78 \\ 16 \end{pmatrix} = \begin{pmatrix} 0.0952 \text{ V} \\ 5.3968 \text{ V} \end{pmatrix}$$

Now that the voltages are known, we can add the currents at terminal a to find I_{Test}.

$$I_{\text{Test}} = i_x + \frac{V_{\text{Test}} - V_1}{5\,\Omega} = 1\text{ A} + \frac{7\text{ V} - 0.0952\text{ V}}{5\,\Omega} = 2.381\text{ A}$$

$$R_{\text{Th}} = \frac{V_{\text{Test}}}{I_{\text{Test}}} = \frac{7\text{ V}}{2.381\text{ A}} = 2.94\,\Omega$$

Lab Hint: When finding the Thevenin or Norton equivalent in the lab, one should never short out any of the terminals nor should one apply a voltage/current to the output terminals of a circuit. Instead, one should first measure the open-circuit voltage to get the Thevenin voltage. Then, a known resistance with a value close to the expected load value should be connected. The voltage across this known load can then be used to find the Thevenin resistance using the basic voltage divider equation.

3.7 Superposition

When analyzing circuits using circuit models, we normally assume that elements behave in a linear fashion resulting in systems of linear equations as we saw in the previous examples. This assumption is only an approximation as real circuit elements are inherently nonlinear. For example, as current passes through a wire, some of the energy is converted into heat raising the temperature of the wire. As the temperature changes, the resistivity of the wire will also change. Therefore, the resistance of the wire will depend on the current flowing in the wire and the current flowing in the wire will depend on the resistance. Normally, however, the impact of the resistance on temperature, as well as other nonlinearities, is small and the linear approximation is sufficiently valid. Under the linear assumption, the voltages at each node and the currents in each branch can be expressed as linear combinations of the independent sources. Therefore, we can also solve a circuit by finding the impact of each source individually and then summing the individual responses to get the total response. Solving for the impact of an individual source requires deactivating the other sources. This means replacing independent voltage sources with short circuits (i.e., setting voltage to zero) and replacing independent current sources with open circuits (i.e., setting currents to zero). Solving circuits in this fashion is called superposition.

Example 3.30 Resolve the circuit from Example 3.14 for V_o using superposition.

Solution:
The circuit from Example 3.14 was given by

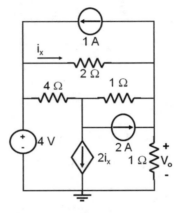

Begin by deactivating all of the independent sources except for the 4 V voltage source. This will give two nodes needing node-voltage equations and one dependent source equation.

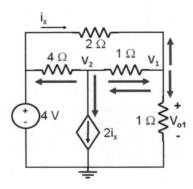

$$i_x = \frac{4 - V_1}{2} \Rightarrow V_1 + 2i_x = 4$$

$$\frac{V_1 - 4}{2} + \frac{V_1 - V_2}{1} + \frac{V_1}{1} = 0 \Rightarrow 5V_1 - 2V_2 = 4$$

$$\frac{V_2 - V_1}{1} + 2i_x + \frac{V_2 - 4}{4} = 0 \Rightarrow -4V_1 + 5V_2 + 8i_x = 4$$

$$\begin{pmatrix} 1 & 0 & 2 \\ 5 & -2 & 0 \\ -4 & 5 & 8 \end{pmatrix} \begin{pmatrix} V_1 \\ V_2 \\ i_x \end{pmatrix} = \begin{pmatrix} 4 \\ 4 \\ 4 \end{pmatrix} \Rightarrow \begin{pmatrix} V_1 \\ V_2 \\ i_x \end{pmatrix} = \begin{pmatrix} -0.8889 & 0.5556 & 0.2222 \\ -2.2222 & 0.8889 & 0.5556 \\ 0.9444 & -0.2778 & -0.1111 \end{pmatrix} \begin{pmatrix} 4 \\ 4 \\ 4 \end{pmatrix}$$

$$= \begin{pmatrix} -0.4444 \text{ V} \\ -3.1111 \text{ V} \\ 2.2222 \text{ A} \end{pmatrix}$$

Therefore, the output voltage due to the 4 V voltage source is given by $V_{o1} = V_1 = -0.4444$ V. Now, solve the circuit again with only the 1 A source.

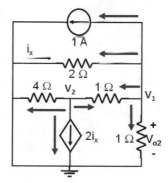

$$i_x = \frac{0 - V_1}{2} \Rightarrow V_1 + 2i_x = 0$$

$$1 \text{ A} + \frac{V_1}{2} + \frac{V_1 - V_2}{1} + \frac{V_1}{1} = 0 \Rightarrow 5V_1 - 2V_2 = -2$$

$$\frac{V_2 - V_1}{1} + 2i_x + \frac{V_2}{4} = 0 \Rightarrow -4V_1 + 5V_2 + 8i_x = 0$$

$$\begin{pmatrix} 1 & 0 & 2 \\ 5 & -2 & 0 \\ -4 & 5 & 8 \end{pmatrix} \begin{pmatrix} V_1 \\ V_2 \\ i_x \end{pmatrix} = \begin{pmatrix} 0 \\ -2 \\ 0 \end{pmatrix} \Rightarrow \begin{pmatrix} V_1 \\ V_2 \\ i_x \end{pmatrix} = \begin{pmatrix} -0.8889 & 0.5556 & 0.2222 \\ -2.2222 & 0.8889 & 0.5556 \\ 0.9444 & -0.2778 & -0.1111 \end{pmatrix} \begin{pmatrix} 0 \\ -2 \\ 0 \end{pmatrix}$$

$$= \begin{pmatrix} -1.1111 \text{ V} \\ -1.7778 \text{ V} \\ 0.5556 \text{ A} \end{pmatrix}$$

Therefore, the output voltage due to the 1 A current source is given by $V_{o2} = V_1 = -1.1111$ V. Now, solve the circuit again with only the 2 A source.

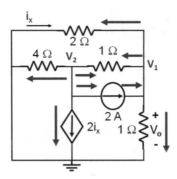

$$i_x = \frac{0 - V_1}{2} \Rightarrow V_1 + 2i_x = 0$$

$$\frac{V_1}{2} + \frac{V_1 - V_2}{1} + \frac{V_1}{1} = 2\,\text{A} \Rightarrow 5V_1 - 2V_2 = 4$$

$$\frac{V_2 - V_1}{1} + 2\,\text{A} + 2i_x + \frac{V_2}{4} = 0 \Rightarrow -4V_1 + 5V_2 + 8i_x = -8$$

$$\begin{pmatrix} 1 & 0 & 2 \\ 5 & -2 & 0 \\ -4 & 5 & 8 \end{pmatrix} \begin{pmatrix} V_1 \\ V_2 \\ i_x \end{pmatrix} = \begin{pmatrix} 0 \\ 4 \\ -8 \end{pmatrix} \Rightarrow \begin{pmatrix} V_1 \\ V_2 \\ i_x \end{pmatrix} = \begin{pmatrix} -0.8889 & 0.5556 & 0.2222 \\ -2.2222 & 0.8889 & 0.5556 \\ 0.9444 & -0.2778 & -0.1111 \end{pmatrix} \begin{pmatrix} 0 \\ 4 \\ -8 \end{pmatrix} = \begin{pmatrix} 0.4444\ \text{V} \\ -0.8889\ \text{V} \\ -0.2222\ \text{A} \end{pmatrix}$$

Therefore, the output voltage due to the 1 A current source is given by $V_{o3} = V_1 = 0.4444$ V. The total output voltage can then be found by adding the contribution from each source $V_o = V_{o1} + V_{o2} + V_{o3} = -1.1111$ V which agrees with our answer from Example 3.14.

Note: Each time we solved the circuit for the different sources, we always had a matrix equation of the form:

$$\begin{pmatrix} 1 & 0 & 2 \\ 5 & -2 & 0 \\ -4 & 5 & 8 \end{pmatrix} \begin{pmatrix} V_1 \\ V_2 \\ i_x \end{pmatrix} = \begin{pmatrix} a \\ b \\ c \end{pmatrix}$$

where the values for a, b, and c varied depending on which sources were active. This result is to be expected for linear circuits because the matrix on the left-hand side of the equation depends on the circuit topology while the values of a, b, and c depend on the source terms.

As illustrated by Example 3.30, solving circuits by superposition is a very tedious process. Therefore, superposition should only be used when we either want to learn more about dependences of a voltage or current on a particular source, or we want to solve our circuit using phasor domain circuit analysis and sources at different frequencies are present. In the second case, the sources could be DC sources that bias the active elements in the circuit while the AC sources are our signal sources from sensors or other inputs. Applying superposition to different frequency sources will be discussed in Chap. 7, so we will focus on the use of superposition to find the dependence of the output on a single source when other sources are present in the following example.

Example 3.31 Find an equation relating the output voltage, V_o, to the current from the senor, I_{sensor}, for the following circuit using superposition (Fig. 3.35).

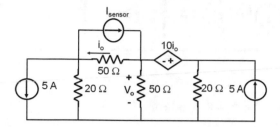

Fig. 3.35 Circuit diagram for Example 3.31

Solution:

We will first find the direct dependence of V_o on I_{sensor} by deactivating the other two independent current sources.

$$i_o = \frac{V_1 - V_2}{50} \Rightarrow V_1 - V_2 - 50i_o = 0$$

$$\frac{V_1 - V_2}{50} + \frac{V_1}{50} + \frac{V_1 + 10i_o}{20} = I_{sensor} \Rightarrow 9V_1 - 2V_2 + 50i_o = 100I_{sensor}$$

$$\frac{V_2 - V_1}{50} + \frac{V_2}{20} + I_{sensor} = 0 \Rightarrow -2V_1 + 7V_2 = -100I_{sensor}$$

$$\begin{pmatrix} 1 & -1 & -50 \\ 9 & -2 & 50 \\ -2 & 7 & 0 \end{pmatrix} \begin{pmatrix} V_1 \\ V_2 \\ i_o \end{pmatrix} = \begin{pmatrix} 0 \\ 100I_{sensor} \\ -100I_{sensor} \end{pmatrix} = \begin{pmatrix} 0 \\ 100 \\ -100 \end{pmatrix} I_{sensor}$$

$$\Rightarrow \begin{pmatrix} V_1 \\ V_2 \\ i_o \end{pmatrix} = \begin{pmatrix} 0.1094 & 0.1094 & 0.0469 \\ 0.0313 & 0.0313 & 0.1563 \\ -0.0184 & 0.0016 & -0.0022 \end{pmatrix} \begin{pmatrix} 0 \\ 100 \\ -100 \end{pmatrix} I_{sensor} = \begin{pmatrix} 6.25 \ V/A \\ -12.5 \ V/A \\ 0.375 \end{pmatrix} I_{sensor}$$

Therefore, the output voltage due to the sensor alone is given by $V_{o1} = V_1 = (6.25 \ V/A)I_{sensor}$. Now, let's find the output voltage due to the other two current sources.

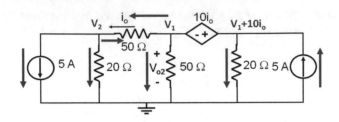

$$i_o = \frac{V_1 - V_2}{50} \Rightarrow V_1 - V_2 - 50i_o = 0$$

$$\frac{V_1 - V_2}{50} + \frac{V_1}{50} + \frac{V_1 + 10i_o}{20} = 5 \text{ A} \Rightarrow 9V_1 - 2V_2 + 50i_o = 500$$

$$\frac{V_2 - V_1}{50} + \frac{V_2}{20} + 5 \text{ A} = 0 \Rightarrow -2V_1 + 7V_2 = -500$$

$$\begin{pmatrix} 1 & -1 & -50 \\ 9 & -2 & 50 \\ -2 & 7 & 0 \end{pmatrix} \begin{pmatrix} V_1 \\ V_2 \\ i_o \end{pmatrix} = \begin{pmatrix} 0 \\ 500 \\ -500 \end{pmatrix}$$

$$\Rightarrow \begin{pmatrix} V_1 \\ V_2 \\ i_o \end{pmatrix} = \begin{pmatrix} 0.1094 & 0.1094 & 0.0469 \\ 0.0313 & 0.0313 & 0.1563 \\ -0.0184 & 0.0016 & -0.0022 \end{pmatrix} \begin{pmatrix} 0 \\ 500 \\ -500 \end{pmatrix} = \begin{pmatrix} 31.25 & V \\ -62.5 & V \\ 1.875 & A \end{pmatrix}$$

Therefore, the output voltage due to the other current sensors alone is given by $V_{o2} = V_1 = 31.25$ V. The total output voltage is the some of both contributions and is given by $V_o = V_{o1} + V_{o2} = (6.25 \ V/A)I_{\text{sensor}} + 31.25$ V.

Hence, using superposition, we are able to know the output for any value of sensor current without needing to resolve the circuit as the sensor input changes.

3.8   Problems

Problem 3.1: Select a ground node and label the other nodes in the following circuit. Also, circle the wires corresponding to each node.

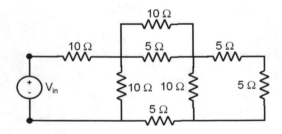

Problem 3.2: Select a ground node and label the other nodes in the following circuit. Also, circle the wires corresponding to each node.

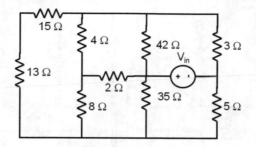

Problem 3.3: Select a ground node and label the other nodes in the following circuit. Also, circle the wires corresponding to each node.

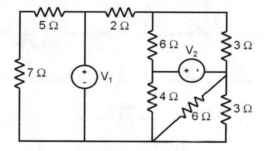

Problem 3.4: Determine all of the meshes and loops for the following circuit.

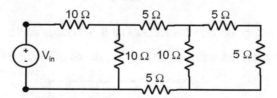

Problem 3.5: How many meshes are in the following circuit?

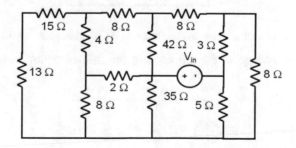

Problem 3.6: Identify each of the following traces as a mesh, loop, or path.

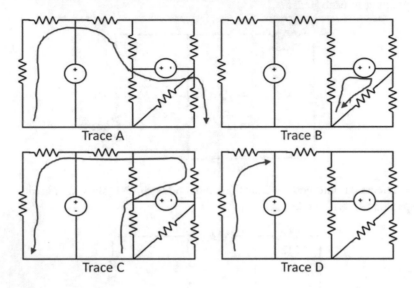

Trace A   Trace B

Trace C   Trace D

Problem 3.7: Find the voltage drop across each resistor for the circuit shown.

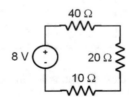

Problem 3.8: Design a voltage divider that reduces an 8 V voltage source to 3 V.

Problem 3.9: Find the current through each resistor for the circuit shown.

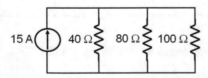

Problem 3.10: Design a current divider that reduces a 10 A current down to 7 A.

Problem 3.11: Find the voltage at the top node in the following circuit using the node-voltage method.

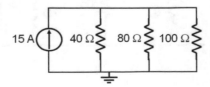

Problem 3.12: Find the voltage at all of the nodes in the following circuit using the node-voltage method as well as the labeled currents I_A and I_B.

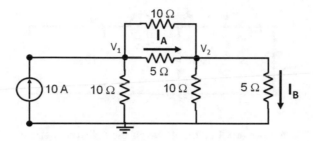

Problem 3.13: Find the voltage at all of the nodes in the following circuit using the node-voltage method as well as the labeled currents I_A and I_B.

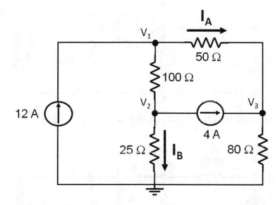

Problem 3.14: Find the voltage at all of the nodes in the following circuit using the node-voltage method.

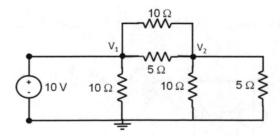

Problem 3.15: Find the voltage at all of the nodes in the following circuit using the node-voltage method.

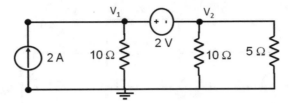

Problem 3.16: Find the voltage at all of the nodes in the following circuit using the node-voltage method.

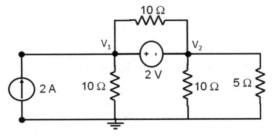

Problem 3.17: Find the voltage at all of the nodes in the following circuit using the node-voltage method.

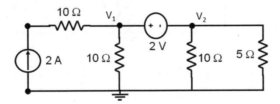

Problem 3.18: Find the voltage at all of the nodes in the following circuit using the node-voltage method.

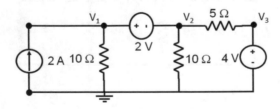

Problem 3.19: Find the voltage at all of the nodes in the following circuit using the node-voltage method.

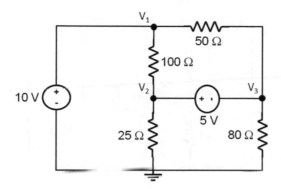

Problem 3.20: Find the voltage at all of the nodes in the following circuit using the node-voltage method.

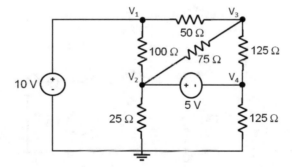

Problem 3.21: Find the voltage at all of the labeled nodes in the following circuit using the node-voltage method.

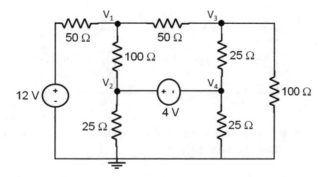

Problem 3.22: Find the voltage at all of the labeled nodes in the following circuit using the node-voltage method.

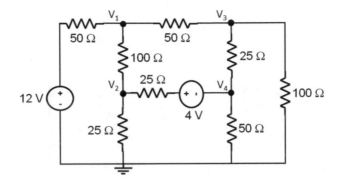

Problem 3.23: Find the voltage at all of the labeled nodes in the following circuit using the node-voltage method.

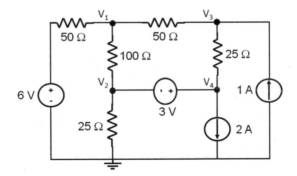

Problem 3.24: Use the node-voltage method to find the voltages at all of the nodes in the circuit.

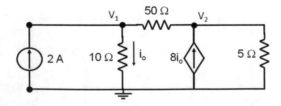

Problem 3.25: Use the node-voltage method to find the voltages at all of the nodes in the circuit as well as the current through the dependent source.

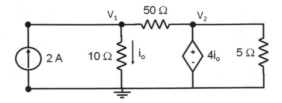

Problem 3.26: Use the node-voltage method to find the voltages at all of the labeled nodes in the circuit as well as the current through the dependent source.

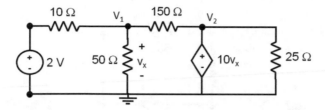

Problem 3.27: Use the node-voltage method to find the voltages at all of the labeled nodes in the circuit.

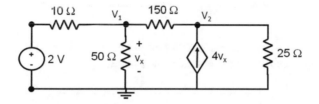

Problem 3.28: Use the node-voltage method to find the voltages at all of the labeled nodes in the circuit.

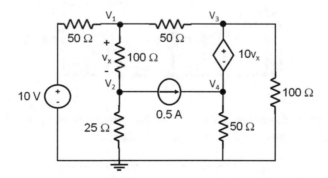

Problem 3.29: Use the node-voltage method to find the voltage across and the current through the 100 Ω resistor in the following circuit.

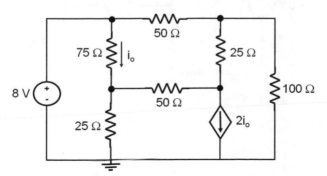

Problem 3.30: Use the node-voltage method to find the voltage across and the current through the 75 Ω resistor in the following circuit.

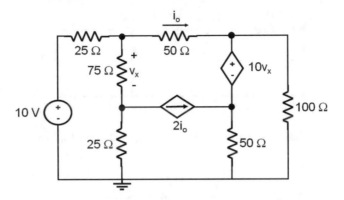

Problem 3.31: Use the mesh-current method to find the currents flowing out of the positive voltage terminal of each voltage source for the circuit shown below.

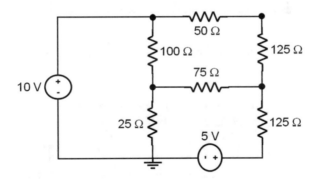

Problem 3.32: Use the mesh-current method to find the currents flowing out of the positive voltage terminal of each voltage source for the circuit shown below.

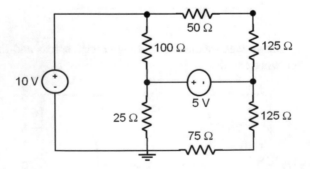

Problem 3.33: Use the mesh-current method to find the currents I_A and I_B for the circuit shown below.

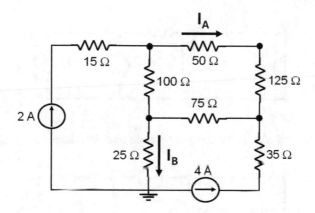

Problem 3.34: Use the mesh-current method to find the currents I_A and I_B for the circuit shown below.

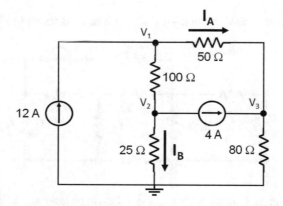

Problem 3.35: Use the mesh-current method to find the currents I_A and I_B as well as the node voltages V_1 and V_2.

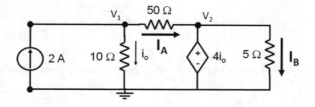

Problem 3.36: Use the mesh-current method to find the currents I_A and I_B as well as the node voltages V_1 and V_2.

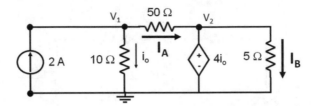

Problem 3.37: Find the voltage V_o by applying source transformations to simplify the circuit.

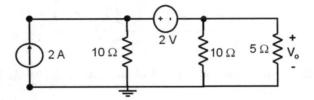

Problem 3.38: Find the voltage V_o by applying source transformations to simplify the circuit.

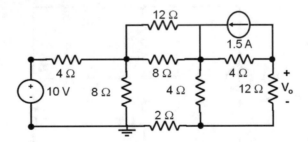

Problem 3.39: Apply source transformation to the following circuit until the circuit has been simplified to where it can be solved by two-node voltage equations and a dependent source equation. After simplification, solve the circuit using the node-voltage method and calculate the current I_A.

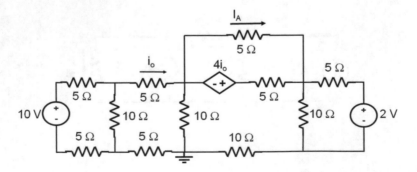

Problem 3.40: Find Thevenin and Norton equivalent for terminals a and b for the following circuit by first finding the open-circuit voltage and the short-circuit current.

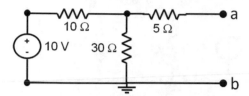

Problem 3.41: Find Thevenin and Norton equivalent for terminals a and b for the following circuit by first finding the open-circuit voltage and the short-circuit current.

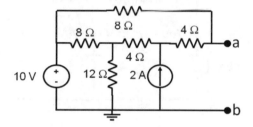

Problem 3.42: Find Thevenin and Norton equivalent for terminals a and b for the following circuit by first finding the open-circuit voltage and the short-circuit current.

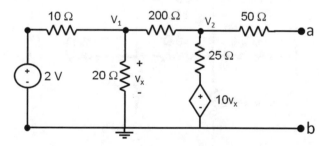

Problem 3.43: Find Thevenin and Norton equivalent for terminals a and b for the following circuit by first finding the open-circuit voltage and the short-circuit current.

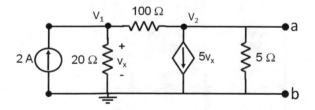

Problem 3.44: Directly find the Thevenin resistance for the following circuit by deactivating the independent sources.

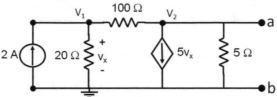

Problem 3.45: Directly find the Thevenin resistance for the following circuit by deactivating the independent sources.

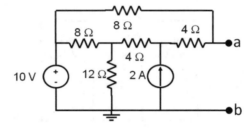

Problem 3.46: Directly find the Thevenin resistance for the following circuit by deactivating the independent sources.

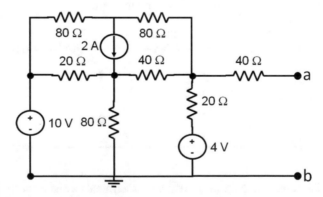

Problem 3.47: Directly find the Thevenin resistance for the following circuit by deactivating the independent sources.

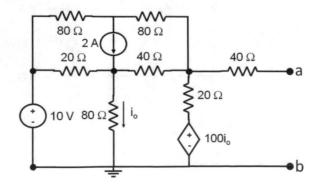

Problem 3.48: Find the voltage, V_o, for the circuit shown below using superposition.

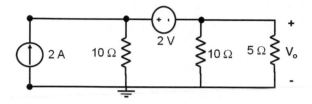

Problem 3.49: Find the voltage, V_o, for the circuit shown below using superposition.

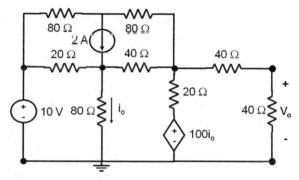

Problem 3.50: Find an equation relating the output voltage, V_o, to the voltage from the senor, V_{sensor}, for the following circuit using superposition.

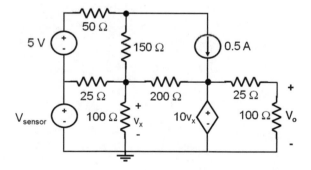

Power and Energy in Electric Circuits

4

Calculating the flow of power and the transfer of energy is vital to the understanding of circuit behavior. Sustainability engineering demands maximum efficiency. Therefore, finding power losses relative to power delivery is vital in the design process. In addition, power losses typically translate into heat, which can pose risks to users of the electric device and lead to temperature-related device failure as well. In this chapter, we will assume that currents and voltages are already known and focus on the calculation of power and energy. In addition, we will use the common convention of a positive sign when power is being absorbed or lost in a circuit element and a negative sign when power is being supplied by the circuit element.

4.1 Power Calculation and Sign Convention

Power is the change in energy, w, per unit time, t.

$$p(t) = \frac{dw}{dt} = i(t)v(t) \tag{4.1}$$

In a circuit, some of the elements act as energy sources that provide power while others act as energy sinks that absorb power. Figure 4.1 shows a circuit element that is providing power as well as a circuit element that is absorbing power. If the charge is flowing out of the positive voltage terminal, then the element is providing power. Otherwise, it is absorbing power or translating power from electrical to some other form.

By convention, energy flowing into a circuit element results in a positive power, and energy flowing out of a circuit element gives a negative power. For example, if 50 J flows into a circuit element over 5 s, the element would have absorbed +10 W of power. Similarly, if 50 J flows out of a circuit element over 5 s, then the element would have supplied 10 W or absorbed −10 W. In circuit analysis, we calculate how much power is absorbed by each circuit element. If the sign of the power is negative, then we know the element is a source instead of a sink. Therefore, power is calculated by multiplying the current flowing into the positive voltage terminal by the voltage as illustrated in Fig. 4.2. A negative sign is included in the formula if the current is instead flowing out of the positive voltage terminal. This standard is known as the Passive Sign Convention in circuit analysis.

© Springer Nature Switzerland AG 2020
T. A. Bigelow, *Electric Circuits, Systems, and Motors*,
https://doi.org/10.1007/978-3-030-31355-5_4

(a) Energy Source

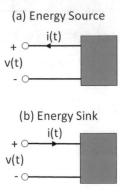

(b) Energy Sink

Fig. 4.1 a Circuit element acting as an energy source. **b** Circuit element acting as an energy sink

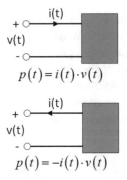

Fig. 4.2 Formulas for power according to the passive sign convention

Example 4.1 Find the power associated with each of the circuit elements given below (Fig. 4.3) based on the passive sign convention if $v(t) = 9$ V and $i(t) = 1$ A.

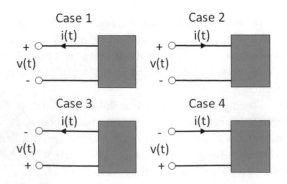

Fig. 4.3 Circuit diagram for Example 4.1

Solution:

Case 1: This circuit has the current flowing out of the positive voltage terminal. Therefore, by the passive sign convention, the correct formula for the power is

$$p(t) = -i(t)v(t) = -(1 \text{ A}) \cdot (9 \text{ V}) = -9 \text{ W}$$

Case 2: This circuit has the current flowing into the positive voltage terminal. Therefore, by the passive sign convention, the correct formula for the power is

$$p(t) = i(t)v(t) = (1 \text{ A}) \cdot (9 \text{ V}) = 9 \text{ W}$$

Case 3: This circuit has the current flowing out of the negative voltage terminal. Since the current leaving the circuit element must also be the same current that enters the element, this means that the current must also be entering the positive voltage terminal. Therefore, by the passive sign convention, the correct formula for the power is

$$p(t) = i(t)v(t) = (1 \text{ A}) \cdot (9 \text{ V}) = 9 \text{ W}$$

Case 4: This circuit has the current flowing into the negative voltage terminal. Since the current entering the circuit element must also be the same current that leaves the element, this means that the current must also be leaving the positive voltage terminal. Therefore, by the passive sign convention, the correct formula for the power is

$$p(t) = -i(t)v(t) = -(1 \text{ A}) \cdot (9 \text{ V}) = -9 \text{ W}.$$

Example 4.2 Find the power associated with each circuit element shown below based on the passive sign convention. The voltages and current shown in the diagram (Fig. 4.4) were solved in Example 3.9.

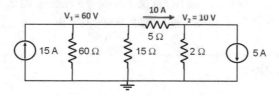

Fig. 4.4 Circuit diagram for Example 4.2

Solution:

15 A Source Power: The voltage across the 15 A source is 60 V with the top node being at a higher voltage then the bottom node. Therefore, the 15 A is flowing out of the positive voltage terminal. Thus, the correct power formula is given by

$$p(t) = -i(t)v(t) = -(15 \text{ A}) \cdot (60 \text{ V}) = -900 \text{ W}$$

60 Ω Resistor Power: The voltage across the resistor is 60 V. Therefore, the current flowing down into the resistors positive voltage terminal is given by $I_{60\Omega} = \frac{V_1 - 0}{60 \ \Omega} = \frac{60 \text{ V}}{60 \ \Omega} = 1 \text{ A}$. Thus, the correct power formula is given by

$$p(t) = i(t)v(t) = (1 \text{ A}) \cdot (60 \text{ V}) = 60 \text{ W}$$

15 Ω Resistor Power: The voltage across the resistor is 60 V. Therefore, the current flowing down into the resistors positive voltage terminal is given by $I_{15\Omega} = \frac{V_1 - 0}{15\ \Omega} = \frac{60\ V}{15\ \Omega} = 4$ A. Thus, the correct power formula is given by

$$p(t) = i(t)v(t) = (4\ \text{A}) \cdot (60\ \text{V}) = 240\ \text{W}$$

5 Ω Resistor Power: The voltage across the resistor is $V_1 - V_2 = 50$ V while the current through the resistor was previously found to be 10 A. Thus, the correct power formula is given by

$$p(t) = i(t)v(t) = (10\ \text{A}) \cdot (50\ \text{V}) = 500\ \text{W}$$

2 Ω Resistor Power: The voltage across the resistor is 10 V. Therefore, the current flowing down into the resistors positive voltage terminal is given by $I_{2\Omega} = \frac{V_2 - 0}{2\ \Omega} = \frac{10\ V}{2\ \Omega} = 5$ A. Thus, the correct power formula is given by

$$p(t) = i(t)v(t) = (5\ \text{A}) \cdot (10\ \text{V}) = 50\ \text{W}$$

5 A Source Power: The voltage across the 5 A source is 10 V with the top node being at a higher voltage then the bottom node. Therefore, the 5 A is flowing into the positive voltage terminal. Thus, the correct power formula is given by

$$p(t) = i(t)v(t) = (5\ \text{A}) \cdot (10\ \text{V}) = 50\ \text{W}$$

Notice that the power associated with the 5 A source is +50 W. Therefore, this source is absorbing power while the 15 A source provides all of the power for the circuit. Also, if we add up all of these power values, the net is 0 W indicating that energy is conserved.

4.2 Energy Calculation

Once the power is known, the energy delivered or supplied over a specific time period can be calculated by integrating with respect to time.

$$w = \int_{t_1}^{t_2} p(t)\mathrm{d}t \tag{4.2}$$

In the United States, those that work in the power generation and delivery industry typically use units of kilowatt-hours (kWh) to quantify the amount of energy rather than the traditional SI units of Joules. One kWh is the amount of energy that would be delivered if 1 kW was supplied for one hour. Therefore,

$$1\ \text{kWh} = (1\ \text{kWh}) \cdot \left(\frac{60\ \text{min}}{1\ \text{h}}\right)\left(\frac{60\ \text{s}}{1\ \text{min}}\right) = 3.6 \times 10^6\ J = 3.6\ \text{MJ} \tag{4.3}$$

Example 4.3 Find the total energy delivered to a circuit element if the voltage applied across its terminals is given by $v(t) = \begin{cases} 0 & t \leq 0 \\ e^{-5t} \text{ V} & t > 0 \end{cases}$ and the current flowing into the positive voltage terminal is given by $i(t) = \begin{cases} 0 & t \leq 0 \\ 7t \text{ A} & t > 0 \end{cases}$.

Solution:

First, we need to find the power delivered to the circuit element. Since the current is flowing into the positive voltage terminal, the correct power formula is given by

$$p(t) = i(t)v(t) = \begin{cases} 0 & t \leq 0 \\ 7te^{-5t} \text{ W} & t > 0 \end{cases}$$

When plotted as a function of time, this power initially increases with time before decaying to zero exponentially.

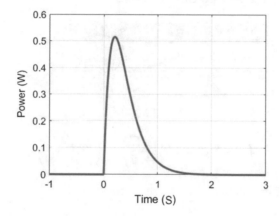

The total energy delivered is then found by integrating as a function of time over all time.

$$w = \int_{-\infty}^{\infty} p(t)dt = \int_{-\infty}^{0} p(t)dt + \int_{0}^{\infty} p(t)dt = \int_{-\infty}^{0} 0 dt + \int_{0}^{\infty} 7te^{-5t} dt = \int_{0}^{\infty} 7te^{-5t} dt$$

This integral can now be solved by integration by parts.

$$\int u \tfrac{dv}{dx}dx = uv - \int v \tfrac{du}{dx}dx$$

$$u = 7t \quad \tfrac{du}{dt} = 7$$

$$\tfrac{dv}{dt} = e^{-5t} \quad v = \int e^{-5t}dt = \frac{e^{-5t}}{-5}$$

$$\int 7te^{-5t}dt = (7t)\left(\frac{e^{-5t}}{-5}\right) - \int 7\left(\frac{e^{-5t}}{-5}\right)dt$$

At this point, we can solve these integrals to obtain the total power OR we could apply integration by parts again to completely remove the integration.

$$u = 7 \quad \frac{du}{dt} = 0$$

$$\frac{dv}{dt} = \left(\frac{e^{-5t}}{-5}\right) \quad v = \int \left(\frac{e^{-5t}}{-5}\right) dt = \frac{e^{-5t}}{25}$$

$$\int 7\left(\frac{e^{-5t}}{-5}\right) dt = (7)\left(\frac{e^{-5t}}{25}\right) - \int 0\left(\frac{e^{-5t}}{25}\right) dt = (7)\left(\frac{e^{-5t}}{25}\right)$$

Therefore,

$$\int 7te^{-5t} dt = (7t)\left(\frac{e^{-5t}}{-5}\right) - \left(\int 7\left(\frac{e^{-5t}}{-5}\right) dt\right) = (7t)\left(\frac{e^{-5t}}{-5}\right) - \left((7)\left(\frac{e^{-5t}}{25}\right)\right)$$

These steps can also be captured in a table and can be performed very quickly for some equations. Specifically, we need some part of the integrand to go to zero after taking the derivative several times while the rest of the integrand remains easy to integrate.

Derivative Column		Integral Column	
7t	Multiply + Sign	exp(-5t)	Add products
7		exp(-5t)/(-5)	
0	Multiply - Sign	exp(-5t)/(25)	

$$\int 7te^{-5t} dt = (7t)\left(\frac{e^{-5t}}{-5}\right) - \left((7)\left(\frac{e^{-5t}}{25}\right)\right)$$

Now, apply the limits of integration.

$$w = \int_0^\infty 7te^{-5t} dt = \left[(7t)\left(\frac{e^{-5t}}{-5}\right) - \left((7)\left(\frac{e^{-5t}}{25}\right)\right)\right]_0^\infty$$

$$= \lim_{t\to\infty} \left(\left[(7t)\left(\frac{e^{-5t}}{-5}\right) - \left((7)\left(\frac{e^{-5t}}{25}\right)\right)\right]\right) - \left[(7\cdot 0)\left(\frac{e^{-5\cdot 0}}{-5}\right) - \left((7)\left(\frac{e^{-5\cdot 0}}{25}\right)\right)\right]$$

$$= 0 - \left[-\left((7)\left(\frac{1}{25}\right)\right)\right] = \frac{7}{25} \ J = 0.28 \ J.$$

Example 4.4 Find the energy delivered to a 3 μF capacitor for time less than zero and for time greater than zero if the voltage applied to the capacitor is given by $v(t) = 2\exp\left(-\left(\frac{t}{200 \ \mu s}\right)^2\right)$ V.

The current flow into this capacitor as a function of time was previously solved to be $i(t) = -300t \cdot \exp\left(-\left(\frac{t}{200 \ \mu s}\right)^2\right)$ A in Example 1.16.

Solution:

The power as a function of time is given by $p(t) = i(t)v(t) = -600t \exp\left(-2\left(\frac{t}{200 \text{ μs}}\right)^2\right)$ W.

For time less than zero, we have

$$w = \int_{-\infty}^{0} p(t)dt = \int_{-\infty}^{0} \left(-600t \exp\left(-2\left(\frac{t}{200 \text{ μs}}\right)^2\right) \text{ W}\right) dt$$

Let's solve by applying a change of variables to get

$$u = 2\left(\frac{t}{200 \text{ μs}}\right)^2 \Rightarrow du = 4\left(\frac{t}{(200 \text{ μs})^2}\right) dt \Rightarrow tdt = \frac{(200 \text{ μs})^2}{4} du$$

$$\int_{-\infty}^{0} \left(-600t \exp\left(-2\left(\frac{t}{200 \text{ μs}}\right)^2\right)\right) dt \Rightarrow \int_{\infty}^{0} (-600 \exp(-u)) \frac{(200 \text{ μs})^2}{4} du$$

$$= \left(6 \times 10^{-6} \text{ J}\right) \int_{0}^{\infty} \exp(-u)du = 6 \text{ μJ}$$

For time greater than zero, we have

$$w = \int_{0}^{\infty} p(t)dt = \int_{0}^{\infty} \left(-600t \exp\left(-2\left(\frac{t}{200 \text{ μs}}\right)^2\right) \text{ W}\right) dt$$

Applying the same change of variables gives

$$u = 2\left(\frac{t}{200 \text{ μs}}\right)^2 \Rightarrow du = 4\left(\frac{t}{(200 \text{ μs})^2}\right) dt \Rightarrow tdt = \frac{(200 \text{ μs})^2}{4} du$$

$$\int_{0}^{\infty} \left(-600t \exp\left(-2\left(\frac{t}{200 \text{ μs}}\right)^2\right)\right) dt \Rightarrow \int_{0}^{\infty} (-600 \exp(-u)) \frac{(200 \text{ μs})^2}{4} du$$

$$= -\left(6 \times 10^{-6} \text{ J}\right) \int_{0}^{\infty} \exp(-u)du = -6 \text{ μJ}$$

Therefore, 6 μJ are stored on the capacitor before t = 0, and this same energy is then removed for t > 0. This can also be determined from the graph of the power as function of time.

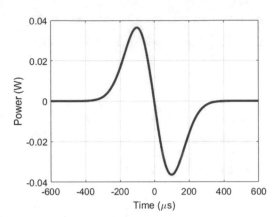

Since this curve has anti-symmetry about t = 0, the total energy must be zero.

Example 4.5 Find the energy delivered in one day to a home for the power flow shown in Fig. 4.5.

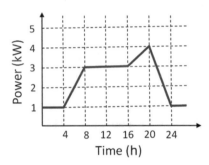

Fig. 4.5 Power as a function of time for Example 4.5

Solution:

The first step in solving this problem is finding a mathematical expression for the power as a function of time. From 0 to 4 h, the power supplied is a constant 1 kW. From 4 to 8 h, the power increases linearly with a change (slope) of 2 kW in 4 h or 0.5 kWh. Therefore, the equation for this line is $\left((0.5 \, \frac{kW}{h}) \cdot t - 1 \, kW\right)$. From 8 to 12 h, the power is constant 3 kW. It then increases to 4 kW over 4 h as given by $\left((0.25 \, \frac{kW}{h}) \cdot t - 1 \, kW\right)$. The power then decreases linearly back to 1 kW over the final 4 h as given by $\left((-0.75 \, \frac{kW}{h}) \cdot t + 19 \, kW\right)$. For each of these lines, the slope is found from the change in power over the change in time while the intercept is found by substituting one of the endpoints into the equation for the line. For example, for the last line segment, the power is 1 kW at 24 h. Therefore, $\left((-0.75 \, \frac{kW}{h}) \cdot 24 \, h + b\right) = 1 \, kW \Rightarrow b = 1 \, kW + \left(0.75 \, \frac{kW}{h}\right) \cdot 24 \, h = 1 \, kW + 18 \, kW = 19 \, kW$.

The power as a function of time can now be written as the following piece-wise function.

$$p(t) = \begin{cases} 1 \, kW & 0 \leq t < 4 \, h \\ \left((0.5 \, \frac{kW}{h}) \cdot t - 1 \, kW\right) & 4 \leq t < 8 \, h \\ 3 \, kW & 8 \leq t < 16 \, h \\ \left((0.25 \, \frac{kW}{h}) \cdot t - 1 \, kW\right) & 16 \leq t < 20 \, h \\ \left((-0.75 \, \frac{kW}{h}) \cdot t + 19 \, kW\right) & 20 \leq t < 24 \, h \end{cases}$$

$$w = \int_{0}^{24\ h} p(t)\mathrm{d}t = \int_{0}^{4\ h} (1\ \mathrm{kW})\mathrm{d}t + \int_{4\ h}^{8\ h} \left(\left(0.5\ \frac{\mathrm{kW}}{\mathrm{h}}\right)\cdot t - 1\ \mathrm{kW}\right)\mathrm{d}t + \int_{8\ h}^{16\ h} (3\ \mathrm{kW})\mathrm{d}t$$

$$+ \int_{16\ h}^{20\ h} \left(\left(0.25\ \frac{\mathrm{kW}}{\mathrm{h}}\right)\cdot t - 1\ \mathrm{kW}\right)\mathrm{d}t + \int_{20\ h}^{24\ h} \left(\left(-0.75\ \frac{\mathrm{kW}}{\mathrm{h}}\right)\cdot t + 19\ \mathrm{kW}\right)\mathrm{d}t$$

$$w = 4\ \mathrm{kWh} + \left(\left(0.5\ \frac{\mathrm{kW}}{\mathrm{h}}\right)\cdot \frac{t^2}{2} - t\ \mathrm{kW}\right)\Bigg|_{4\ h}^{8\ h} + (3\ \mathrm{kW})\cdot(16\ \mathrm{h} - 8\ \mathrm{h})$$

$$+ \left(\left(0.25\ \frac{\mathrm{kW}}{\mathrm{h}}\right)\cdot \frac{t^2}{2} - t\ \mathrm{kW}\right)\Bigg|_{16\ h}^{20\ h} + \left(-\left(0.75\ \frac{\mathrm{kW}}{\mathrm{h}}\right)\cdot \frac{t^2}{2} + 19t\ \mathrm{kW}\right)\Bigg|_{20\ h}^{24\ h}$$

$$w = 4\ \mathrm{kWh} + \left((0.25\ \mathrm{kWh})\cdot(8^2 - 4^2) - 4\ \mathrm{kWh}\right) + 24\ \mathrm{kWh}$$
$$+ \left((0.125\ \mathrm{kWh})\cdot(20^2 - 16^2) - 4\ \mathrm{kWh}\right)$$
$$+ \left(-(0.375\ \mathrm{kWh})\cdot(24^2 - 20^2) + (19\ \mathrm{kWh})\cdot(24 - 20)\right)$$
$$w = 4\ \mathrm{kWh} + 8\ \mathrm{kWh} + 24\ \mathrm{kWh} + 14\ \mathrm{kWh} + 10\ \mathrm{kWh} = 60\ \mathrm{kWh}$$

Practical Point of Reference: Many new and exciting advances are being made in the area of electrical vehicles. However, it is easy for the novice to lose sight of the power required to move such vehicles relative to the power stored in traditional gasoline. For example, one gallon of gasoline has approximately 130 MJ or 36 kWh of energy. On the other hand, a traditional car battery has about 0.7 kWh of energy. Therefore, it would take approximately 50 traditional batteries to provide the same energy as one gallon of gas neglecting the difference in energy conversion efficiency. This is why so much effort is going into increasing the energy storage capacity of batteries.

4.3   Power in Resistive Circuits

The power absorbed by resistors can be directly found from either the voltage or the current since the relationship between voltage and current for a resistor is given by Ohm's Law (i.e., $v(t) = i(t) \cdot R$).

$$p(t) = v(t) \cdot i(t) \Rightarrow \begin{cases} p(t) = (i(t) \cdot R) \cdot i(t) = (i(t))^2 \cdot R \\ p(t) = v(t) \cdot \left(\frac{v(t)}{R}\right) = \frac{(v(t))^2}{R} \end{cases} \tag{4.4}$$

Notice that the power absorbed by the resistor is always proportional to the square of the voltage or the square of the current. This means that resistors can never supply power. They can only act as power sinks. This makes sense as resistors transfer energy from electrical to some other form (usually heat).

Example 4.6 Find the power absorbed in each of the resistors, as well as the power, supplied by the source for the circuit shown in Fig. 4.6. The voltages and current in the 20 Ω resistor for this circuit were originally solved in Example 3.7.

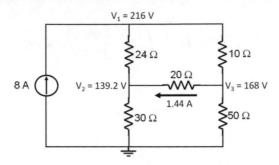

Fig. 4.6 Circuit diagram for Example 4.6

Solution:
Let's first find the power from the source. The source provides 8 A of current while the voltage across the source is $V_1 = 216$ V. Therefore, the power aborobed by the source is

$$p(t) = -i(t)v(t) = -1728 \text{ W}$$

If -1728 W are being absorbed, then this means the source is supplying $+1728$ W to the circuit. We can now find the power absorbed by the 20 Ω resistor from the current and the resistance.

$$p_{20\Omega}(t) = (i(t))^2 R = (1.44 \text{ A})^2 \cdot 20 \text{ } \Omega = 41.472 \text{ W}$$

The power absorbed by the other resistors can be found from the voltage drop across the resistor and the resistance.

$$p_{30\Omega}(t) = \frac{(v(t))^2}{R} = \frac{(139.2 \text{ V})^2}{30 \text{ } \Omega} = 645.888 \text{ W}$$

$$p_{50\Omega}(t) = \frac{(v(t))^2}{R} = \frac{(168 \text{ V})^2}{50 \text{ } \Omega} = 564.48 \text{ W}$$

$$p_{24\Omega}(t) = \frac{(v(t))^2}{R} = \frac{(216 - 139.2 \text{ V})^2}{24 \text{ } \Omega} = 245.76 \text{ W}$$

$$p_{10\Omega}(t) = \frac{(v(t))^2}{R} = \frac{(216 - 168 \text{ V})^2}{10 \text{ } \Omega} = 230.4 \text{ W}$$

Therefore, the 20 Ω resistor is absorbing the least amount of power while the 30 Ω resistor is absorbing the most.

Example 4.7 Find the voltage across the 2 A source using conservation of energy for the circuit shown in Fig. 4.7. The mesh currents shown in the figure were previously solved in Example 3.16.

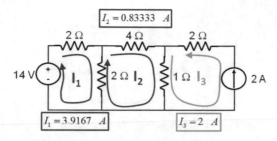

Fig. 4.7 Circuit diagram for Example 4.7

Solution:

First, find the power associated with the 14 V source I_1 flows out of the positive voltage terminal so the power is given by

$$P_{14V} = -I_1 \cdot (14 \text{ V}) = -54.8333 \text{ W}$$

Now, find the power absorbed by all of the resistors.

$$P_{\text{resistors}} = I_1^2 \cdot (2 \text{ }\Omega) + I_2^2 \cdot (4 \text{ }\Omega) + I_3^2 \cdot (2 \text{ }\Omega) + (I_1 - I_2)^2 \cdot (2 \text{ }\Omega) + (I_2 + I_3)^2 \cdot (1 \text{ }\Omega)$$
$$= 30.6806 \text{ W} + 2.7778 \text{ W} + 8 \text{ W} + 19.0139 \text{ W} + 8.0278 \text{ W} = 68.5 \text{ W}$$

In order to conserve energy, the sum of all of the powers must be zero.

$$P_{14V} + P_{\text{resistors}} + P_{2A} = 0 \Rightarrow P_{2A} = -P_{14V} - P_{\text{resistors}} = -13.6667 \text{ W}$$

The negative sign indicates that the 2 A source is supplying power. Therefore, the current must be flowing out of the positive voltage terminal with the magnitude of the voltage given by 13.6667 W/2 A = 6.8333 V.

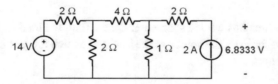

4.4   Maximum Power Transfer

When working with power in circuit analysis, designing circuits for the maximum transfer of power is often desirable. The goal may be to maximize the efficiency of power delivery to a device or to maximize the retrieval of power form a sensor. Power not delivered to a specific load is often lost into an unusable form such as waste heat. In this section, we will discuss maximum power transfer in purely resistive circuits. Maximum power transfer in circuits with inductors and capacitors will be saved for Chap. 8 when we have the benefit of phasor domain circuit analysis.

To begin, consider the Thevenin Equivalent circuit shown in Fig. 4.8. Since any circuit can be translated into a Thevenin equivalent, this simple circuit allows us to generalize our analysis to any circuit.

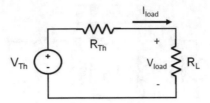

Fig. 4.8 Thevenin Equivalent circuit for maximum power transfer analysis

For this circuit, the power delivered to the load is given by

$$P_{\text{load}} = V_{\text{load}} \cdot I_{\text{load}} = \overbrace{\left(V_{\text{Th}} \frac{R_L}{R_{\text{Th}} + R_L} \right)}^{\substack{\text{Voltage from} \\ \text{Voltage Divider}}} \overbrace{\left(\frac{V_{\text{Th}}}{R_{\text{Th}} + R_L} \right)}^{\substack{\text{Current from} \\ \text{Ohm's Law}}} = \frac{V_{\text{Th}}^2 R_L}{(R_{\text{Th}} + R_L)^2} \tag{4.5}$$

In order to find the resistor, R_L, that will receive the most possible power from the source, we need to take the derivative of P_{load} with respect to R_L and set this derivative to zero.

$$\begin{aligned}
\frac{\partial P_{\text{load}}}{\partial R_L} &= \frac{\partial}{\partial R_L} \left(\frac{V_{\text{Th}}^2 R_L}{(R_{\text{Th}} + R_L)^2} \right) = V_{\text{Th}}^2 \frac{\partial}{\partial R_L} \left(\frac{R_L}{(R_{\text{Th}} + R_L)^2} \right) = \\
&= V_{\text{Th}}^2 \left(\frac{(R_{\text{Th}} + R_L)^2 - 2R_L(R_{\text{Th}} + R_L)}{(R_{\text{Th}} + R_L)^4} \right) = 0
\end{aligned} \tag{4.6}$$

$$\Rightarrow (R_{\text{Th}} + R_L)^2 - 2R_L(R_{\text{Th}} + R_L) = 0 \Rightarrow (R_{\text{Th}} + R_L) - 2R_L = 0 \Rightarrow R_{\text{Th}} = R_L$$

Therefore, the largest amount of power is delivered to the load when the load resistance, R_L, is the same as the Thevenin or source resistance, R_{Th}. Any other values of R_L will result in less power being delivered to the load.

Example 4.8 A 10 V source has a 50 Ω source resistance. Find the load resistance that will result in the maximum power transfer from the source to the load and sketch the power delivered to the load as a function of load resistance.

Solution:
The maximum power transfer will occur when the load resistance is matched to the source. Therefore, $R_L = 50\ \Omega$. The power delivered to the load as a function of load resistance for this circuit is shown below.

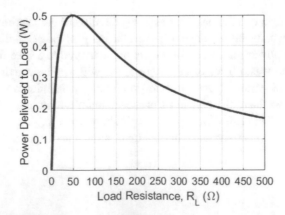

Practical Hint: While designing the load to achieve maximum power transfer is common in multiple applications ranging from power delivery to sensing and measurement, not all systems focus on power. For example, oscilloscopes and the analog to digital converters used to digitize measured signals capture the voltage and not the power of the waveform. Therefore, for these applications, matching the load resistance to the source resistance is not the optimal solution. Instead, the load resistance should be made as large as possible to maximize the voltage across the load. Likewise, if the goal is to maximize the current flowing into the load, then the load resistance should be made as small as possible. The needs of the load always dictate the optimal load resistance for a given circuit.

One common mistake made when working with maximum power transfer is to change the source resistance to match the load resistance rather than changing the load resistance to match source resistance. In other words, you should never increase the source resistance to match the load resistance. Instead, you should lower the load resistance to match the source resistance. The equation, $R_{\text{Th}} = R_{\text{L}}$, does give the maximum power transfer for a given value of R_{Th}, but it was derived assuming R_{L} is changing. If we have control over R_{Th}, then the smallest possible value of R_{Th} will give the most efficient power delivery to the load. Increasing, R_{Th} by adding a resistor in series with the output will only result in additional power being absorbed and less efficient overall power delivery.

Example 4.9 Three power supplies are available to deliver power to a 50 Ω load. All three power supplies have a voltage of 10 V, but the source resistance varies as 10 Ω, 50 Ω, and 100 Ω for each supply as shown in Fig. 4.9. What power supply would maximize the power delivered to the load? What is the power delivered to the load for each case?

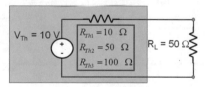

Fig. 4.9 Circuit diagram for Example 4.9

Solution:
The most efficient case will be when there is the least amount of power lost in the source. This corresponds to the 10 Ω source resistance. This can be confirmed by looking at the power delivered to the load for each case.

10 Ω Source Resistance:

$$P_{\text{load}} = \frac{V_{\text{Th}}^2 R_{\text{L}}}{(R_{\text{Th}} + R_{\text{L}})^2} = \frac{100 \cdot 50}{(10 + 50)^2} = 1.3889 \text{ W}$$

50 Ω Source Resistance:

$$P_{\text{load}} = \frac{V_{\text{Th}}^2 R_{\text{L}}}{(R_{\text{Th}} + R_{\text{L}})^2} = \frac{100 \cdot 50}{(50 + 50)^2} = 0.5 \text{ W}$$

100 Ω Source Resistance:

$$P_{load} = \frac{V_{Th}^2 R_L}{(R_{Th} + R_L)^2} = \frac{100 \cdot 50}{(100 + 50)^2} = 0.2222 \text{ W}$$

Clearly, the power delivered for the 10 Ω case is much greater than when the source resistance is made to match the load resistance. The best case, however, would be if we could also lower the load resistance to match the 10 Ω source resistance.

10 Ω Source AND Load Resistances:

$$P_{load} = \frac{V_{Th}^2 R_L}{(R_{Th} + R_L)^2} = \frac{100 \cdot 10}{(10 + 10)^2} = 2.5 \text{ W}$$

4.5   Problems

Problem 4.1: Find the power associated with each of the circuit elements given below based on the passive sign convention if $v(t) = 20$ V and $i(t) = 0.5$ A.

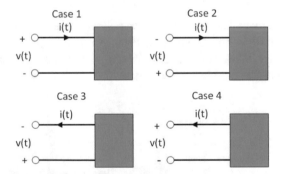

Problem 4.2: Find the power associated with each of the circuit elements given below based on the passive sign convention.

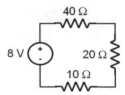

Problem 4.3: Find the power associated with each of the circuit elements given below based on the passive sign convention.

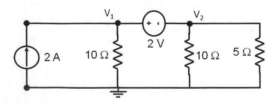

Problem 4.4: Find the total energy delivered to a circuit element if the voltage applied across its terminals is given by $v(t) = \begin{cases} 0 & t \le 0 \\ e^{-2t} \text{ V} & t > 0 \end{cases}$ and the current flowing into the positive voltage terminal is given by $i(t) = \begin{cases} 0 & t \le 0 \\ 0.4 \text{ A} & t > 0 \end{cases}$.

Problem 4.5: Find the total energy delivered to a circuit element if the voltage applied across its terminals is given by $v(t) = \begin{cases} 0 & t \le 0 \\ 2te^{-t} \text{ V} & t > 0 \end{cases}$ and the current flowing into the positive voltage terminal is given by $i(t) = \begin{cases} 0 & t \le 0 \\ e^{-t} \text{ A} & t > 0 \end{cases}$.

Problem 4.6: Find the total energy delivered to a circuit element for time less than zero and time greater than zero if the voltage applied across its terminals is given by $v(t) = 5\exp\left(-(2t)^2\right)$ V and the current flowing into the positive voltage terminal is given by $i(t) = -4t \cdot \exp\left(-(2t)^2\right)$ A.

Problem 4.7: Find the energy delivered in one day in kWh to a home for the power flow shown below.

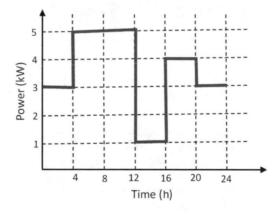

Problem 4.8: Find the energy delivered in one day in kWh to a home for the power flow shown below.

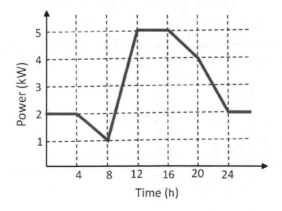

Problem 4.9: Find the power absorbed in each of the resistors, as well as the power, supplied by the source for the circuit shown below.

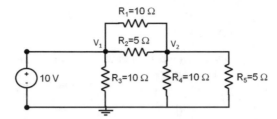

Problem 4.10: Find the power absorbed in each of the resistors, as well as the power, supplied by the sources for the circuit shown below.

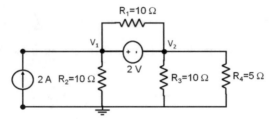

Problem 4.11: Find the power associated with each of the sources for the circuit shown below, as well as the total power, absorbed by the resistors.

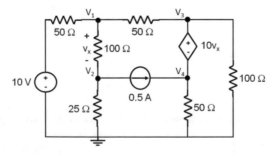

Problem 4.12: Use the conservation of energy to find the unknown resistance value in the circuit shown below if $V_1 = 7.2$ V and $V_2 = 8$ V.

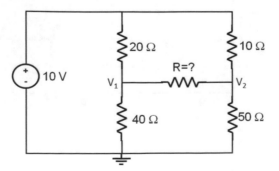

Problem 4.13: If the source resistance of a voltage source is 33 Ω, what is the load resistance that will result in the maximum power transfer to the load if possible values range from 10 Ω to 100 Ω?

Problem 4.14: If the load resistance connected to a voltage source is 25 Ω, what is the best Thevenin resistance of the source that will result in the maximum power transfer to the load if possible values range from 10 Ω to 100 Ω?

Problem 4.15: Three power supplies are available to deliver power to a 40 Ω load. All three power supplies have a voltage of 20 V, but the source resistance varies as 20 Ω, 40 Ω, and 80 Ω for each supply. What power supply would maximize the power delivered to the load? What is the power delivered to the load for each case?

Problem 4.16: Three loads are available to connect to a power supply with a voltage of 15 V and a source resistance of 100 Ω. The available load resistances are 50 Ω, 100 Ω, and 200 Ω. What load resistance would maximize the power delivered to the load? What is the power delivered to the load for each case?

Problem 4.17: Three loads are available to connect to a power supply with a voltage of 100 V and a source resistance of 40 Ω. The available load resistances are 10 Ω, 30 Ω, and 60 Ω. What load resistance would maximize the power delivered to the load? What is the power delivered to the load for each case?

DC Motors and Generators

5

One of the earliest areas of electrical engineering is the translation of electrical energy into motion (motors) and the transfer of motion into electrical energy (generators). The first usable motors and generators got their start in the mid to late 1800s and have served as the backbone of the modern industry ever since. In this chapter, we will cover the basics of direct current (DC) motors and generators. The analysis will assume steady-state operation leaving the transient response of the motor to a more advanced book. We will also discuss stepper motors and brushless DC motors due to their importance in modern industry, robotics, and control. However, stepper motors and brushless DC motors rely on carefully controlled time-varying pulse sequences to enable the motion, so it is a bit of a misnomer to label them as DC motors.

5.1 Review of Torque for Rotating Machines

The basic function of motors is to provide a sufficient torque to move a mechanical load at the desired speed. While rotational speed is relatively simple to understand, the concept of torque is difficult to quantify. Most students have not seen the concept of torque since it was first presented in their physics classes early in their engineering studies. Therefore, the goal of this section is to review the concept of torque and relate it to the motion of real objects enabling the student to design better mechatronic systems in the future.

By definition, torque is the rotational force acting on a body. For example, consider the impact of the downward force on the axis shown in Fig. 5.1. This downward force will cause the "shaft" to spin in the counterclockwise direction. The torque, $\vec{T}$, resulting from this force will be given by

$$\vec{T} = \vec{r} \times \vec{F} \tag{5.1}$$

where $\vec{F}$ is the force and $\vec{r}$ is the position vector drawn from the axis of rotation. Since the force is perpendicular to the position vector, Eq. (5.1) simplifies to

$$T = |\vec{F}| \cdot r = F \cdot r \tag{5.2}$$

where r is just the distance from the location of the force to the axis of rotation (i.e., radius of the shaft).

© Springer Nature Switzerland AG 2020
T. A. Bigelow, *Electric Circuits, Systems, and Motors*,
https://doi.org/10.1007/978-3-030-31355-5_5

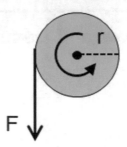

Fig. 5.1 Relating torque and force on a rotating body

Now, let us assume that the net force, F, acting on the shaft is greater than zero. Under this condition, we would expect that the rotational speed to increase due to this unbalanced force just as the linear speed would increase if a force was applied to an object under linear motion. Recall that for linear motion,

$$F = m \cdot a = m \cdot \frac{dv}{dt} \tag{5.3}$$

where m is the mass of the object, a is the linear acceleration, and v is the linear velocity of the object. For rotational motion, however, we do not use the mass. Instead, the rate of change in rotational speed, ω_m, with torque is limited by the moment of inertia, J_m, of the object.

$$T = J_m \cdot \frac{d\omega_m}{dt} \tag{5.4}$$

For a motor, the J_m would need to include the equivalent inertia for the entire mechanical load attached to the shaft including the impact of any gears connecting the motor shaft to the load. For example, if the load had an inertia of J_{load}, and it was connected to the motor shaft via a gear ratio of N, then the equivalent inertia seen by the motor shaft would be

$$J_m = J_{load} \cdot N^2 = J_{load} \cdot \left(\frac{\text{rotational speed of load}}{\text{rotational speed of motor}} \right)^2 \tag{5.5}$$

This is exactly equivalent to how the load on a transformer can be reflected to the source to find the total load on the source as will be discussed in Chap. 10. Obviously, the lower the inertia, the easier it will be to accelerate the mechanical system.

5.2   Common Speed–Torque Demands for Mechanical Loads

In our introduction to the study of motors, we are primarily interested in motors operating at some steady state. In steady state, the net torque acting on the shaft should be zero, and the motor should be spinning at a constant speed $(d\omega_m/dt = 0)$. Under these conditions, the torque supplied by the motor should exactly balance the torque demanded by the load. If the torque from the motor is greater than the torque demanded by the load, the motor speed, ω_m, will increase. Likewise, if the mechanical load is demanding more torque than the motor can provide, the motor will slow down until equilibrium is achieved. For very high torque demands, the motor will stall or stop spinning completely. However, the torque demanded by the load can also depend on the speed depending on the type of load. Some common load types are given below.

5.2.1 Constant Torque Loads

The simplest mechanical load would be one where the torque is a constant independent of speed.

$$T = Constant \tag{5.6}$$

Hoists and elevators would primarily have this type of torque due to the load being predominantly due to gravity. In addition, loads dominated by Coulomb friction would also require a constant torque with respect to speed. Coulomb friction is the friction between two dry surfaces while an object is in motion. For example, a box sliding along the floor or automobile brakes would both experience Coulomb friction. As a reminder, the Coulomb friction should not be confused with the sticking friction. Sticking friction would be the frictional force that would need to be overcome to get a stationary object moving. Sticking friction is critical for getting a motor started, but does not play a role during steady-state operation which is the focus of this text.

5.2.2 Fluid Loads

Another common load for a mechanical system would be fluid loads. Any object moving through a fluid, air, or water, would experience a drag force due to the fluid. Assuming laminar flow, the torque load resulting from the movement through the fluid is proportional to the square of the velocity.

$$T_{\text{fluid}} \propto \omega_m^2 \tag{5.7}$$

Examples of fluid loads would include fans and blowers as well as wind resistance. For example, the wheels of a car on the road would need to be able to provide both a constant torque due to the interaction between the tires and the pavement as well as a torque that increased with the square of the speed due to the flow of the air around the vehicle. In high winds or at high speeds, the fluid load will have a noticeable impact on the energy efficiency of the vehicle.

5.2.3 Viscous Friction Loads

In addition to fluid loads, viscous loads might dominate some applications where the required torque would vary linearly with speed.

$$T_{\text{viscous}} \propto \omega_m \tag{5.8}$$

Viscous friction is usually confined to the movement of bearings embedded in a lubricant. For most mechanical systems, the viscous load is relatively small except when the motor is very cold resulting in high viscosity of the lubricant. Some electrical motors may have trouble starting in cold weather due to this "unexpected" change in the motor loading conditions.

5.3   Linear Machines

The most basic electrical machine is a linear motor. While linear motors are rarely found in practical applications, they serve to simply illustrate the basic operation of electrical motors and can lead to a more intuitive understanding of their operation. Figure 5.2 shows a simple linear machine consisting of a voltage source, a resistor to model the conductive losses inside the machine, and a current-carrying bar in a magnetic field. The bar is free to move along two parallel rails. The current flow in the bar is perpendicular to the magnetic field and flows in the $\vec{L}$ direction where L is the length of the bar between the rails. The X's on the page show that the magnetic field is directed into the page (like you see the fletching on an arrow as it moves away from you).

The motion of the charges (i.e., current) in the presence of the magnetic field generates a force on the bar due to the Lorentz force that is given by

$$\vec{f} = I_A \vec{L} \times \vec{B} \tag{5.9}$$

The direction of the force is given by the cross product between the current direction and the magnetic field direction and can thus be solved by a right-hand rule. In this case, if you point your fingers in the direction of the magnetic field and your thumb in the direction of the current, then the force on the bar will be in the direction of your palm as illustrated in Fig. 5.3. Given that the current flow and magnetic field are perpendicular, Eq. (5.9) can be simplified as

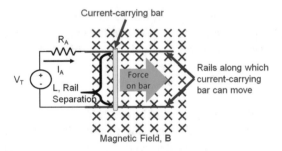

Fig. 5.2 Diagram of a simple linear motor

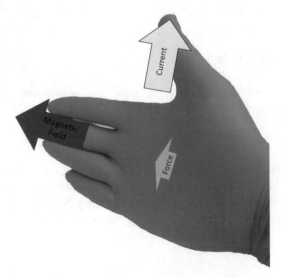

Fig. 5.3 Right-hand rule to find the direction of force for linear motor

$$f = I_A LB \quad \text{(To Right)} \tag{5.10}$$

to remove the cross product.

Example 5.1 Find the direction of the force on the bar for the linear motors shown in Fig. 5.4.

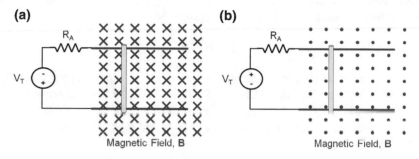

Fig. 5.4 Linear motors for Example 5.1

Solution:

Figure 5.4a: For this motor, the current will be flowing up through the bar since the polarity of the voltage is reversed. The magnetic field, however, is still going into the page. Therefore, by the right-hand rule, the force will be to the left.

Figure 5.4b: For this motor, the current will be flowing down through the bar as the top node is at a higher potential. This time, however, the magnetic field is coming out of the page as is indicated by the points rather than the crosses. The points are like the tip of an arrow as it moves toward you. Given the new direction of the magnetic field, the force will also be to the left.

Since the current flowing in the bar produces a force on the bar, the bar will begin to accelerate in the direction of that force. However, anytime a conductor cuts across magnetic field lines a voltage is induced across the conductor. The induced voltage, or back electromotive force (emf), can be found by first determining the force, $\vec{F}_m$, acting on a charged particle, Q, that is moving in the presence of a magnetic field at some velocity, $\vec{u}$.

$$\vec{F}_m = Q\vec{u} \times \vec{B} \tag{5.11}$$

Therefore, any charge particles (i.e., electrons) in the bar will experience this force as the bar moves. From basic physics, we know that forces acting on charge particles can be translated into electric field intensity, $\vec{E}$, by

$$\vec{E} = \frac{\vec{F}_m}{Q} \tag{5.12}$$

Also, once we know the electric field, we can find the voltage between two points, by integrating the electric field along a line connecting the two points.

$$V_{AB} = -\int_A^B \vec{E} \cdot d\vec{l} \tag{5.13}$$

Applying these equations to the case of the bar of the linear machine shown in Fig. 5.5 gives

$$E_{\mathrm{A}} = -\int_{L}^{0} \left(\vec{u} \times \vec{B} \right) \cdot \hat{x} dx = \int_{0}^{L} \left(\vec{u} \times \vec{B} \right) \cdot \hat{x} dx = \left(\vec{u} \times \vec{B} \right) \cdot \hat{x} L \qquad (5.14)$$

where we have recognized that $\vec{u}$ and $\vec{B}$ are constant along the bar. Also, since $\vec{u}$, $\vec{B}$, and $\hat{x}$ are all perpendicular to each other, Eq. (5.14) can be simplified as

$$E_{\mathrm{A}} = uLB \qquad (5.15)$$

Notice, however, that if the direction of the magnetic field was to flip or the bar was to be moving to the left instead of to the right, then the polarity of E_a shown in Fig. 5.5 would reverse. Also, while the force from the bar is directly related to current, the speed of the bar is directly related to the voltage across the bar. The dependence of the speed on the voltage and the force on the current is true for all motors as we will see in the future.

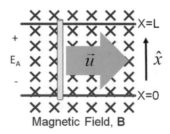

Fig. 5.5 Electromotive force (emf) induced due to motion of conducting bar in the presence of a magnetic field

The induced emf due to the motion of the bar will oppose the current, I_{A}, flowing into the bar with the resulting current being given by

$$I_{\mathrm{A}} = \frac{V_{\mathrm{T}} - E_{\mathrm{A}}}{R_{\mathrm{A}}} = \frac{V_{\mathrm{T}} - uLB}{R_{\mathrm{A}}} \qquad (5.16)$$

Therefore, as the bar accelerates due to the Lorentz force, the induced emf will increase reducing the current flow. As the current decreases, the Lorentz force will also decrease. Eventually, all of the forces will balance, the bar will cease to accelerate and will continue to travel down the rails at some velocity, u. The final equivalent circuit for the linear machine is thus given by Fig. 5.6.

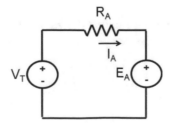

Fig. 5.6 Equivalent circuit for a linear machine

Example 5.2 For the linear machine shown in Fig. 5.2, find the initial starting force and the final speed if the length of the bar is 4 cm, V_T is 10 V, R_A is 0.5 Ω, and the magnitude of the magnetic field is 5 T. Assume that there are no frictional losses associated with the bars motion.

Solution:
The initial force on the bar will occur before the bar starts moving. Therefore, $u = 0$ m/s and $E_A = 0$ V. Therefore, the current flow in the bar is given by $I_A = \frac{V_T - 0}{R_A} = 20$ A and the force on the bar is given by $f = I_A LB = 4$ N   (To Right).

The final speed of the bar will occur after all of the forces have balanced. Since there are no frictional losses, the final Lorentz force must be zero otherwise the bar would continue to accelerate. In order for the Lorentz force to be zero, the current, I_A, must be zero. Therefore, $V_T = E_A = uLB$. Hence, the final speed on the bar is $u = \frac{V_T}{LB} = 50$ m/s.

Example 5.3 For the linear machine shown in Fig. 5.2, find the speed of the bar, power flowing into the bar, and power supplied by the voltage source if a mechanical load of 1.5 N is applied to the left. This load could be due to friction or some other drag force. The length of the bar is 4 cm, V_T is 10 V, R_A is 0.5 Ω, and the magnitude of the magnetic field is 5 T.

Solution:
This example differs from Example 5.2 in that now we have included some losses in the system. At steady state, all of the forces must still balance. Since the mechanical load in this case is 1.5 N to the left, the Lorentz force on the bar must be 1.5 N to the right in order to be at equilibrium. A larger force to the right would cause the bar to accelerate, and a smaller force would cause the bar to slow down. In order to provide this force, the current flowing in the bar must be

$$I_A = \frac{f}{LB} = \frac{(1.5 \text{ N})}{(4 \text{ cm})(5 \text{ T})} = 7.5 \text{ A}$$

Hence, E_A would be given by

$$I_A = \frac{V_T - E_A}{R_A} \Rightarrow E_A = V_T - I_A R_A = (10 \text{ V}) - (7.5 \text{ A})(0.5 \text{ } \Omega) = 6.25 \text{ V}$$

The speed of the bar would then be given by

$$u = \frac{E_A}{LB} = \frac{(6.25 \text{ V})}{(4 \text{ cm})(5 \text{ T})} = 31.25 \text{ m/s}$$

Therefore, the bar is moving slower as a result of the applied force. The power flowing into the bar is given by the voltage across the bar, E_A, multiplied by the current flowing into the bar, I_A.

$$P_{bar} = E_A I_A = (6.25 \text{ V})(7.5 \text{ A}) = 46.875 \text{ W}$$

Similarly, the power for the voltage source is given by the negative of the source voltage, V_T, multiplied by the current I_A with the negative sign resulting from the current leaving the positive voltage terminal (i.e., passive sign convention).

$$P_{\text{source}} = -V_T I_A = -(10 \text{ V})(7.5 \text{ A}) = -75 \text{ W}$$

Since this value is negative, the source is supplying 75 W of power to the linear motor.

Example 5.4 For the linear machine shown in Fig. 5.2, find the force on the bar, power flowing into the bar, and power for the voltage source if a pulling force is applied that causes the motor to move at (a) 25 m/s and (b) 75 m/s. The length of the bar is 4 cm, V_T is 10 V, R_A is 0.5 Ω, and the magnitude of the magnetic field is 5 T.

Solution:
(a) 25 m/s *bar speed*:
A velocity of 25 m/s is slower than the "no-load" speed of 50 m/s from Example 5.2. Therefore, an external load or friction must be drawing power from the linear motor. Since we know the speed, we can find the back emf.

$$E_A = uLB = (25 \text{ m/s})(4 \text{ cm})(5 \text{ T}) = 5 \text{ V}$$

Therefore, the current flowing in the motor will be given by

$$I_A = \frac{V_T - E_A}{R_A} = \frac{(10 \text{ V}) - (5 \text{ V})}{(0.5 \text{ }\Omega)} = 10 \text{ A}$$

so the force acting on the bar by the load would be

$$f = I_A LB = (10 \text{ A})(4 \text{ cm})(5 \text{ T}) = 2 \text{ N} \quad \text{(To Left)}$$

as the Lorentz force is 2 N to the right. The power flowing into the bar is given by

$$P_{\text{bar}} = E_A I_A = (5 \text{ V})(10 \text{ A}) = 50 \text{ W}$$

while the power for the voltage source is given by

$$P_{\text{source}} = -V_T I_A = -(10 \text{ V})(10 \text{ A}) = -100 \text{ W}$$

Since this value is negative, the source is supplying 100 W of power to the linear motor.
(b) 75 m/s *bar speed*:
A velocity of 75 m/s is faster than the "no-load" speed of 50 m/s from Example 5.2. Therefore, an external force is pulling the bar faster than is expected. This external force will be feeding power back into the linear machine turning it into a generator. To see this more clearly, let's complete the voltage, current, and power calculations. The back emf is once again given by

$$E_A = uLB = (75 \text{ m/s})(4 \text{ cm})(5 \text{ T}) = 15 \text{ V}$$

This voltage is greater than V_T, once again showing that the bar is providing power back to V_T and thus acting as a generator.

From the voltage, E_A, we can find the current I_A.

$$I_A = \frac{V_T - E_A}{R_A} = \frac{(10\ V) - (15\ V)}{(0.5\ \Omega)} = -10\ A$$

The sign of the current is negative indicating that the current is actually flowing in the opposite direction or back into V_T.

The power flowing into the bar is given by

$$P_{bar} = E_A I_A = (15\ V)(-10\ A) = -150\ W$$

The negative sign on the power once again shows that the bar is supplying the power and not absorbing the power.

The power for the source is given by

$$P_{source} = -V_T I_A = -(10\ V)(-10\ A) = 100\ W$$

The positive power for the source says that the source is receiving/dissipating 100 W of power from the motion of the bar. The source that is supplying this power can be found from

$$f = I_A LB = (-10\ A)(4\ cm)(5\ T) = -2\ N \quad (\text{To Left}) = 2\ N \quad (\text{To Right})$$

Therefore, an external mechanical load is pulling on the bar with a net force of 2 N to the right.

As can be seen from Example 5.4, an external force acting on the bar can cause the bar to act as a generator that provides power rather than a load which absorbs power. When acting as a generator, this force is known as the prime mover. As a generator, the machine can be directly connected to an electrical load (i.e., light bulb, secondary machine, etc.) without the need for the voltage source as is illustrated by Example 5.5.

Example 5.5 A DC linear generator is connected directly to a load as shown in Fig. 5.7. If the magnetic field is 2 T and the prime mover pushes on the 0.1 m long bar so that the velocity of the bar is 25 m/s, find the force on the bar and the power delivered to the electrical load, R_{load}.

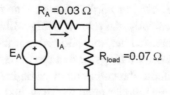

Fig. 5.7 Circuit diagram for Example 5.5

Solution:
Since the speed of the bar is known, the emf generated by the bar's motion can be determined.

$$E_A = uLB = (25\ m/s)(0.1\ m)(2\ T) = 5\ V$$

With the voltage known, we can find the current, I_A, flowing in the circuit. The current can be found from Ohm's law as

$$I_A = \frac{E_A}{R_A + R_{load}} = \frac{(5\ V)}{(0.03\ \Omega) + (0.07\ \Omega)} = 50\ A$$

The direction of the current is defined to be in the opposite direction from the current shown in Fig. 5.2. Therefore, the direction of the force would be reversed, but this will not affect the magnitude of the force that the prime mover needs to provide.

$$f = I_A LB = (50 \text{ A})(0.1 \text{ m})(2 \text{ T}) = 10 \text{ N}$$

Clearly, as the current demand increases, the force needed from the prime mover will also increase. If the prime mover cannot provide the necessary force, the speed of the bar will decrease lowering the induced emf. Eventually, the current resulting from the induced emf and the force available from the prime mover will balance. The power delivered to R_{load} can be found from

$$P_{\text{load}} = R_{\text{load}}(I_A)^2 = (0.07 \text{ } \Omega)(50 \text{ A})^2 = 175 \text{ W}$$

5.4 Basic Operation of Rotating DC Motors

The basic structure of a DC motor is shown in Fig. 5.8. The motor consists of a series of wires (i.e., windings) wrapped around a core that is usually made of laminated iron. The rotor is also connected to a shaft that allows the mechanical motion to be translated outside of the motor housing. The spinning rotor is encased by the stator which in addition to providing support also generates a magnetic field inside of the motor. The operation of the motor is easier to understand if we remove the core and focus on a single set of rotor windings as shown in Fig. 5.9.

In Fig. 5.9, the orientation of the magnetic field is shown relative to the rotor windings. This magnetic field is either established by additional wires on the stator (i.e., stator or field windings) or by the placement of permanent magnets. Current is allowed to flow in the rotor windings as the commutator provides an electrical connection to the external circuit via the brushes. In normal operation, the commutator slides passed the brushes during rotation while maintaining an electrical connection. Therefore, only the rotor windings and commutator from Fig. 5.9 will spin during motor operation. When a positive voltage is applied to the terminal on the right, as shown in Fig. 5.10, there will be a Lorentz force on the wire in the upward direction. This force will cause the rotor to turn in the counterclockwise direction. Eventually, the rotor winding will have rotated by approximately 90°. At this point, the brushes will connect to the opposite pads on the commutator. This will cause the current to change direction in the rotor windings so that the Lorentz force, and hence the torque, can continue in the same direction. Without the commutator changing the direction of the current in the windings, the DC motor would stop spinning once the rotor had turned 90°.

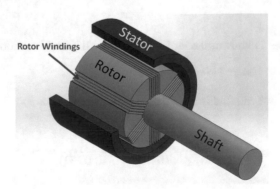

Fig. 5.8 Basic structure of DC motor

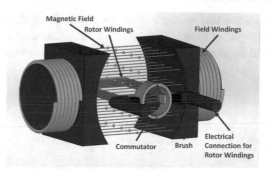

Fig. 5.9 Simplified diagram for DC motor focusing on a single rotor winding to illustrate the motor operation

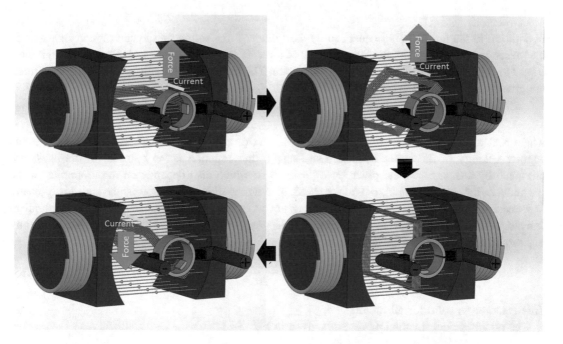

Fig. 5.10 Diagram illustrating basic motor operation emphasizing the importance of the commutator to change current direction so that the Lorentz force, and hence the torque, can always be in the same direction as the motor turns

The torque on the rotor due to the Lorentz force is given by

$$T_{\mathrm{dev}} = K\phi I_{\mathrm{A}} \tag{5.17}$$

where K is a machine constant, ϕ is the magnetic flux produced by each stator pole, and I_{A} is the current flowing in the rotor windings. This equation is very similar to the force produced by a linear machine given by $f = L \cdot B \cdot I_{\mathrm{A}}$ which also depends on the properties of the machine, L, the magnetic field, B, and the current, I_{A}. The developed torque, T_{dev}, is not the same as the output torque on the shaft of the motor, T_{out}, due to mechanical losses inside of the machine. Therefore, the output torque is given by

$$T_{\mathrm{out}} = T_{\mathrm{dev}} - T_{\mathrm{loss}} \tag{5.18}$$

where T_{loss} is the loss of torque due to mechanical losses inside of the machine. Often, it is more convenient to express the output in terms of power rather than torque or speed. Given that the power is just the torque multiplied by the speed, Eq. (5.18) can also be written as

$$P_{\text{out}} = P_{\text{dev}} - P_{\text{loss}} = T_{\text{out}} \cdot \omega_m \tag{5.19}$$

where

$$P_{\text{dev}} = T_{\text{dev}} \cdot \omega_m \tag{5.20}$$

and

$$P_{\text{loss}} = T_{\text{loss}} \cdot \omega_m \tag{5.21}$$

In addition to the torque being generated by the Lorentz force acting on the current-carrying wire in the presence of the magnetic field, the passage of the wire through the magnetic field will also induce a back emf just as was observed for the linear motor. The faster the rotor cuts the magnetic field lines, the greater the generated back emf. The back emf, E_A, is given by

$$E_A = K\phi\omega_m \tag{5.22}$$

where K is a machine constant, ϕ is the magnetic flux produced by each stator pole, and ω_m is the angular velocity of the motor in rad/s. Once again, this equation is very similar to the back emf produced by a linear machine given by $E_A = L \cdot B \cdot u$ which also depends on the properties of the machine, L, the magnetic field, B, and the speed, u. When working with motors, the speed is typically expressed in revolutions per minute (rpm) and not in radians per second. However, it is trivial to convert from rad/s to rpm as one revolution is equivalent to 2π radians. Therefore, the speed in rpm, n_m, can be found from the speed in rad/s, ω_m, from,

$$n_m = \frac{60\omega_m}{2\pi} \tag{5.23}$$

This equation is valid for all motors.

The circuit model for the DC motor is given in Fig. 5.11.

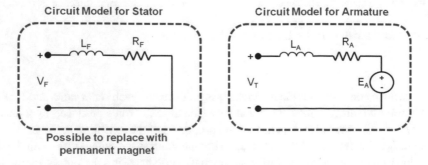

Fig. 5.11 Circuit model for DC motor

In this model, L_F is the inductance of the field windings, R_F is the resistance of the field windings, and V_F is the voltage applied to the field windings to establish the magnetic field. The field windings can also be replaced by a permanent magnet as was previously noted. For the armature, V_T is the voltage applied to the armature, L_A is the inductance of the armature windings, and R_A is the resistance that includes the combined resistances of the windings, commutator, and brushes. Since we are only interested in the DC steady-state analysis for the motor, the inductors can be treated as short circuits. The circuit model allows us to calculate the torque and speed from the current, I_A, and voltage, E_A, in the armature. In addition, the developed power can be found from

$$P_{dev} = E_A I_A \tag{5.24}$$

5.5   Shunt-Connected DC Motors

As was seen previously, DC motors require current flow in the field windings on the stator to establish a magnetic field, and current flow on the armature windings to provide the necessary torque. However, it is rare to find a motor where the field and armature windings are separately excited. Instead, a single voltage source is normally used to power both windings. Therefore, when connecting to the voltage source, the windings can be connected in either series or in parallel. When connected in parallel, the same voltage is applied across both windings as shown in Fig. 5.12, and this configuration is known as the shunt-connected DC motor. Often a variable resistor, R_{adj}, is included in the field branch to provide simple speed control as will be discussed later.

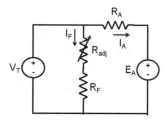

Fig. 5.12 Circuit model for shunt-connected DC motor. The inductors have been replaced by short circuits as we are assuming steady-state DC excitation

The relationship between speed and torque for the shunt-connected DC motor can be derived from the circuit diagram shown in Fig. 5.12 and Eqs. (5.17) and (5.22). The current I_A can be found by the voltage drop across R_A divided by the resistance. Therefore, the torque is related to the speed by

$$I_A = \frac{V_T - E_A}{R_A} = \frac{V_T - K\phi\omega_m}{R_A}$$
$$T_{dev} = K\phi I_A = \frac{K\phi}{R_A}(V_T - K\phi\omega_m) \tag{5.25}$$

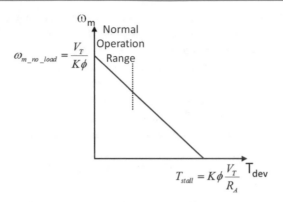

Fig. 5.13 Speed/torque relationship for shunt-connected DC motor

Likewise, the speed can be written as a function of the torque as

$$K\phi\omega_m = \left(V_T - \frac{R_A}{K\phi} T_{dev} \right)$$

$$\omega_m = \frac{V_T}{K\phi} - \frac{R_A}{(K\phi)^2} T_{dev} \qquad (5.26)$$

Therefore, there is a linear relationship between speed and torque for the shunt-connected DC motor as shown in Fig. 5.13. When no load is applied, T_{dev} of approximately 0 Nm, the motor will spin at its fastest speed of

$$\omega_{m\_no\_load} = \frac{V_T}{K\phi} \qquad (5.27)$$

Conversely, the maximum torque that will stop the motor or stall torque, T_{stall}, when ω_m is 0 rad/s is given by

$$T_{stall} = K\phi \frac{V_T}{R_A} \qquad (5.28)$$

Normally, the shunt-connected DC motor is designed to operate near its no-load speed at relatively low torques.

When solving shunt-connected DC motor problems, the first step is often to find the value of $K\phi$. Recall that $K\phi$ depends on the specific machine and the magnetic field generated by the field windings. Also, the magnetic field will strongly depend on the current in the field coils, I_F. As the field current increases, the magnetic field will also increase resulting in an increase of $K\phi$. We can find the value of $K\phi$ for a specific motor from the field current using the magnetization curve. The magnetization curve gives the back emf, E_A, as a function of field current, I_F, for a specific motor speed and is experimentally determined by the motor manufacturer. An example magnetization curve for a DC motor is shown in Fig. 5.14. The curve is approximately linear for low field currents, but it begins to saturate as the field current is increased. Since the magnetization curve provides the back emf, E_A, as a function of field current for a given speed, $K\phi$ can be calculated from (5.22) as is illustrated in the next examples.

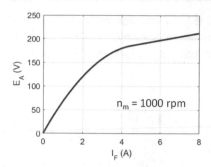

Fig. 5.14 Example magnetization curve

Example 5.6 Find the value of $K\phi$ for a shunt-connected DC motor with the magnetization curve shown in Fig. 5.14 if the field current is 4 A.

Solution:

While we DO NOT KNOW the speed of our shunt-connected DC motor currently, we can still use the information from the magnetization curve to get the $K\phi$ value. By reading off the chart, we can see that the E_A value would be 180 V when the field current is 4 A IF AND ONLY IF the speed is 1000 rpm.

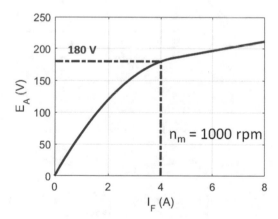

Therefore,

$$(E_A)_{\text{chart}} = 180 \text{ V}$$

$$(\omega_m)_{\text{chart}} = (n_m)_{\text{chart}} \cdot \frac{2\pi}{60} = 104.72 \text{ rad/s}$$

This speed is just the speed from the specification sheet measured by the manufacturer to facilitate the calculation of $K\phi$. It DOES NOT correspond to the speed of the motor when not being used to generate the magnetization curve. From these values,

$$(E_A)_{\text{chart}} = K\phi(\omega_m)_{\text{chart}} \Rightarrow K\phi = \frac{(E_A)_{\text{chart}}}{(\omega_m)_{\text{chart}}} = 1.7189 \text{ Wb}$$

where Wb is the abbreviation for the units of Webers.

Example 5.7 A shunt-connected DC motor has an R_{adj} of 0 Ω, an R_{F} of 20 Ω, a R_{A} of 3 Ω, and a V_{T} of 40 V. The motor also has the magnetization curve shown in Fig. 5.14. (a) Find the no-load speed for the motor in revolutions per minute. (b) Find the developed torque that drops the speed to 20% less than the no-load speed.

Solution:
(a) The first step is once again to find $K\phi$. In order to find $K\phi$, we need to know the field current.

$$I_{\text{F}} = \frac{V_{\text{T}}}{R_{\text{F}} + R_{\text{adj}}} = \frac{(40\ \text{V})}{(20\ \Omega) + (0\ \Omega)} = 2\ \text{A}$$

We can now use the chart to find $K\phi$.

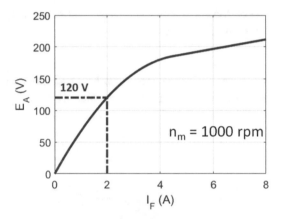

Therefore,

$$(E_{\text{A}})_{\text{chart}} = 120\ \text{V}$$

$$(\omega_m)_{\text{chart}} = (n_m)_{\text{chart}} \cdot \frac{2\pi}{60} = 104.72\ \text{rad/s}$$

$$(E_{\text{A}})_{\text{chart}} = K\phi(\omega_m)_{\text{chart}} \Rightarrow K\phi = \frac{(E_{\text{A}})_{\text{chart}}}{(\omega_m)_{\text{chart}}} = 1.146\ \text{Wb}$$

The no-load speed is then given by

$$\omega_{\text{m\_no\_load}} = \frac{V_{\text{T}}}{K\phi} = 34.91\ \text{rad/s}$$

$$n_{\text{m\_no\_load}} = \frac{60\omega_{\text{m\_no\_load}}}{2\pi} = 333.3\ \text{rpm}$$

Notice that the speed of the motor is not related in any way to the speed given on the magnetization curve.
(b) To find the torque that drops the speed by 20%, we need to first find the new speed.

$$\omega_m = 0.8\omega_{m\_no\_load} = 27.93 \text{ rad/s}$$

The developed torque can then be found from

$$T_{dev} = \frac{K\phi}{R_A}(V_T - K\phi\omega_m) = 3.0558 \text{ Nm}$$

Example 5.8 A shunt-connected DC motor has the magnetization curve shown in Fig. 5.15 and a R_{adj} of 0 Ω, an R_F of 30 Ω, a R_A of 5 Ω, and a V_T of 285 V. Find the speed and torque when the power delivered by the motor is 3000 W and there are no rotational losses given that the motor is operating in its normal operating range.

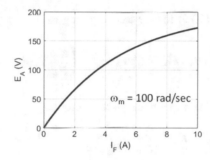

Fig. 5.15 Magnetization curve for Example 5.8

Solution:
The first step is to find $K\phi$ from the magnetization curve. The field current can be found from

$$I_F = \frac{V_T}{R_F + R_{adj}} = 9.5 \text{ A}$$

We can now use the chart to find $K\phi$.

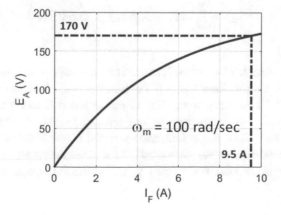

Therefore,

$$(E_A)_{chart} = 170 \text{ V}$$

$$(\omega_m)_{chart} = 100 \text{ rad/s}$$

$$(E_A)_{chart} = K\phi(\omega_m)_{chart} \Rightarrow K\phi = \frac{(E_A)_{chart}}{(\omega_m)_{chart}} = 1.7 \text{ Wb}$$

Now, we need to translate the given output power into a torque and a speed for the motor. Since there are no rotational losses in the motor, the output power is the same as the developed power. The developed power is related to the speed and torque by

$$P_{dev} = T_{dev} \cdot \omega_m$$

We can now substitute in Eq. (5.25) for the torque giving us

$$P_{dev} = \left(\frac{K\phi}{R_A} (V_T - K\phi\omega_m) \right) \cdot \omega_m = \left(\frac{K\phi}{R_A} V_T \omega_m - \frac{(K\phi)^2}{R_A} \omega_m^2 \right)$$

This expression can be written as a quadratic with two roots.

$$\frac{(K\phi)^2}{R_A} \omega_m^2 - \frac{K\phi}{R_A} V_T \omega_m + P_{dev} = 0$$

$$0.5780\omega_m^2 - 96.9\omega_m + 3000 = 0 \Rightarrow \begin{cases} \omega_m = 126.7 \text{ rad/s} \\ \omega_m = 40.97 \text{ rad/s} \end{cases}$$

Of these two possible solutions, we need to pick the higher speed as we are told that the motor is operating in its normal operating range. For a shunt-connected DC motor, this means near the no-load speed for the motor. The no-load speed is the highest speed at which the motor can spin. For this example, the no-load speed is 167.6 rad/s. Notice that no-load speed is NOT THE SAME as the motor speed and both also differ from the speed on the magnetization curve. Once the speed is known, the torque can be found from

$$T_{dev} = \frac{K\phi}{R_A} (V_T - K\phi\omega_m) = 23.68 \text{ Nm}$$

Example 5.9 A shunt-connected DC motor is needed that can supply 50 Nm of torque to a load after rotational losses. The internal rotational losses of the motor are 0.2 Nm when no gearbox is connected and the equivalent of 0.5 Nm when a gearbox is connected that has a ratio of 10 (i.e., output torque = 10× input torque). The motor has the magnetization curve shown in Fig. 5.16. Also, we know that V_T = 200 V, R_A = 2 Ω, R_{adj} = 0 Ω, and R_F = 200 Ω. (a) What is the developed torque on the motor with and without the gearbox connected? (b) What is the speed of the shaft coming out of the motor in rad/s, speed at the load, and efficiency in % of the motor with and without the gear box connected?

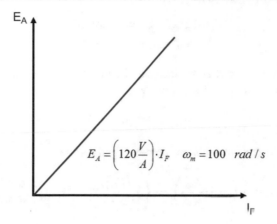

Fig. 5.16 Magnetization curve for Example 5.9

Solution:
(a) With no gearbox connected, the developed torque is just the torque needed plus the rotational losses.

$$T_{\text{dev}} = 50 \text{ Nm} + 0.2 \text{ Nm} = 50.2 \text{ Nm}$$

When the gearbox is connected, the 50 Nm of torque is reduced by a factor of 10 by the gearbox. However, the rotational losses are increased due to the increased friction. The developed torque is thus given by

$$T_{\text{dev}} = \frac{50 \text{ Nm}}{10} + 0.5 \text{ Nm} = 5.5 \text{ Nm}$$

(b) In order to find the speed, we must first find $K\phi$. For this example, the field current is given by

$$I_{\text{F}} = \frac{V_{\text{T}}}{R_{\text{F}} + R_{\text{adj}}} = 1 \text{ A}$$

Since the magnetization curve is just a line, $(E_A)_{\text{chart}}$ can be found as

$$(E_{\text{A}})_{\text{chart}} = \left(120 \ \frac{V}{A} \right) \cdot I_{\text{F}} = 120 \text{ V}$$

$$(\omega_m)_{\text{chart}} = 100 \text{ rad/s}$$

$$(E_{\text{A}})_{\text{chart}} = K\phi(\omega_m)_{\text{chart}} \Rightarrow K\phi = \frac{(E_{\text{A}})_{\text{chart}}}{(\omega_m)_{\text{chart}}} = 1.2 \text{ Wb}$$

Once $K\phi$ is known, the shaft speed can be found from

$$\omega_m = \frac{V_{\text{T}}}{K\phi} - \frac{R_{\text{A}}}{(K\phi)^2} T_{\text{dev}} \Rightarrow \begin{cases} (\omega_m)_{\text{Gear\_Box}} = 159.0 \text{ rad/s} \\ (\omega_m)_{\text{No\_Gear\_Box}} = 96.94 \text{ rad/s} \end{cases}$$

The speed at the load without the gear box will be the same as the speed at the output shaft of the motor. However, with the gearbox, the speed is reduced by a factor of 10. Therefore,

$$(\omega_{\text{load}})_{\text{Gear\_Box}} = 15.90 \text{ rad/s}$$
$$(\omega_{\text{load}})_{\text{No\_Gear\_Box}} = 96.94 \text{ rad/s}$$

In order to find the efficiency, we need to find both the input power and the output power. The input power is from the voltage source and is given by

$$P_{\text{in}} = V_{\text{T}} \cdot (I_{\text{F}} + I_{\text{A}})$$

However, the armature current depends on the developed torque.

$$P_{\text{in}} = V_{\text{T}} \cdot \left(I_{\text{F}} + \frac{T_{\text{dev}}}{K\phi}\right) \Rightarrow \begin{cases} (P_{in})_{\text{Gear\_Box}} = 1116.7 \text{ W} \\ (P_{\text{in}})_{\text{No\_Gear\_Box}} = 8566.7 \text{ W} \end{cases}$$

The output power is given by the output torque multiplied by the speed at the load.

$$P_{\text{out}} = \omega_{\text{load}} \cdot T_{\text{load}} = \begin{cases} (P_{\text{out}})_{\text{Gear\_Box}} = 795.1 \text{ W} \\ (P_{\text{out}})_{\text{No\_Gear\_Box}} = 4847.2 \text{ W} \end{cases}$$

The output power could also be found from the shaft speed multiplied by the output torque corrected for rotational losses.

$$P_{\text{out}} = (T_{\text{dev}} - T_{\text{rotational\_losses}}) \cdot \omega_m = \begin{cases} (P_{\text{out}})_{\text{Gear\_Box}} = 795.1 \text{ W} \\ (P_{\text{out}})_{\text{No\_Gear\_Box}} = 4847.2 \text{ W} \end{cases}$$

The efficiency is then found from

$$\eta = 100\frac{P_{\text{out}}}{P_{\text{in}}} \Rightarrow \begin{cases} (\eta)_{\text{Gear\_Box}} = 71.21\% \\ (\eta)_{\text{No\_Gear\_Box}} = 56.58\% \end{cases}$$

Operating motors in an efficient manner is critical. In Example 5.9, we saw that adding a gearbox significantly improved the efficiency of the motor. The lower developed torque needed with the gearbox improved the efficiency by reducing the conductive losses in the armature branch of the circuit. When optimizing efficiency, understanding where and how power is lost is critically important. Figure 5.17 shows the power flow in a DC shunt-connected motor. This diagram can be

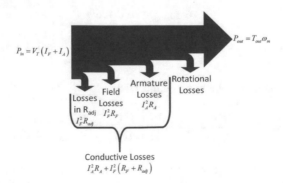

Fig. 5.17 Power flow diagram for shunt-connected DC motor

useful in quantifying and optimizing power efficiency for the motor. Examples 5.10 and 5.11 further illustrate the importance of operating the shunt-connected motor at high speed/low torque to maximize the efficiency.

Example 5.10 $V_T = 100$ V is applied to a shunt-connected DC motor with $R_{adj} = 5$ Ω. For the motor, $R_A = 0.5$ Ω and $R_F = 45$ Ω. The motor has 0.5 Nm of rotational losses and the magnetization curve shown in Fig. 5.18. The motor needs to supply a torque of 4 Nm. What is the motor speed and efficiency?

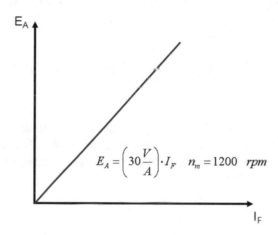

$$E_A = \left(30 \frac{V}{A} \right) \cdot I_F \quad n_m = 1200 \quad rpm$$

Fig. 5.18 Magnetization curve for Example 5.10

Solution:
Begin by finding $K\phi$.

$$I_F = \frac{V_T}{R_F + R_{adj}} = 2 \text{ A}$$

$$(E_A)_{chart} = \left(30 \frac{V}{A} \right) \cdot I_F = 60 \text{ V}$$

$$(\omega_m)_{chart} = (n_m)_{chart} \cdot \frac{2\pi}{60} = 125.664 \text{ rad/s}$$

$$K\phi = \frac{(E_A)_{chart}}{(\omega_m)_{chart}} = 0.4775 \text{ Wb}$$

Now, we need to find the developed torque.

$$T_{dev} = T_{out} + T_{loss} = 4.5 \text{ Nm}$$

From the developed torque, we can find the speed.

$$\omega_m = \frac{V_T}{K\phi} - \frac{R_A}{(K\phi)^2} T_{dev} = 199.57 \text{ rad/s}$$

To find the efficiency, we need to find the ratio of the output power to the input power. The output power is given by

$$P_{\text{out}} = T_{\text{out}}\omega_m = 798.28 \text{ W}$$

This time to find the input power, we will sum up the field losses, armature losses, rotational losses, and losses due to R_{adj}.

$$P_{\text{conductive\_losses}} = I_A^2 R_A + I_F^2(R_F + R_{\text{adj}}) = \left(\frac{T_{\text{dev}}}{K\phi}\right)^2 R_A + I_F^2(R_F + R_{\text{adj}}) = 44.41 \text{ W} + 200 \text{ W}$$

$$P_{\text{total\_losses}} = P_{\text{conductive\_losses}} + T_{\text{loss}}\omega_m = 244.41 \text{ W} + 99.79 \text{ W} = 344.20 \text{ W}$$

We can now find the efficiency from

$$\eta = 100\frac{P_{\text{out}}}{P_{\text{in}}} = 100\frac{P_{\text{out}}}{P_{\text{out}} + P_{\text{total\_losses}}} = 69.87\%$$

Example 5.11 $V_T = 100$ V is applied to a shunt-connected DC motor with $R_{\text{adj}} = 5$ Ω. For the motor, $R_A = 0.5$ Ω and $R_F = 20$ Ω. The motor has negligible rotational losses and the magnetization curve shown in Fig. 5.18. Plot the efficiency versus torque curve. What is the most efficient torque? What is the efficiency at this torque?

Solution:
Begin by finding $K\phi$.

$$I_F = \frac{V_T}{R_F + R_{\text{adj}}} = 4 \text{ A}$$

Since the magnetization curve is just a line, $(E_A)_{\text{chart}}$ can be found as

$$(E_A)_{\text{chart}} = \left(30 \frac{V}{A}\right) \cdot I_F = 120 \text{ V}$$

$$(\omega_m)_{\text{chart}} = (n_m)_{\text{chart}} \cdot \frac{2\pi}{60} = 125.664 \text{ rad/s}$$

$$K\phi = \frac{(E_A)_{\text{chart}}}{(\omega_m)_{\text{chart}}} = 0.9549 \text{ Wb}$$

Next we need to find an expression for the efficiency in terms of the torque. Thus, we need an expression for both the output power and the input power in terms of the torque. We know that

$$P_{\text{out}} = \omega_m \cdot T_{\text{out}} = \omega_m \cdot T_{\text{dev}}$$

because the motor had negligible rotational losses. Substituting Eq. (5.26) in for ω_m gives

$$P_{\text{out}} = \left(\frac{V_T}{K\phi} - \frac{R_A}{(K\phi)^2}T_{\text{dev}}\right)T_{\text{dev}}$$

The only unknown in this equation is T_{dev}. The input power is given by

$$P_{\text{in}} = V_{\text{T}}(I_{\text{A}} + I_{\text{F}}) = V_{\text{T}}\left(\frac{T_{\text{dev}}}{K\phi} + I_{\text{F}}\right)$$

where once again, the only unknown is T_{dev}. Thus, the efficiency in terms of T_{dev} is given by

$$\eta = 100\frac{P_{\text{out}}}{P_{\text{in}}} = \frac{\left(\frac{V_{\text{T}}}{K\phi}T_{\text{dev}} - \frac{R_{\text{A}}}{(K\phi)^2}T_{\text{dev}}^2\right)}{V_{\text{T}}\left(\frac{T_{\text{dev}}}{K\phi} + I_{\text{F}}\right)}$$

A plot of the efficiency versus torque is shown below. Clearly, the optimal efficiency occurs at relatively low torque.

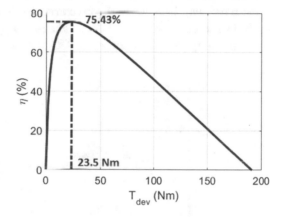

The peak in the efficiency can be found by taking the derivative of η with respect to T_{dev} and setting the result equal to zero to find the maximum point.

$$\frac{\partial\eta}{\partial T_{\text{dev}}} = \frac{\left(\frac{V_{\text{T}}}{K\phi} - \frac{2R_{\text{A}}T_{\text{dev}}}{(K\phi)^2}\right)V_{\text{T}}\left(\frac{T_{\text{dev}}}{K\phi} + I_{\text{F}}\right) - \left(\frac{V_{\text{T}}}{K\phi}T_{\text{dev}} - \frac{R_{\text{A}}}{(K\phi)^2}T_{\text{dev}}^2\right)\left(\frac{V_{\text{T}}}{K\phi}\right)}{V_{\text{T}}^2\left(\frac{T_{\text{dev}}}{K\phi} + I_{\text{F}}\right)^2} = 0$$

$$\Rightarrow \left(\frac{V_{\text{T}}}{K\phi} - \frac{2R_{\text{A}}T_{\text{dev}}}{(K\phi)^2}\right)V_{\text{T}}\left(\frac{T_{\text{dev}}}{K\phi} + I_{\text{F}}\right) = \left(\frac{V_{\text{T}}}{K\phi}T_{\text{dev}} - \frac{R_{\text{A}}}{(K\phi)^2}T_{\text{dev}}^2\right)\left(\frac{V_{\text{T}}}{K\phi}\right)$$

$$\Rightarrow \left(\frac{V_{\text{T}}}{K\phi} - \frac{2R_{\text{A}}T_{\text{dev}}}{(K\phi)^2}\right)\left(\frac{T_{\text{dev}}}{K\phi} + I_{\text{F}}\right) = \left(\frac{V_{\text{T}}}{(K\phi)^2}T_{\text{dev}} - \frac{R_{\text{A}}}{(K\phi)^3}T_{\text{dev}}^2\right)$$

$$\Rightarrow \left(\frac{V_{\text{T}}}{(K\phi)^2}T_{\text{dev}} - \frac{2R_{\text{A}}}{(K\phi)^3}T_{\text{dev}}^2 + \frac{V_{\text{T}}I_{\text{F}}}{K\phi} - \frac{2R_{\text{A}}I_{\text{F}}}{(K\phi)^2}T_{\text{dev}}\right) = \left(\frac{V_{\text{T}}}{(K\phi)^2}T_{\text{dev}} - \frac{R_{\text{A}}}{(K\phi)^3}T_{\text{dev}}^2\right)$$

$$\Rightarrow \frac{R_{\text{A}}}{(K\phi)^2}T_{\text{dev}}^2 + \frac{2R_{\text{A}}I_{\text{F}}}{(K\phi)}T_{\text{dev}} - V_{\text{T}}I_{\text{F}} = 0$$

This expression has two roots.

$$T_{dev} = -31.0980 \text{ Nm} \quad or \quad 23.4585 \text{ Nm}$$

However, since negative torque is meaningless in this context, the most efficient torque is 23.4585 Nm which has a corresponding efficiency of 75.43%.

Previously we stated that the variable resistor, R_{adj}, could be used to control the speed of the shunt-connected DC motor. If the resistance, R_{adj}, is increased, then this will result in a decrease in the field current, I_F. As the field current decreases, the magnetic field generated by the field coils will decrease resulting in a decrease in $K\phi$. As $K\phi$ decreases, the no-load speed $\omega_{m\_no\_load} = \frac{V_T}{K\phi}$ will increase and the slope of the speed/torque curve given by $\omega_m = \frac{V_T}{K\phi} - \frac{R_A}{(K\phi)^2} T_{dev}$ will sharpen as illustrated in Fig. 5.19. Therefore, if we are operating the motor in its normal operating range near the no-load speed, the speed of the motor will increase as R_{adj} increases. Notice that this type of motor control only works at relatively low torques. At higher torques, the motor speed may decrease with increasing R_{adj}. However, it is rare to operate shunt-connected DC motors at high torques.

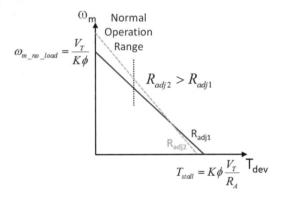

Fig. 5.19 Change in speed/torque relationship for shunt-connected DC motor when R_{adj} is increased

Example 5.12 Now we want to increase the speed of the motor described in Example 5.10 (developed torque still at 4.5 Nm) by 10%. What should be the new value of R_{adj}? Sketch the speed torque curve for the motor with both values or R_{adj}.

Solution:
In order to solve for R_{adj}, we need to first find the new desired speed. Since the original speed was 199.57 rad/s, an increase of 10% would give us a speed of 219.53 rad/s. We can now use this speed and the given torque to find an equation for $K\phi$ from

$$T_{dev} = \frac{K\phi}{R_A}(V_T - K\phi\omega_m) = (K\phi)\frac{V_T}{R_A} - (K\phi)^2\frac{\omega_m}{R_A}$$

$$\Rightarrow (K\phi)^2\frac{\omega_m}{R_A} - (K\phi)\frac{V_T}{R_A} + T_{dev} = 0$$

$$\Rightarrow (K\phi)^2 439.05 - (K\phi)200 + 4.5 = 0$$

This quadratic yields two possible values for $K\phi$.

$$(K\phi) = 0.4318 \text{ Wb} \quad OR \quad 0.0237 \text{ Wb}$$

The best choice is the value that is closest to the original $K\phi$ value as it is reasonable to assume that only a small change in the field current was needed to achieve the small change in speed.

$$(K\phi) = 0.4318 \text{ Wb}$$

With $K\phi$ known, we can find the E_A value from the magnetization curve that would give this $K\phi$ value.

$$(E_A)_{\text{chart}} = K\phi(\omega_m)_{\text{chart}} = 54.26 \text{ V}$$

We can then read the needed field current value from the magnetization curve and use this value to calculate the needed R_{adj}.

$$I_F = \frac{(E_A)_{\text{chart}}}{\left(30\ \frac{V}{A}\right)} = 1.809 \text{ A} = \frac{V_T}{R_F + R_{\text{adj}}} \Rightarrow R_{\text{adj}} = 10.29 \ \Omega$$

A sketch of the speed/torque curves for both values of R_{adj} is shown below.

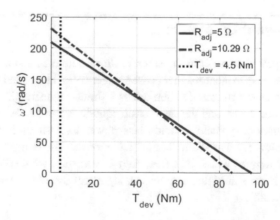

Clearly, as R_{adj} increased, the change in slope of the curve resulted in the higher speed.

The increase in speed as $K\phi$ decreases can lead to dangerous motor operation if a shunt-connected DC motor is not properly designed or utilized. For example, if the field windings should suddenly become disconnected from the voltage source (i.e., $R_{\text{adj}} \rightarrow \infty$), $K\phi$ will fall toward zero and the motor speed will increase rapidly. $K\phi$ will never exactly reach zero due to residual magnetization. However, at high speed, the motor could tear itself apart.

In addition to a decrease in $K\phi$ due to a sudden loss of the field windings, $K\phi$ can also be reduced by flux weakening. Flux weakening is a reduction of the average magnetic flux due to the magnetic field generated by current flow in the armature. Flux weakening is the most significant when operating near the saturation region of the magnetization curve as illustrated in Fig. 5.20. Since the armature current depends on torque and an increasing armature current will result in greater flux weakening, the speed of the shunt-connected DC motor can actually increase with increasing torque. Also, as was discussed previously, the torque demand for many loads increases with speed.

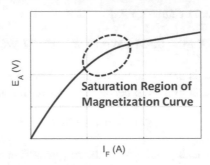

Fig. 5.20 Saturation region of the magnetization curve

Therefore, a speed increase due to flux weakening can result in a demand for more torque by the load which will result in even more significant flux weakening. As the flux weakening continues, the motor speed can increase out of control. To compensate for flux weakening, it is common to add several turns of field windings in series with the armature in addition to the field windings that are in parallel with the armature (i.e., type of compound DC motor). The flux from the field windings in series with the armature will increase as the torque/armature current increases compensating for the impact on the flux of the armature currents. A motor with these extra windings is called a stabilized shunt motor.

5.6 Permanent Magnet DC Motors

Permanent magnet DC motors operate on the same basic principle as a shunt-connected DC motor. The only difference is that a permanent magnet is used to generate the needed flux rather than the field windings. Therefore, permanent magnet DC motors and shunt-connected DC motors have the same linearly shaped speed/torque curve, and the same basic equations can be used to analyze both motors. Permanent magnet DC motors typically provide less power than shunt-connected DC motors and most small/cheap DC motors are permanent magnet DC motors. The motors are typically operated at higher speeds and lower torques as the flux (i.e., $K\phi$) is less than what can be achieved with field windings. Operating the motors at higher torques can significantly reduce the life of the brushes due to the higher current demands.

5.7 Series-Connected DC Motors

In addition to having the field coils and the armature windings connected in parallel, the two sets of coils can be connected in series. The circuit model for the series connected motor is shown in Fig. 5.21. For the series-connected motor, the current that provides the torque by flowing in the

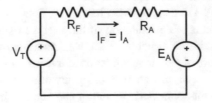

Fig. 5.21 Circuit model for series-connected DC motor

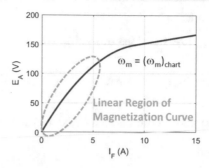

Fig. 5.22 Linear region of the magnetization curve

armature windings is the same current that generates the magnetic field produced by the field windings. Therefore, changes in torque on the motor will change the field current resulting in a change in $K\phi$.

Since $K\phi$ depends on the motor torque, finding $K\phi$ is rarely the first step when analyzing series-connected DC motors. Instead, we assume that the motor is running in the linear region of the magnetization curve as illustrated in Fig. 5.22. In this region, $(E_A)_{chart}$ is directly proportional to the field current I_F as given by

$$(E_A)_{chart} = \overbrace{\left(KK_F \cdot (\omega_m)_{chart}\right)}^{\text{Proportionality Constant}} I_F \tag{5.29}$$

Therefore,

$$K\phi = \frac{(E_A)_{chart}}{(\omega_m)_{chart}} = \frac{\left(KK_F \cdot (\omega_m)_{chart}\right)I_F}{(\omega_m)_{chart}} = KK_F I_F \tag{5.30}$$

Hence, finding KK_F is normally the first step in solving problems with series-connected DC motors.

Example 5.13 Find KK_F for the magnetization curves shown in Figs. 5.16 and 5.18.

Solution:
Figure 5.13: The magnetization curve from Fig. 5.13 is shown below.

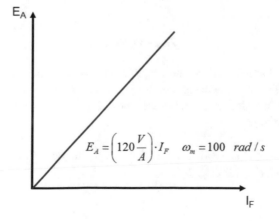

$\overbrace{}^{\text{Proportionality Constant}}$

Since $(E_A)_{\text{chart}} = \overbrace{(KK_F \cdot (\omega_m)_{\text{chart}})}^{\text{Proportionality Constant}} I_F$, the slope of the E_A versus I_F curve must be given by

$$Slope = \left(KK_F \cdot (\omega_m)_{\text{chart}}\right) \Rightarrow KK_F = \frac{Slope}{(\omega_m)_{\text{chart}}} = \frac{120}{100} = 1.2 \text{ Wb/A}$$

Figure 5.15: The magnetization curve from Fig. 5.13 is shown below.

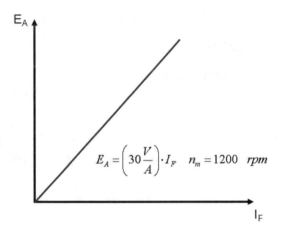

For this magnetization curve, the speed from the chart first needs to be converted to rad/s.

$$(\omega_m)_{\text{chart}} = (n_m)_{\text{chart}} \cdot \frac{2\pi}{60} = 125.664 \text{ rad/s}$$

$$KK_F = \frac{Slope}{(\omega_m)_{\text{chart}}} = \frac{30}{125.664} = 0.2387 \text{ Wb/A}$$

Assuming we are working in the linear region of the magnetization curve, we can derive equations relating speed, torque, voltage, and current for the series-connected DC motor. From Eq. (5.17), we know that $T_{\text{dev}} = K\phi I_A$. Since $K\phi = KK_F I_F$ and $I_F = I_A$, the developed torque is related to the current by

$$T_{\text{dev}} = KK_F I_A^2 \tag{5.31}$$

Hence, for series-connected DC motors, the torque increases with the square of the current making them very useful in applications requiring high torques such as drills, starter motors in cars, elevator motors, and the motors in trains. Also, since we know $E_A = K\phi\omega_m$ from Eq. (5.22), the terminal voltage, V_T, can be written as

$$V_T = I_A(R_F + R_A) + E_A = I_A(R_F + R_A) + K\phi\omega_m = I_A(R_F + R_A) + KK_F I_A \omega_m$$
$$= ((R_F + R_A) + KK_F \omega_m)I_A \tag{5.32}$$

Therefore,

$$T_{\text{dev}} = KK_F I_A^2 = KK_F \left(\frac{V_T}{(R_F + R_A) + KK_F \omega_m} \right)^2 = \frac{KK_F V_T^2}{\left((R_F + R_A) + KK_F \omega_m \right)^2} \quad (5.33)$$

Notice that the speed will go to zero when

$$T_{\text{dev}} = T_{\text{stall}} = \frac{KK_F V_T^2}{(R_F + R_A)^2} \quad (5.34)$$

We can also write the speed as a function of the torque as

$$\omega_m = \frac{V_T}{\sqrt{T_{\text{dev}}} \sqrt{KK_F}} - \frac{(R_F + R_A)}{KK_F} \quad (5.35)$$

Unlike a shunt connected or permanent magnet DC motor, speed and torque are not linearly related for a series-connected DC motor.

Example 5.14 A series-connected DC motor has $V_T = 200$ V, $R_A = 1$ Ω, $R_F = 14$ Ω, and the magnetization curve shown in Fig. 5.23. Sketch the speed/torque characteristics for this motor.

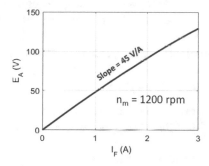

Fig. 5.23 Magnetization curve for Example 5.14

Solution:
The first step is to find KK_F from the slope of the magnetization curve and the speed shown on the magnetization curve.

$$(\omega_m)_{chart} = (n_m)_{chart} \cdot \frac{2\pi}{60} = 125.664 \text{ rad/s}$$

$$KK_F = \frac{Slope}{(\omega_m)_{chart}} = \frac{45}{125.664} = 0.3581 \text{ Wb/A}$$

The maximum torque in the plot is the stalling torque where the speed would drop to zero.

$$T_{\text{dev}} = T_{\text{stall}} = \frac{KK_F V_T^2}{(R_F + R_A)^2} = 63.6619 \text{ Nm}$$

The minimum torque is more complicated as a torque of zero would result in an infinite speed according to Eq. (5.35). Therefore, for our plot, we will limit the minimum torque to 1.5 Nm. For real series-connected DC motors, the minimum developed torque will always be limited by the rotational losses of the system. From these torque values, the speed can be found from (5.35) as shown below.

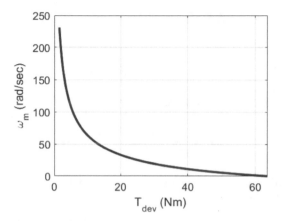

The speed/torque curve found in Example 5.14 is relatively typical for series-connected DC motor. At low loads, the speed of the motor increases dramatically. Therefore, there is a danger of the motor ripping itself apart should the motor be run with no-load attached. Often safety circuitry needs to be enabled to protect series-connected DC motors should the load suddenly drop-off to prevent the motor from running too fast. One shortcoming of Example 5.14 is that the magnetization curve was assumed to be linear over the entire operating range of the motor. This is normally not the case at higher loads/load currents as is illustrated by Example 5.15.

Example 5.15 A series-connected DC motor has $V_T = 200$ V, $R_A = 1$ Ω, $R_F = 14$ Ω, and the magnetization curve shown in Fig. 5.24. Sketch the speed/torque characteristics for this motor.

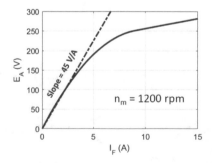

Fig. 5.24 Magnetization curve for Example 5.15

Solution:

The simplest solution is to vary the field current over a specified range and calculate the speed and torque for every value of the field current. From Example 5.14, we found that the field current varied from 2.0467 to 13.3333 A when the torque varied from 1.5 Nm to T_{stall}. Therefore, we will limit our field current values to this range. From the values of field current, we can find appropriate values of $K\phi$ from the magnetization curve. Each field current value will have a unique $K\phi$ value given by

$$K\phi = \frac{(E_A)_{\text{chart}}}{(\omega_m)_{\text{chart}}}$$

Of course, this requires the magnetization curve to be specified by values in a computer file to be done efficiently. Once the $K\phi$ value for each current value is known, the developed torque can be found from $T_{\text{dev}} = K\phi I_F$ because $I_A = I_F$. Also, the back emf, E_A, can be found from $E_A = V_T - I_F(R_A + R_F)$. This E_A is NOT the same as the E_A from the chart used to find $K\phi$. Once E_A is known, ω_m can be found from $E_A = K\phi\omega_m$. The resulting speed/torque curve is shown below.

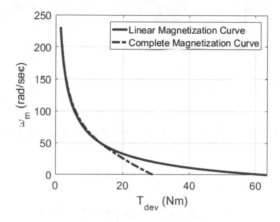

The linear approximation of the magnetization curve is fine at low torques. However, at high torques, the speed is reduced resulting in a lower stalling torque.

When solving for series-connected DC motors, it is often useful to know how power is flowing in the circuit. The power flow for a series-connected DC motor is summarized in Fig. 5.25. Examples 5.16 and 5.17 illustrate how solving for power flow can help solve for the operating point of the motor.

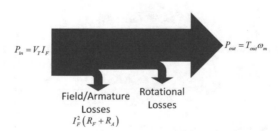

Fig. 5.25 Power flow diagram for series-connected DC motor

Example 5.16 A series-connected DC motor has $V_T = 800$ V, $R_A = 2.5$ Ω, $R_F = 155$ Ω, and the magnetization curve shown in Fig. 5.26. What is the rotational speed for the motor, the torque supplied by the motor, and the efficiency of power delivery to the mechanical load if the output power is 695 W and the rotational losses are negligible?

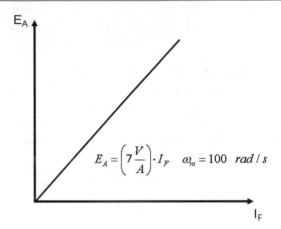

Fig. 5.26 Magnetization curve for Example 5.16

Solution:

The first step is to find KK_F from the slope of the magnetization curve and the speed shown on the magnetization curve.

$$(\omega_m)_{\text{chart}} = 100 \text{ rad/s}$$

$$KK_F = \frac{Slope}{(\omega_m)_{\text{chart}}} = \frac{7}{100} = 0.07 \text{ Wb/A}$$

The next step is to find a relationship between the developed power which is given and either the speed or the torque. Since,

$$T_{\text{dev}} = \frac{KK_F V_{\text{T}}^2}{\left((R_{\text{F}} + R_{\text{A}}) + KK_F\omega_m\right)^2}$$

and

$$P_{\text{dev}} = T_{\text{dev}} \cdot \omega_m$$

The developed power can be written as

$$P_{\text{dev}} = \frac{\omega_m KK_F V_{\text{T}}^2}{\left((R_{\text{F}} + R_{\text{A}}) + KK_F\omega_m\right)^2}$$

This expression can now be written as a quadratic in ω_m.

$$\Rightarrow P_{\text{dev}}\left((R_{\text{F}} + R_{\text{A}}) + KK_F\omega_m\right)^2 = \omega_m KK_F V_{\text{T}}^2$$

$$P_{\text{dev}}(R_{\text{F}} + R_{\text{A}})^2 + P_{\text{dev}}(KK_F)^2\omega_m^2 + 2P_{\text{dev}}(R_{\text{F}} + R_{\text{A}})KK_F\omega_m = \omega_m KK_F V_{\text{T}}^2$$

$$\left[P_{\text{dev}}(KK_F)^2\right]\omega_m^2 + \omega_m\left[2P_{\text{dev}}(R_{\text{F}} + R_{\text{A}})KK_F - KK_F V_{\text{T}}^2\right] + P_{\text{dev}}(R_{\text{F}} + R_{\text{A}})^2 = 0$$

This expression has two possible roots.

$$\omega_m = 8017.4 \text{ rad/s} \quad OR \quad 631.4 \text{ rad/s}$$

With the speed known, the torque can be found from $T_{dev} = P_{dev}/\omega_m$.
Hence, $T_{dev} = 0.0867$ Nm $\quad OR \quad 1.1012$ Nm

With the torque known, the field current can be found from $I_F = I_A = \sqrt{\frac{T_{dev}}{KK_F}}$ and used with $P_{in} = V_T I_F$ to get the input powers of

$$P_{in} = 890.5 \text{ W} \quad OR \quad 3173 \text{ W}$$

Hence the efficiency of the motor is given by

$$\eta = 100 \frac{P_{out}}{P_{in}} = 78.0861\% \quad OR \quad 21.9139\%$$

Therefore, the higher speed solution is the most efficient. However, a motor spinning at 8017 rad/s (76,560 rpm) is unlikely. Therefore, if this were a real motor, the second solution is more reasonable.

Example 5.17 A series-connected DC motor needs to provide 330 W of power to a mechanical load. The motor has a KK_F value of 0.04 Wb/A. If the input power is 430 W, what is the speed of the motor, torque on the motor, and voltage V_T applied to the motor assuming negligible rotational losses given $R_A + R_F = 7\ \Omega$?

Solution:
Since KK_F is given, the first step is to find an equation that can relate the provided power quantities to the motor's voltage, current, torque, or speed. Since rotational losses are negligible, the only losses will be the conductive losses in R_A and R_F. Therefore,

$$P_{conductive\_losses} = P_{in} - P_{out} = I_F^2(R_F + R_A)$$
$$430 \text{ W} - 330 \text{ W} = I_F^2(7\ \Omega) \Rightarrow I_F = 3.7796 \text{ A}$$

Also, since $P_{in} = V_T I_F \Rightarrow V_T = 113.8$ V
Likewise, $T_{dev} = KK_F I_A^2 = 0.5714$ Nm and

$$\omega_m = \frac{V_T}{\sqrt{KK_F}} \frac{1}{\sqrt{T_{dev}}} - \frac{R_A + R_F}{KK_F} = 577.5 \text{ rad/s}$$

5.8   Universal Motors

Universal motors are a special class of series-connected DC motors that can be powered by single-phase AC voltage sources. Notice that for series-connected DC motors, the torque is independent of the direction of the current or the polarity of the voltage in Eqs. (5.31) and (5.33). Therefore, the polarity can be changing with time and the torque will continue to be in the same direction. Usually, the only difference in construction between a traditional series-connected DC motor and a universal motor is the lamination of the stator to avoid eddy current losses. Actually, if you examine an "AC motor" and find that it has brushes and a commutator, it is likely a universal motor.

For a given weight, universal motors produce more power than other types making them advantages for handheld tools and small appliances. Also, when the load torque increases, the universal motor slows down. Other motors (shunt DC, AC induction) tend to run at relatively constant speed and thus could potentially draw excessive current. Therefore, the universal motor is more suitable for loads that experience varying torques (i.e., drills, blenders, etc.). Universal motors can also be designed to operate at much higher speeds than other AC motors. Unfortunately, universal motors and DC machines, in general, are not as robust as brushless motors as the brushes and commutators tend to wear out relatively quickly.

5.9   Compound DC Motors

As was mentioned when discussing the impact of flux weakening in shunt-connected DC motors, it is possible to have field windings connected both in series and in parallel with the armature windings as shown in Fig. 5.27. Motors with windings in both series and in parallel are called Compound DC motors. Compound motors combine some of the best features of series and shunt-connected DC motors and the properties can be tailored by adjusting the number of windings in serial and in parallel. Compound motors will have a higher starting torque than shunt-connected DC motors while avoiding the high speeds that can occur when no load is connected as happens for the series-connected motor.

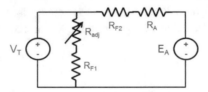

Fig. 5.27   Circuit diagram for compound DC motor

5.10   DC Generators

Recall that for a DC motor, the motion of the wires in the presence of the magnetic field from the stator resulted in a back emf, E_A. Balancing E_A with respect to the other currents and voltages in the circuit allowed us to calculate the speed/torque curve for the motor. However, all that is required to generate E_A is to have wires cutting through a magnetic field. Therefore, just as we saw for linear machines, kinetic energy from a prime mover can also be used to generate E_A and subsequently electrical energy. All that is needed is to have a magnetic field from the stator to generate the necessary Kϕ. The back emf is then still given by $E_A = K\phi\omega_m$.

5.10.1   Separately Excited or Permanent Magnet DC Generators

The first type of DC generator uses either a separate voltage source connected to the field windings or a permanent magnet to generate the required magnetic field as illustrated in Fig. 5.28. For the separately excited generator, a variable resistor is often included in series with the field windings to allow control of the magnetic field. Changes in the magnetic field will change Kϕ resulting in a controlled change in E_A. Permanent magnet generators replace the field windings with a permanent magnet similar to permanent magnet DC motors. In fact, a permanent magnet DC motor can be used as a simple generator by disconnecting it from it power source and applying a torque to the shaft. For the generator, the terminal voltage V_T refers to the voltage at the output of the generator being applied across the electrical load, R_{Load}, which is being powered by the generator.

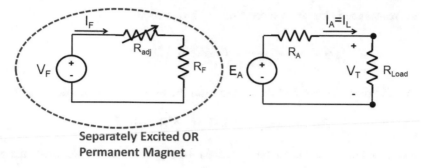

Separately Excited OR
Permanent Magnet

Fig. 5.28 Circuit model for separately excited/permanent magnet DC generator

Example 5.18 A permanent magnet DC generator has a $K\phi$ value of 3.2 Wb and an R_A of 5 Ω. The prime mover rotates the armature at a speed of 1370 rpm. Determine the no-load voltage, the full-load voltage, and the torque required by the prime mover at full load if the full-load power required by the load is 300 W. Assume that most of the power is delivered to the load and rotational losses are negligible.

Solution:
Since we know the speed, we can directly find the value for E_A.

$$E_A = K\phi\omega_m = K\phi \cdot n_m \cdot \frac{2\pi}{60} = 459.09 \text{ V}$$

With no load connected, there would be no current flow through R_A. This would mean no current drop across R_A, so the terminal voltage would still be E_A.

$$V_{T\_no\_load} = E_A = 459.09 \text{ V}$$

When the full load is connected, we know the power going into the load.

$$P_{\text{load}} = 300 \text{ W} = V_T I_A$$

Both the terminal voltage, V_T, and the load current, I_A, can be written in terms of E_A and the resistor values.

$$I_A = \frac{E_A}{R_{\text{Load}} + R_A} \Rightarrow V_T = I_A R_{\text{load}} = \frac{R_{\text{Load}} E_A}{R_{\text{Load}} + R_A}$$

Therefore,

$$P_{\text{load}} = 300 \text{ W} = \frac{R_{\text{Load}} E_A^2}{(R_{\text{Load}} + R_A)^2} = \frac{R_{\text{Load}} E_A^2}{(R_{\text{Load}}^2 + 2R_{\text{Load}} R_A + R_A^2)}$$

$$R_{\text{Load}}^2 + 10R_{\text{Load}} + 25 = 702.545 R_{\text{Load}}$$

$$R_{\text{Load}}^2 - 692.545 R_{\text{Load}} + 25 = 0$$

$$R_{\text{Load}} = 692.51 \text{ Ω} \quad OR \quad 0.0361 \text{ Ω}$$

Once the load resistance is known, we can find the load current.

$$I_A = \frac{E_A}{R_{\text{Load}} + R_A} = 0.6582 \text{ A} \quad OR \quad 91.16 \text{ A}$$

With the current known, we can find the power from the prime mover and the terminal voltage.

$$P_{\text{dev}} = E_A I_A = 302.17 \text{ W} \quad OR \quad 41.85 \text{ kW}$$

Cleary, if most of the power is lost in the load, then the higher value of load resistance provides the correct solution.

$$R_{\text{Load}} = 692.51 \text{ } \Omega$$
$$I_A = 0.6582 \text{ A}$$
$$P_{\text{dev}} = 302.17 \text{ W}$$
$$V_{\text{T\_full\_load}} = I_A R_{\text{Load}} = 455.81 \text{ V}$$

From either the power or the current, we can find the torque needed from the prime mover.

$$T_{\text{dev}} = K\phi I_A = \frac{P_{\text{dev}}}{\omega_m} = 2.1062 \text{ Nm}$$

Example 5.19 For the generator described in Example 5.18, plot the torque required by the prime mover as a function of current and power delivered to the load. The generator has a $K\phi$ value of 3.2 Wb and an R_A of 5 Ω. The prime mover rotates the armature at a speed of 1370 rpm. The load current and load power are changed by varying R_{Load} from 50 to 50,000 Ω.

Solution:
The torque demanded by the generator is given by

$$T_{\text{dev}} = \frac{P_{\text{dev}}}{\omega_m} = K\phi I_A = 3.2 I_A$$

$$I_A = \frac{E_A}{R_A + R_{\text{Load}}} = \frac{K\phi\omega_m}{R_A + R_{\text{Load}}} \Rightarrow T_{\text{dev}} = 3.2 \frac{K\phi\omega_m}{R_A + R_{\text{Load}}}$$

Similarly, the torque relationship to load power is obtained by first translating the power into a current.

$$P_{\text{load}} = V_T I_A = (I_A R_{\text{Load}}) I_A = R_{\text{Load}} I_A^2$$

Therefore, the torque required as a function of current and power can be plotted as shown below.

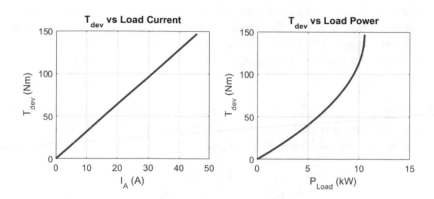

The torque demanded from the prime mover increases as the current/power demanded by the load increases. If no current is needed from the generator, the shaft can turn very easily. As more current is needed, it becomes harder and harder for the prime mover to turn the shaft. If the prime mover cannot supply the needed torque, the generator will stall.

When working with generators, the voltage across the load will depend on the load due to the voltage drop across the armature windings. The amount of drop for a particular generator is quantified by the percent voltage regulation given by

$$Voltage\ Regulation = 100\frac{V_{T\_no\_load} - V_{T\_full\_load}}{V_{T\_full\_load}} \tag{5.36}$$

where $V_{T\_no\_load}$ is the terminal voltage when no load is connected and $V_{T\_full\_load}$ is the terminal voltage when the maximum rated load current (i.e., minimum resistance) is connected. For most applications, the voltage regulation should be within a few percent.

Example 5.20 A separately excited DC generator has $V_F = 200$ V, R_{adj} variable from 0 to 60 Ω, $R_F = 40$ Ω, $R_A = 2$ Ω and the magnetization curve shown in Fig. 5.29. The prime mover is rotating the shaft of the generator at a speed of 1500 rpm.

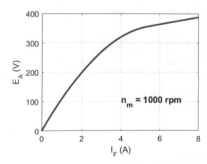

Fig. 5.29 Magnetization curve for Example 5.20

(a) What is the maximum and minimum no-load terminal voltages from the generator?

(b) If R_{adj} was 60 Ω, what would be the terminal voltage, voltage regulation, and torque demanded from the prime mover at full load if the load resistance for full load was 50 Ω?

(c) What value of R_{adj} would return the terminal voltage to the no-load value?

Solution:

Part (a) Changing the resistance R_{adj} in the field branch will change the field current. As the field current changes, $K\phi$ will also change. Therefore, the first step is to find the values of $K\phi$ corresponding to the largest and smallest values of R_{adj}.

$$I_F = \frac{V_F}{R_F + R_{adj}} = \overbrace{2\ A}^{R_{adj}=60\ \Omega} \quad OR \quad \overbrace{5\ A}^{R_{adj}=0\ \Omega}$$

Using the field current and the magnetization curve, $(E_A)_{chart}$ can be found as

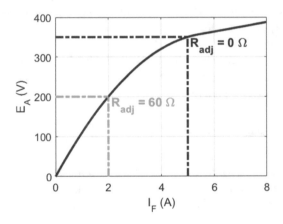

$$(E_A)_{chart} = 200\ V \quad OR \quad 350\ V$$

$(E_A)_{chart}$ can then be used to find $K\phi$ using the motor speed from the magnetization curve.

$$(\omega_m)_{chart} = (n_m)_{chart} \cdot \frac{2\pi}{60} = 104.72\ rad/s$$

$$K\phi = \frac{(E_A)_{chart}}{(\omega_m)_{chart}} = \overbrace{1.9099\ Wb}^{R_{adj}=60\ \Omega} \quad OR \quad \overbrace{3.3432\ Wb}^{R_{adj}=0\ \Omega}$$

Once $K\phi$ is known, $E_A = K\phi\omega_m$ can be found for the generator given the shaft rotation speed of 1500 rpm. This E_A is the no-load terminal voltage for the generator.

$$V_{T\_no\_load} = \overbrace{300\ V}^{R_{adj}=60\ \Omega} \quad OR \quad \overbrace{525\ V}^{R_{adj}=0\ \Omega}$$

Part (b) With R_{adj} = 60 Ω, we know that

$$V_{T\_no\_load} = E_A = 300\ V$$

This E_A is across both R_A and the load resistance as shown below.

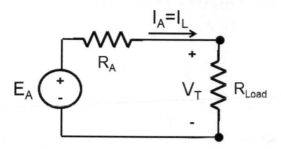

Therefore,

$$I_A = \frac{E_A}{R_A + R_{\text{Load}}} = 5.7692 \text{ A}$$

$$T_{\text{dev}} = K\phi I_A = (1.9099)I_A = 11.0184 \text{ Nm}$$

$$V_{\text{T\_full\_load}} = I_A R_{\text{Load}} = 288.4615 \text{ V}$$

$$Voltage\ Regulation = 100\frac{V_{\text{T\_no\_load}} - V_{\text{T\_full\_load}}}{V_{\text{T\_full\_load}}} = 100\frac{300 - 288.4615}{288.4615} = 4\%$$

Part (c) In order to increase the terminal voltage at full load, we need to increase E_A. To increase, E_A we need to increase $K\phi$. To increase $K\phi$, we need to increase the current in the field windings which will require a decrease in R_{adj}. First, however, we need to find the desired E_A.

$$(V_{\text{T\_full\_load}})_{\text{desired}} = 300 \text{ V} \Rightarrow I_A = \frac{300 \text{ V}}{R_{\text{Load}}} = 6 \text{ A}$$
$$(E_A)_{\text{desired}} = I_A(R_A + R_{\text{Load}}) = 312 \text{ V}$$

We can use this desired E_A to find the desired $K\phi$ and the desired $(E_A)_{\text{chart}}$.

$$(K\phi)_{\text{desired}} = \frac{(E_A)_{\text{desired}}}{\omega_m} = 1.9863 \text{ Wb}$$

$$\left((E_A)_{\text{chart}}\right)_{\text{desired}} = (K\phi)_{\text{desired}}(\omega_m)_{\text{chart}} = 208 \text{ V}$$

The magnetization curve can then be used in reverse to find the desired field current.

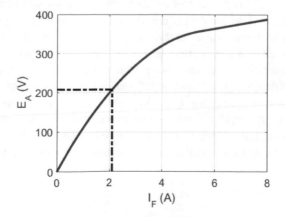

$$(I_F)_{desired} = 2.1 \text{ A} \Rightarrow (R_F + R_{adj}) = \frac{V_F}{(I_F)_{desired}} = 95.2381 \ \Omega$$

Therefore,

$$(R_{adj})_{desired} = 55.2381 \ \Omega$$

5.10.2 Shunt DC Generators

Using a separate voltage source to excite the field windings is often not practical. Therefore, another type of DC generator connects the field windings in parallel with the electrical load so that the generator can be used to power the coils and the load as shown in Fig. 5.30. For this generator, the load current, I_L, is not the same as the current through the armature windings, I_A, or the current through the field windings, I_F. Since $I_A = I_F + I_L$, the voltage drop across R_A will be greater than for a separately excited or permanent magnet generator for the same load current. As a result, a shunt-connected DC generator will have poorer voltage regulation.

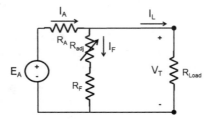

Fig. 5.30 Shunt-connected DC generator

At first, it may seem as if the shunt-connected DC generator could not possibly work. The field current used to generate the magnetic field, $K\phi$, must come from the voltage supplied by E_A. However, in order to generate E_A, a magnetic field is needed as $E_A = K\phi\omega_m$. How then does the magnetic field get established in the first place? The answer is that there is usually a residual magnetic field remaining on the poles of the generator (i.e., small "permanent" magnet). The same thing can happen when you leave a screwdriver in the presence of a magnetic field. This residual magnetic field will allow a small value of E_A to be established when the generator starts to turn. The E_A will then cause a larger field current to flow increasing the magnetic field. As the magnetic field increases, the induced voltage will increase creating a stronger magnetic field. The process will continue until a stable operating point is reached usually near where the magnetization curve saturates.

There are times when the shunt DC generator is started but no voltage builds up on the terminals. This can occur if the residual magnetic field is too small or if the generator is rotating in the wrong direction. If spinning in the wrong direction, the field current will be attempting to produce a magnetic field that opposes the residual field. In both cases, the issue can be resolved by "flashing the field" which involves temporarily connecting the generator to a DC source. The DC source will allow current to flow in the field windings establishing the residual magnetic field with the proper direction.

5.11 Stepper Motors and Brushless DC Motors

While there are some differences in construction, stepper motors and brushless DC motors operate on the same basic principle. A microcontroller is used to apply a DC voltage to different field windings based on the position of the rotor. The rotor, which normally consists of one or more permanent magnets, aligns with the magnetic field following each excitation causing the motor to spin. Stepper motors and brushless DC motors do not have brushes or a commutator significantly improving their reliability over traditional DC motors. Stepper motors also allow for very precise control of position making them ideally suited for many robotics applications. Stepper motors and brushless DC motors are also able to spin at very high speeds (greater than 50,000 rpm). Such speeds are not achievable with other motor designs. However, stepper motors and brushless DC motors tend to be more expensive than their traditional DC motor counterparts and they are restricted to lower power applications.

In order to understand the basic operation of these motors, consider the artistic rendering of a 3-phase/2-Pole stepper motor shown in Fig. 5.31. For a stepper motor, the number of poles (P) must be an even number and references the number of "North" and "South" poles on the rotor. The number of phases (N), gives the number of stator coils that can be independently excited to control the direction of the magnetic field. In Fig. 5.31, the 2 poles are labeled N and S corresponding to the North and South pole of a single magnet. Likewise, the 3 phases are denoted a, b, and c and are positioned symmetrically about the stator.

When a positive voltage is applied to a, the current will flow into the page establishing a magnetic field pointing up according to the right-hand rule relating current and magnetic field. The poles on the rotor will rotate to align with this magnetic field. If a positive voltage is then applied to c′ with no voltage applied to a, the direction of the magnetic field will change by 60° and the rotor will rotate to align with the new magnetic field. If a positive voltage is then applied to b, the direction of the magnetic field will once again change by 60° resulting in another 60° step for the rotor. Applying a voltage to a′, c, and b′ will step the magnetic field and the rotor by an additional 60° step for each voltage pulse. Therefore, the rotor will complete one revolution after 6 voltage pulses. The number of degrees per step is given by

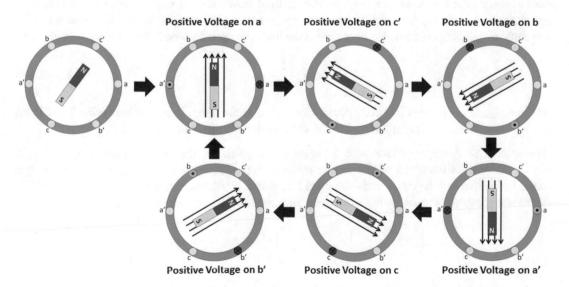

Fig. 5.31 Basic operation of a 3-Phase/2-Pole stepper motor

$$\theta_{step} = \frac{360°}{NP} \tag{5.37}$$

Example 5.21 What is the step size in degrees for a stepper motor with 3 phases and 4 poles?

Solution:
Based on the formula, the step size is given by

$$\theta_{step} = \frac{360°}{NP} = \frac{360°}{3 \cdot 4} = 30°$$

This can also be solved by considering the artistic rendering of a 3-Phase/4-Pole motor shown below.

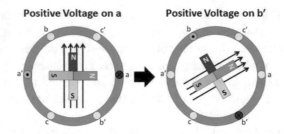

When a positive voltage is applied to a, one set of poles will align with this field. If we then apply a positive voltage to b', the other set of poles is closer and will, therefore, align with this new field location. Since the second set of poles are rotated 90° with respect to the first set of poles, the rotor will only move by 30° for this new field location. Obviously, the motor controller must know the number of poles on the rotor to provide the correct pulse sequence.

The speed of the stepper motor and brushless DC motor is set by the rate at which the pulses are sent and not by balancing the speed/torque characteristics of the motor. In fact, the motor can spin continuously or it can pause for long periods of time between each step. Stepper motors do have speed/torque curves, but they give the maximum speed for a given torque and not the exact speed for a specific torque. If the torque does not exceed the limit at a specific speed, then the speed is given by

$$n_m = \frac{1}{NP} n_{pulses} \tag{5.38}$$

where n_{pulses} is the number of pulses per minute and n_m is the speed in revolutions per minute. If the torque exceeds the maximum torque for a given speed, then the motor will stall.

Example 5.22 A stepper motor with 5 phases and 8 poles has the speed–torque curve shown in Fig. 5.32. What is the speed of the stepper motor in rpm if the developed torque from the stepper motor is 0.4 Nm and the motor is driven by pulses sent at 30,000 pulses per minute? What would be the motor speed if the number of pulses were increased to 80,000 pulses per minute?

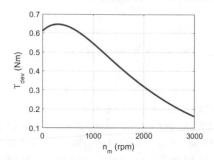

Fig. 5.32 Speed-torque curve for stepper motor in Example 5.22

Solution:

The speed of the motor is given by

$$n_m = \frac{1}{NP} n_{\text{pulses}}$$

provided the motor does not stall. We can determine if the motor will stall by plotting the operating point on the same graph as the speed versus torque curve. For the 30,000 and 80,000 pulses per minute, the motor will try to spin at 750 rpm and 2000 rpm, respectively, while supplying a torque of 0.4 Nm. When plotted below, we see that the 750 rpm case is below the speed–torque curve so the motor will not stall and will thus spin at 750 rpm. However, for the 2000 rpm case, the desired operating point is above the speed-torque curve. Therefore, the motor would stall and the speed would be 0 rpm.

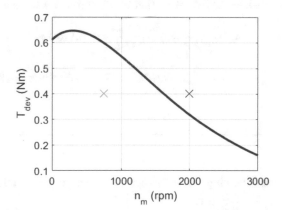

Example 5.23 A 3-phase stepper motor must be able to move in steps no bigger than 8°. How many poles are needed? How many pulses are needed from the controller every second to operate at 500 rpm?

Solution:

Number of Poles:

$$\theta_{\text{step}} = \frac{360°}{NP} \Rightarrow P = \frac{360°}{N\theta_{\text{step}}} = 15$$

However, the number of poles must be an even number. Therefore, 16 poles are needed if the step size is to be no bigger than 8°. With 16 poles, the step size would be 7.5°.

Pulse Rate:

$$n_m = \frac{1}{NP}n_{\text{pulses}} \Rightarrow n_{\text{pulses}} = NPn_m = 3 \cdot 16 \cdot \frac{500 \text{ rev}}{\text{min}} \cdot \frac{1 \text{ min}}{60 \text{ s}} = 400/\text{s}$$

Example 5.24 A 3-phase stepping motor is to be used to drive a linear axis for a robot. The motor output shaft will be connected to a screw thread with a screw pitch of 1.5 mm. We want to be able to have a spatial control of at least 0.05 mm. How many poles are needed? How many pulses are needed from the controller every second to move the linear axis at a rate of 90 mm/s?

Solution:
Number of Poles:
We first need to translate the spatial step into an angular step using the screw pitch. The screw pitch is the spacing between the threads. Therefore, one complete rotation will move the screw the length of the screw pitch.

$$\theta_{\text{step}} < (0.05 \text{ mm})\frac{360°}{1.5 \text{ mm}} = 12°$$

$$\theta_{\text{step}} = \frac{360°}{NP} \Rightarrow P \geq \frac{360°}{N\theta_{\text{step}}} = 10$$

Pulse Rate:
Once again, we need to translate the lateral motion into rotation via the screw pitch.

$$n_m = \frac{v}{pitch} = \frac{90 \text{ mm/s}}{1.5 \text{ mm/rev}} = 60 \text{ rev/s}$$

$$n_m = \frac{1}{NP}n_{\text{pulses}} \Rightarrow n_{\text{pulses}} = NPn_m = 3 \cdot 10 \cdot 60 \text{ rev/s} = 1800/\text{s}$$

5.12   Problems

Problem 5.1: Which of the following linear motors will have the bar moving to the left?

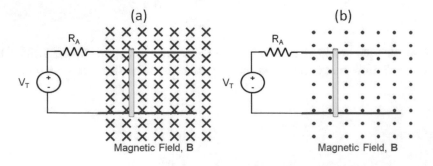

Problem 5.2: A linear machine has a bar length of 3 cm, V_T is 10 V, R_A is 0.25 Ω, and the magnitude of the magnetic field is 4 T. Find the initial starting force and the final speed. Assume that there are no frictional losses associated with the bars motion.

Problem 5.3: For the linear machine shown below, find the speed of the bar, power flowing into the bar, and power supplied by the voltage source if a mechanical load of 0.25 N is applied to the left. This load could be due to friction or some other drag force. The length of the bar is 2 cm, V_T is 10 V, R_A is 0.5 Ω, and the magnitude of the magnetic field is 2 T.

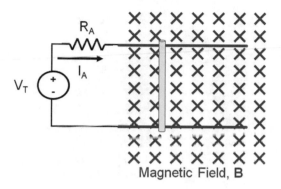

Magnetic Field, **B**

Problem 5.4: For the linear machine shown below, find the speed of the bar, power flowing into the bar, and power supplied by the voltage source if a mechanical load of 0.25 N is applied to the left. The length of the bar is 2 cm, V_T is 10 V, R_A is 0.5 Ω, and the magnitude of the magnetic field is 2 T.

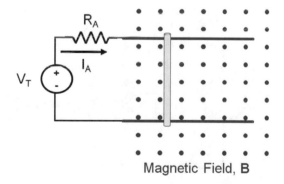

Magnetic Field, **B**

Problem 5.5: For the linear machine shown below, find the speed of the bar, power flowing into the bar, and power supplied by the voltage source if a mechanical load of 1.5 N is applied to the right. The length of the bar is 5 cm, V_T is 2 V, R_A is 0.25 Ω, and the magnitude of the magnetic field is 8 T.

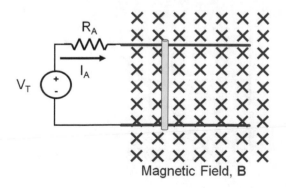

Magnetic Field, **B**

Problem 5.6: Repeat problem 5 assuming no load is applied to the bar.

Problem 5.7: For the linear machine shown below, find the force on the bar, power flowing into the bar, and power for the voltage source if a pulling force is applied that causes the motor to move to the right at (a) 30 m/s and (b) 90 m/s. The length of the bar is 10 cm, V_T is 10 V, R_A is 0.2 Ω, and the magnitude of the magnetic field is 1.5 T.

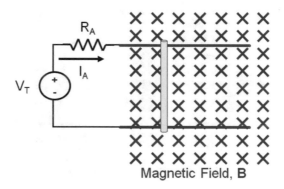

Magnetic Field, **B**

Problem 5.8: A DC linear generator is connected directly to a load as shown below. If the magnetic field is 2 T and the prime mover pushes on the 0.2 m long bar so that the velocity of the bar is 10 m/s, find the magnitude of the force on the bar and the power delivered to the electrical load, R_{load}.

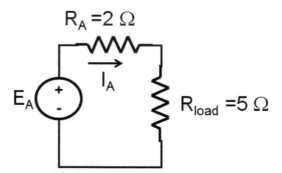

Problem 5.9: A shunt-connected DC motor has the magnetization curve shown with $V_T = 130$ V, $R_A = 3$ Ω, and $R_F + R_{adj} = 160$ Ω. (a) Find the no-load speed for the motor in rad/s. (b) Find the developed torque that drops the speed to 5% less than the no-load speed.

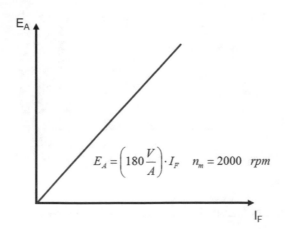

Problem 5.10: $V_T = 100$ V is applied to a shunt-connected DC motor with $R_{adj} = 50$ Ω and the magnetization curve shown. For the motor, $R_A = 0.5$ Ω and $R_F = 20$ Ω. The motor has negligible rotational losses. Plot the speed versus torque curve. What is the maximum speed and maximum torque available from the motor?

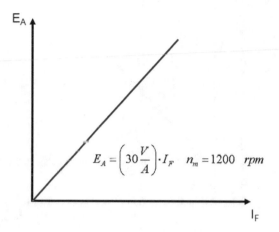

$$E_A = \left(30\frac{V}{A}\right) \cdot I_F \quad n_m = 1200 \quad rpm$$

Problem 5.11: Assume that you have a DC motor with the following magnetization curve over the operation range of interest and with $R_F + R_{adj} = 25$ Ω and $R_A = 2$ Ω. Assume that the motor is in the shunt configuration with $V_T = 95$ V as shown below. Find the speed in rad/s and torque developed in Nm when the power delivered by the motor is 100 W and there are no rotational losses given the motor is operating in its normal operating range.

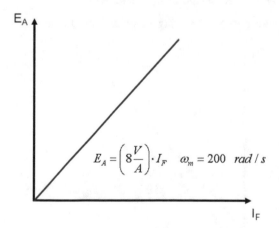

$$E_A = \left(8\frac{V}{A}\right) \cdot I_F \quad \omega_m = 200 \quad rad/s$$

Problem 5.12: A $V_T = 180$ V is applied to a shunt-connected DC motor with R_{adj} set so that $I_F = 3$ A and the magnetization curve shown. For the motor, $R_A = 0.5\ \Omega$ and $R_F = 20\ \Omega$. The motor needs to supply a torque of 50 Nm and has rotational losses equivalent to 5 Nm of torque. Determine the motor speed in rad/s and value of R_{adj}.

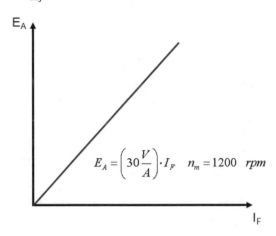

$$E_A = \left(30\frac{V}{A}\right)\cdot I_F \quad n_m = 1200 \quad rpm$$

Problem 5.13: For the motor described in Problem 12, what is the motor speed in rad/s if the new developed torque demanded from the motor is 80 Nm?

Problem 5.14: For the motor described in Problem 12, what should be the new value of R_{adj} to reduce the speed by 3% from the value found when solving Problem 12. Assume that the torque values remain the same and only a small change in R_{adj} is needed.

Problem 5.15: A DC motor has the following magnetization curve.

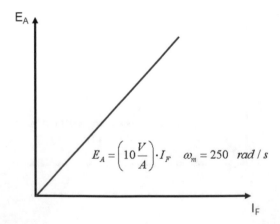

$$E_A = \left(10\frac{V}{A}\right)\cdot I_F \quad \omega_m = 250 \quad rad/s$$

Given that the motor is connected in the Shunt configuration with $V_T = 200$ V, $R_{adj} = 5\ \Omega$, $R_A = 0.5\ \Omega$, and $R_F = 10\ \Omega$. Find the following:

(a) The speed of the motor if the torque supplied by the motor is 25 Nm (assume no rotational losses).

(b) Now you want the motor to spin at 300 rad/s with the same load and same voltage source. Find the new value for R_{adj}.

Problem 5.16: Assume that you have a shunt-connected DC motor with a motor controller where you DO NOT know the magnetization curve, the resistances or any of the applied voltages. However, you can accurately measure the speed of the motor for varying loads. Specifically, you know the following.

- When a load of T_{dev} = 50 Nm, the motor stops spinning. (i.e., T_{stall} = 50 Nm).
- When a load of T_{dev} = 5 Nm is connected to the motor, the motor spins at a speed of 380 rad/s
- You may assume rotational losses are 0 W.

What would be the speed of the motor if the load connected to the motor were 10 Nm?

Problem 5.17: A shunt-connected DC motor has the following magnetization curve. If R_F = 30 Ω, R_A = 5 Ω, R_{adj} = 0 Ω, and V_T = 285 V. Find the speed and torque developed when the power delivered by the motor is 3000 W and there are no rotational losses given the motor is operating in its normal operating range.

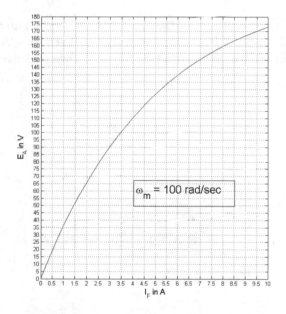

Problem 5.18: A shunt-connected DC motor is needed that can supply 12 Nm of torque to a load after rotational losses. The internal rotational losses of the motor are 0.1 Nm when no gearbox is connected and the equivalent of 0.4 Nm when a gearbox is connected that has a ratio of 10 (i.e., output torque = 10 × input torque). The motor has the magnetization curve shown below. Also, we know that V_T = 150 V, R_A = 1 Ω, R_{adj} = 0 Ω, and R_F = 200 Ω. (a) What is the developed torque on the motor with and without the gearbox connected? (b) What is the speed of the shaft coming out of the motor in rad/s, speed at the load, and efficiency in % of the motor with and without the gear box connected?

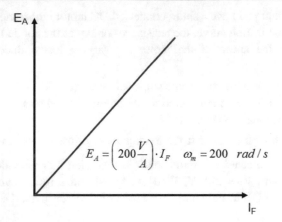

Problem 5.19: $V_T = 200$ V is applied to a shunt-connected DC motor with the magnetization curve shown with R_{adj} set so that $I_F = 1.4$ A. For the motor, $R_A = 2\ \Omega$ and $R_F = 75\ \Omega$. The motor needs to supply a torque of 6.5 Nm and has 0.5 Nm or rotational losses. Determine the motor speed, efficiency, and value of R_{adj}.

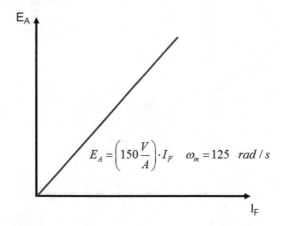

Problem 5.20: $V_T = 100$ V is applied to a shunt-connected DC motor with $R_{adj} = 50\ \Omega$. For the motor, $R_A = 0.5\ \Omega$ and $R_F = 20\ \Omega$. The motor has negligible rotational losses and the magnetization curve shown below. Plot the efficiency versus torque curve assuming no rotational losses. What is the most efficient torque? What is the efficiency at this torque?

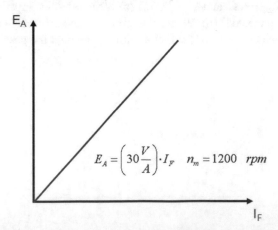

Problem 5.21: Assume that you have a shunt-connected DC motor with the following magnetization curve over the operation range of interest and with $R_F = 4\ \Omega$, $R_{adj} = 0\ \Omega$, $L_F = 500$ mH, and $R_A = 0.25\ \Omega$. Also, you want to connect this to a battery with a $V_T = 36$ V.

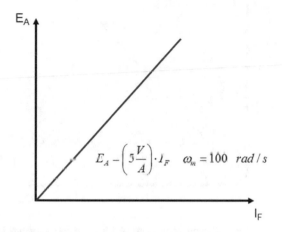

$$E_A = \left(5\frac{V}{A}\right) \cdot I_F \quad \omega_m = 100 \quad rad/s$$

(a) The motor is connected to a shaft with a diameter of 10 cm that is used to wind a cable and lift a load straight up off of the ground. If the torque needed by the motor is 20 Nm what is the mass of the load assuming no rotational losses?

(b) Now assume that a different load is attached that demands a torque of 10 Nm. What is the speed of the motor in revolutions per minute when lifting this new load?

(c) How high would the load from part b get after 10 s?

Problem 5.22: Assume that you have a DC motor with the following magnetization curve over the operation range of interest and with $R_F = 2\ \Omega$, $R_{adj} = 0\ \Omega$, $L_F = 500$ mH, and $R_A = 0.25\ \Omega$. Also, you want to connect this to a battery with a $V_T = 36$ V and the motor needs to provide 1 Nm of torque with negligible rotational losses. Find the speed of the motor and the power from the battery if the motor is connected in (a) the shunt and (b) the series configuration.

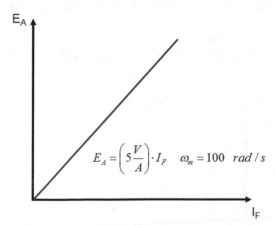

$$E_A = \left(5\frac{V}{A}\right) \cdot I_F \quad \omega_m = 100 \quad rad/s$$

Problem 5.23: A series-connected DC motor has $R_F + R_A = 3\ \Omega$, and the following magnetization curve. You may neglect rotational losses.

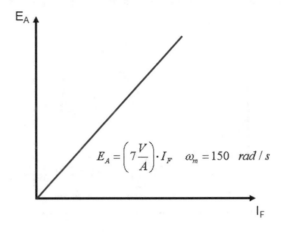

$$E_A = \left(7\frac{V}{A}\right)\cdot I_F \quad \omega_m = 150 \quad rad/s$$

(a) Find the torque on the motor if the current from the DC voltage source is 20 A and $V_T = 200$ V.
(b) Find the speed **in rpm** if the current from the DC voltage source is 20 A and $V_T = 200$ V.
(c) Find the new voltage, V_T, that needs to be applied to the motor when a 50 Nm load is connected so that the speed is 200 rad/s.

Problem 5.24: If a series-connected motor has $R_F + R_A = 5\ \Omega$ and draws 2 A of current from a DC voltage source of $V_T = 120$ V when the torque is 2 Nm with negligible rotational losses, find the speed when the torque is 5 Nm

Problem 5.25: A series-connected DC motor needs to provide an output torque of 15 Nm with negligible rotational losses. The motor has resistance values of $R_A = 3\ \Omega$ and $R_F = 22\ \Omega$ while the magnetization curve for the motor is shown below. The motor is connected to a motor controller that can provide a V_T from 50 to 350 V.

(a) What is the smallest value of V_T for which the motor will still be able to turn for the desired torque?
(b) Derive an expression for the efficiency in terms of V_T.
(c) What is the most efficient V_T and the motor speed corresponding to this V_T?

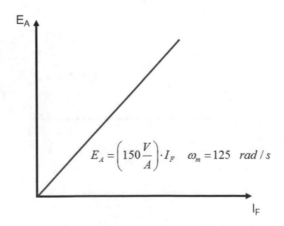

$$E_A = \left(150\frac{V}{A}\right)\cdot I_F \quad \omega_m = 125 \quad rad/s$$

Problem 5.26: A series-connected DC motor needs to provide $P_{out} = 200$ W of power to a mechanical load. The magnetization curve for the motor is shown below. If the input power is $P_{in} = 250$ W, what is the speed of the motor, torque on the motor, and voltage V_T applied to the motor assuming negligible rotational losses given $R_A + R_F = 12$ Ω.

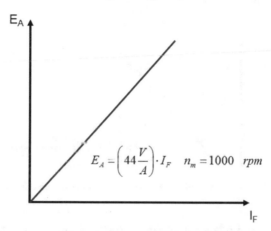

$$E_A = \left(44\frac{V}{A}\right)\cdot I_F \quad n_m = 1000 \ rpm$$

Problem 5.27: A series-connected DC motor has $R_A = 5$ Ω, and $R_F = 25$ Ω. When driving an unknown load at 1100 rpm, the current is $I_A = 2$ A from a voltage source of $V_T = 300$ V with a rotational loss of 1 Nm.

(a) Find the output power and the developed torque.
(b) What fraction of the input power makes it to the output?
(c) Find the maximum allowed developed torque and corresponding motor speed that will ensure that no more than 25% of the power is lost in conductive losses in the armature and field windings.

Problem 5.28: Derive an expression for the efficiency of a series-connected DC motor in terms of the required torque (T_{dev}), the armature resistance (R_A), the field resistance (R_F), the applied voltage (V_T), and KK_F assuming negligible rotational losses.

Problem 5.29: A series-connected DC motor has $R_F + R_A = 2$ Ω and demands 1 kW of power from the DC voltage source of $V_T = 200$ V when running at 100 rad/s. Neglecting rotational losses:

(a) Find the torque on the motor.
(b) Find the new speed in rpm if the new torque is 25 Nm.
(c) Find the new voltage, V_T, that needs to be applied to the motor with the 25 Nm load so that the speed returns to 100 rad/s.

Problem 5.30: A permanent magnet DC generator has a kϕ value of 2 Wb and an R_A of 5 Ω. The prime mover rotates the armature at a speed of 1200 rpm. Determine the no-load voltage, the full-load voltage, and the torque required by the prime mover at full load if the full-load power required by the load is 120 W (assuming most of the power is delivered to the load) neglecting rotational losses.

Problem 5.31: A separately excited DC generator has $V_F = 80$ V, R_{adj} variable from 0 to 50 Ω, $R_F = 20$ Ω, $R_A = 0.5$ Ω and the magnetization curve shown. The prime mover is rotating the shaft of the generator at a speed of 1800 rpm.

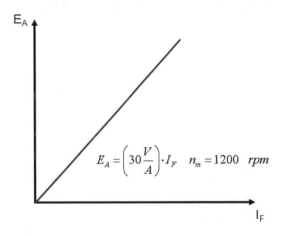

$$E_A = \left(30\frac{V}{A}\right)\cdot I_F \quad n_m = 1200 \quad rpm$$

(a) What is the maximum and minimum no-load terminal voltages from the generator?
(b) If R_{adj} was 50 Ω, what would be the terminal voltage, voltage regulation, and torque demanded from the prime mover at full load if the load resistance for full load was 8 Ω?
(c) What value of R_{adj} would return the terminal voltage to the no-load value?

Problem 5.32: A shunt-connected DC motor has the following magnetization curve taken at $\omega_m = 95$ rad/s.

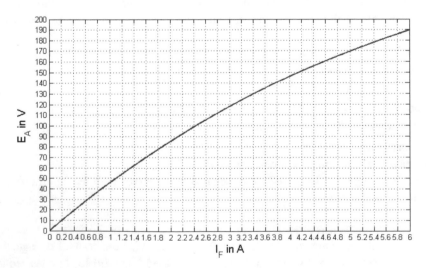

For the motor, $R_F = 80$ Ω, $R_A = 17$ Ω, $R_{adj} = 0$ Ω, and $V_T = 128$ V.

(a) Find the speed of the motor in rad/s if the required torque is 1.25 Nm.
(b) If this motor with the speed/torque determined is being used to drive a permanent magnet DC generator with a has a $k\phi$ value of 0.5 Wb **for the generator** and an R_A of 12 Ω **for the generator**. What is the load resistance, R_{load}, connected to the generator?

Problem 5.33: A series-connected DC motor is connected to a permenant magnet DC generator as shown via the rotating shaft.

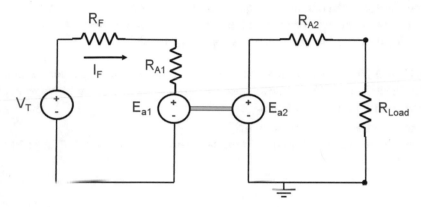

We know that **rotational losses along the shaft correspond to a torque of 5 Nm**, $R_F + R_{A1} = 9\ \Omega$, and R_{Load} is $2\ \Omega$. Also, we know:

- From the specification sheet, we know that if the shaft of the generator were to spin at 130 rad/s with no load connected, $R_{Load} = \infty$, the voltage at the output of the generator would be 180 V.
- When the series-connected DC motor is connected, the shaft spins at a speed of 200 rad/s, the voltage across the load, R_{Load}, is 200 V, and the current from voltage source supplying the DC motor, I_F, is 50 A.

Determine the following:

(a) What is the armature resistance of the generator, R_{A2}?
(b) What is the value of KK_F for the motor?
(c) What terminal voltage is being applied to the series-connected DC motor?

Problem 5.34: A stepper motor with 3 phases and 6 poles has the speed-torque curve shown.

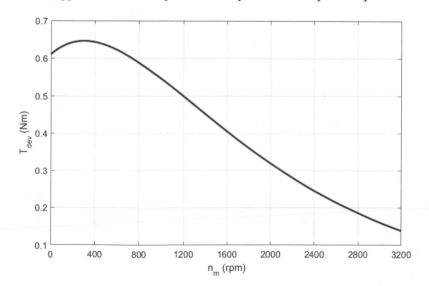

(a) What is the step size for the stepper motor in degrees?
(b) What is the speed of the stepper motor in rpm if the motor is driven by pulses sent at 35,000 pulses per minute if the developed torque from the stepper motor is 0.4 Nm?
(c) What would be the motor speed in rpm if the motor is driven by pulses sent at 15,000 pulses per minute if the developed torque from the stepper motor is 0.5 Nm?

Problem 5.35: A 3-phase stepping motor is to be used to drive a linear axis for a robot. The motor output shaft will be connected to a screw thread with a screw pitch of 4.5 mm. We want to be able to have a spatial control of at least 0.1 mm.

(a) How many poles should the motor have?
(b) How many pulses are needed from the controller every second to move the linear axis at a rate of 100 mm/s?

Reactive Circuit Transient Response

The circuit analysis techniques that we have presented thus far work very well when only resistors are present in the circuit. However, reactive elements such as inductors and capacitors replace the simple algebraic relationships between voltage and current with differential equations. If you recall, current and voltage for a capacitor are related by

$$i(t) = C \cdot \frac{dv(t)}{dt} \tag{6.1}$$

while current and voltage for an inductor are given by

$$v(t) = L \cdot \frac{di(t)}{dt} \tag{6.2}$$

If the voltages or currents are not changing with time, or are no longer changing with time, then $dv/dt = 0$ and $di/dt = 0$. As a result, there would be no current flow into the capacitor, and no voltage drop across the inductor. In this case, the inductors could be replaced by short circuits (wires), and the capacitors could be replaced by open circuits (leads not connected to anything). The circuit could then be solved using the DC circuit analysis technique introduced in the previous chapters. If the voltages or currents are changing with time, however, the derivatives in the capacitor and inductor equations will require solving a differential equation to get a proper solution.

6.1 DC Step Response of Resistor–Capacitor (RC) Circuits

We will begin our analysis by considering the circuit shown in Fig. 6.1 which consists of a single resistor and capacitor in series. At time equal to zero, the switch is closed connecting the DC voltage source, V_s, to the rest of the circuit. V_s is the driving term for the differential equation. Also, before closing the switch, the voltage across the capacitor is given as V_o. From differential equations, this is known as the initial condition for $v(t)$. The voltage across the capacitor cannot change instantaneously (i.e., $v(t = 0^-) = v(t = 0^+)$) as this would require an infinite current to flow due to the derivative in Eq. (6.1). Closing the switch will cause the capacitor to charge or discharge depending on the relative sizes of the initial voltage, V_o, and source voltage, V_s.

© Springer Nature Switzerland AG 2020
T. A. Bigelow, *Electric Circuits, Systems, and Motors*,
https://doi.org/10.1007/978-3-030-31355-5_6

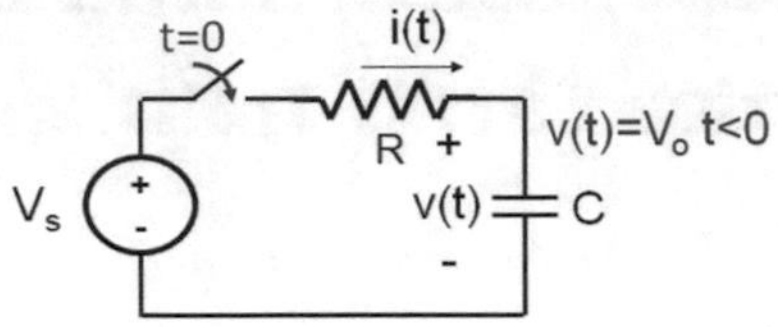

Fig. 6.1 Simple RC circuit

To find the differential equation for the circuit, we can use the mesh current method to sum the voltages around the loop.

$$-V_s + R \cdot i(t) + v(t) = 0 \tag{6.3}$$

However, the current through the capacitor, $i(t)$, must be given by Eq. (6.1), therefore

$$-V_s + R \cdot C \cdot \frac{dv(t)}{dt} + v(t) = 0 \Rightarrow \frac{dv(t)}{dt} + \frac{1}{RC} v(t) = \frac{V_s}{RC} \tag{6.4}$$

In order to solve this differential equation, we first need to find the homogeneous solution (i.e., solution without the driving term). We then include the contribution of the driving term and match the initial conditions. The homogeneous solution is given by the solution to

$$\frac{dv_h(t)}{dt} + \frac{1}{RC} v_h(t) = 0 \tag{6.5}$$

The homogeneous solution to this form of the differential equation is given by $v_h(t) = K_h e^{-at}$ where K_h and a are constants. Substituting this form into Eq. (6.5) gives

$$\begin{aligned} \frac{d(K_h e^{-at})}{dt} + \frac{1}{RC}(K_h e^{-at}) = 0 \\ -aK_h e^{-at} + \frac{1}{RC}(K_h e^{-at}) = 0 \Rightarrow a = \frac{1}{RC} \end{aligned} \tag{6.6}$$

The quantity RC has units of seconds and is known as the time constant, or τ, for the circuit. It governs how fast the capacitor will charge (or discharge). In one time constant, the voltage across the capacitor will change by a factor of e^{-1} or 0.3679. Therefore, the homogeneous solution to the differential equation is given by

$$v_h(t) = K_h e^{-\frac{t}{RC}} \tag{6.7}$$

Once the homogenous solution has been found, we can find the particular solution resulting from the driving term V_s. Since V_s is a constant for time greater than zero, the particular solution for time greater than zero will also be a constant given by $v_p(t) = K_p$. The particular solution must also satisfy the differential equation with the driving term included.

$$\begin{aligned} \frac{d(K_p)}{dt} + \frac{1}{RC}(K_p) = \frac{V_s}{RC} \\ \frac{1}{RC}(K_p) = \frac{V_s}{RC} \Rightarrow K_p = V_s \end{aligned} \tag{6.8}$$

The total solution is then thus the sum of the homogeneous and particular solutions and is given by

$$v(t) = v_h(t) + v_p(t) = V_s + K_h e^{-\frac{t}{RC}} \qquad (6.9)$$

In this equation, all of the terms are known except for the constant K_h which can be found from the initial conditions.

$$v(0) = V_s + K_h e^{-\frac{0}{RC}} = V_o \Rightarrow K_h = V_o - V_s \qquad (6.10)$$

Therefore, the complete solution to the differential equation is given by

$$v(t) = \begin{cases} V_o & t < 0 \\ V_s - (V_s - V_o)e^{-\frac{t}{RC}} & t \geq 0 \end{cases} \qquad (6.11)$$

Once the voltage across the capacitor is known, we can find the current flowing in the circuit and the voltage across the resistor.

$$i(t) = C\frac{dv(t)}{dt} = \begin{cases} 0 & t < 0 \\ \frac{1}{R}(V_s - V_o)e^{-\frac{t}{RC}} & t \geq 0 \end{cases} \qquad (6.12)$$

$$v_{resistor}(t) = R \cdot i(t) = \begin{cases} 0 & t < 0 \\ (V_s - V_o)e^{-\frac{t}{RC}} & t \geq 0 \end{cases} \qquad (6.13)$$

Notice that while the voltage across the capacitor cannot change instantaneously, the voltage across the resistor and the current both can and do change instantaneously after the switch is flipped.

Notice that the final voltage across the capacitor as $t \to \infty$ is V_s. This makes sense as the voltage will stop changing for a long time (many time constants) after the switch has been flipped. As a result, the capacitor will eventually behave as an open circuit with no current flow. If there is no current flow, there would be no voltage drop across the resistor, and all of the source voltage would be across the capacitor.

Example 6.1 The initial voltage across a 10 µF capacitor is 5 V. At time, t, equal to zero, the capacitor is discharged through a 500 Ω resistor as shown in Fig. 6.2. Plot the voltage across the resistor and across the capacitor as a function of time.

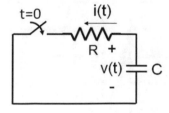

Fig. 6.2 RC circuit for Example 6.1

Solution:

The voltage across the capacitor is given by

$$v(t) = \begin{cases} V_o & t < 0 \\ V_s - (V_s - V_o)e^{-\frac{t}{RC}} & t \geq 0 \end{cases}$$

In this equation, V_o is the initial voltage across the capacitor given by 5 V, R is 500 Ω, and C is 10 μF. The time constant, τ, is given by $RC = 0.005$ s $= (1/200)$ s. Also, since there is no other voltage after the switch is flipped, $V_s = 0$ V. Therefore, the voltage across the capacitor is given by

$$v(t) = \begin{cases} 5 \text{ V} & t < 0 \\ 5e^{-200t} \text{ V} & t \geq 0 \end{cases}$$

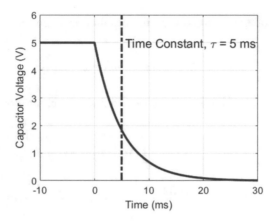

To find the voltage across the resistor, it may be useful to first find the current flowing in the resistor. From Ohm's Law, the current in the resistor is given by $i(t) = \frac{v(t)}{R}$ for $t \geq 0$ and 0 for $t < 0$. The current is 0 for time less than zero because the switch is open, and therefore, the current has no path.

$$i(t) = \begin{cases} 0 \text{ A} & t < 0 \\ 10e^{-200t} \text{ mA} & t \geq 0 \end{cases}$$

$$v_{resistor}(t) = R \cdot i(t) = \begin{cases} 0 \text{ V} & t < 0 \\ 5e^{-200t} \text{ V} & t \geq 0 \end{cases}$$

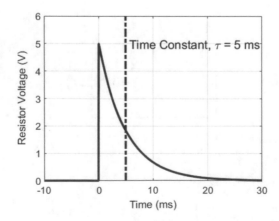

Notice that the polarity of the voltage is reversed from the expression given by Eq. (6.13) because the direction of the current has also been reversed.

Example 6.2 The initial voltage across a 4 mF capacitor is 0 V. At time, t, equal to zero, the capacitor is connected to a 10 V source through a 100 Ω resistor. Plot the voltage across the capacitor as a function of time.

Solution:
The voltage across the capacitor is given by

$$v(t) = \begin{cases} V_o & t < 0 \\ V_s - (V_s - V_o)e^{-\frac{t}{RC}} & t \ge 0 \end{cases}$$

Since the voltage is initially zero, $V_o = 0$ V. The time constant, τ, is given by $RC = 0.4\,s = (1/2.5)\,s$, and the final voltage is given by $V_s = 10$ V. Therefore,

$$v(t) = \begin{cases} 0 & t < 0 \\ 10 - 10e^{-2.5t}\ \text{V} & t \ge 0 \end{cases}$$

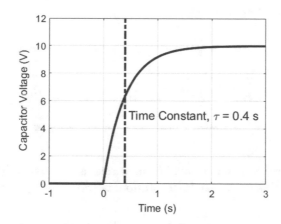

Example 6.3 Find the voltage across the capacitor, v(t) as a function of time for the circuit shown in Fig. 6.3 assuming that the switch has been opened for a very long period of time prior to closing the switch.

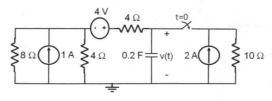

Fig. 6.3 RC circuit for Example 6.3

Solution:

The first step to solving this circuit is to find the voltage across the capacitor before the switch has been flipped.

The circuit prior to closing the switch is given by

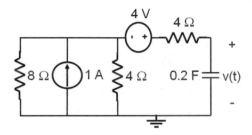

The switch has been open for a long time. Therefore, the capacitor acts like an open circuit with the voltage across the capacitor being given by V_o. This will also be the initial voltage, V_o, for the capacitor circuit after the switch is flipped.

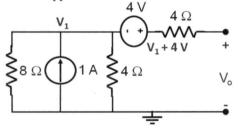

V_o is then given by $V_1 + 4$ V as there would be no current flow through the top 4 Ω resistor when finding the open-circuit voltage. V_1 can be found from the node-voltage method.

$$1\,\text{A} = \frac{V_1}{8\,\Omega} + \frac{V_1}{4\,\Omega} \Rightarrow 8 = V_1 + 2V_1 \Rightarrow V_1 = 2.6667\,\text{V}$$
$$\Rightarrow V_o = V_1 + 4\,\text{V} = 6.6667\,\text{V}$$

Flipping the switch now gives the new circuit shown below.

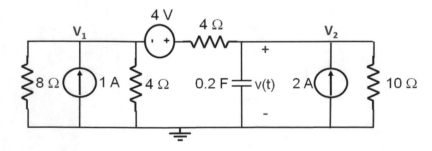

We now need to convert the circuit connected to the capacitor into its Thevenin equivalent so that we can use our previously derived formulas.

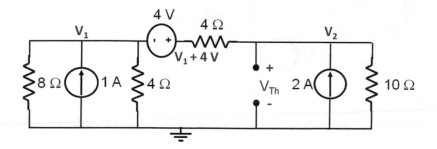

Applying node-voltage to this circuit gives

$$1\,\text{A} = \frac{V_1}{8\,\Omega} + \frac{V_1}{4\,\Omega} + \frac{(V_1 + 4\,\text{V}) - V_2}{4\,\Omega}$$
$$\Rightarrow 8 = V_1 + 2V_1 + 2V_1 + 8\,\text{V} - 2V_2 \Rightarrow V_1 = 0.4V_2$$

$$2\,\text{A} = \frac{V_2 - (V_1 + 4\,\text{V})}{4\,\Omega} + \frac{V_2}{10\,\Omega}$$
$$\Rightarrow 40 = 5V_2 - 5V_1 - 20 + 2V_2 \Rightarrow 60 = 7V_2 - 5V_1$$
$$\Rightarrow 60 = 7V_2 - 2V_2 \Rightarrow V_2 = V_{\text{Th}} = 12\,\text{V}$$

Likewise, we can find the Thevenin resistance, R_{Th}, by deactivating the sources and applying a test voltage to the terminals.

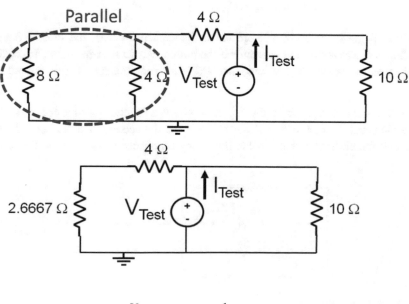

$$R_{\text{Th}} = \frac{V_{\text{Test}}}{I_{\text{Test}}} = \frac{1}{\left(\frac{1}{2.6667\,\Omega + 4\,\Omega}\right) + \frac{1}{10\,\Omega}} = 4\,\Omega$$

Therefore, the circuit after flipping the switch is given by

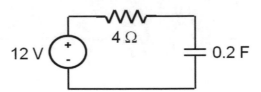

which has the same form as the circuit in Fig. 6.1. Therefore, the voltage across the capacitor is given by

$$v(t) = \begin{cases} V_o & t<0 \\ V_s - (V_s - V_o)e^{-\frac{t}{RC}} & t\geq0 \end{cases} = \begin{cases} 6.6667 \ V & t<0 \\ (12 - 5.3333e^{-\frac{t}{0.8}})V & t\geq0 \end{cases}$$

Up to this point, we have assumed that the switch is always flipped at $t = 0$. However, choosing the time of the switch activation as our reference time may not always be possible or desirable. For example, a circuit may consist of multiple switches that are activated at different times. In this scenario, not all of the switches can be referenced to $t = 0$. Suppose that we have a switch that is flipped at $t = t_o$. The process, in this case, is to first find the voltage across the capacitor immediately prior to when the switch is flipped usually using the Thevenin equivalent circuit. This will serve as the initial voltage for the RC circuit after the switch is flipped. We then find the new Thevenin equivalent circuit for the new switch position. The voltage across the capacitor after the switch is flipped is then given by

$$v(t) = V_{\text{final}} - (V_{\text{final}} - V_{\text{initial}})e^{-\frac{t-t_o}{RC}} \quad t\geq t_o \tag{6.14}$$

In this equation, V_{initial} is the initial voltage across the capacitor before the switch changes position, V_{final} is the final voltage across the capacitor if the circuit was to remain in its current form (i.e., no additional changes in the switch positions), and RC is the time constant for the circuit in its current form.

Example 6.4 Find the voltage across the capacitor, $v(t)$ as a function of time for the circuit shown in Fig. 6.4 assuming that the first switch closes at $t = 0$ and the second switch opens at $t = 0.3$ s. You may assume that the circuit has been stable for a long time prior to closing the first switch.

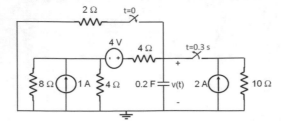

Fig. 6.4 RC circuit for Example 6.4

Solution:
Once again, the first step is to find the Thevenin equivalent circuit for each configuration of the switches. For $t < 0$, the circuit has the same form as Example 6.3 with the switch closed. Therefore, we will refer to that example and simply provide the Thevenin equivalent circuit for $t < 0$.

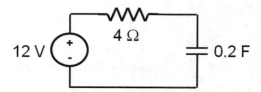

From this circuit, it is clear that $v(t = 0) = 12$ V as the circuit has been stable for a long time prior to closing the first switch. This would have allowed the voltage across the capacitor to reach its maximum value for this circuit. For $0 \leq t < 0.3$ s, the circuit will have the form

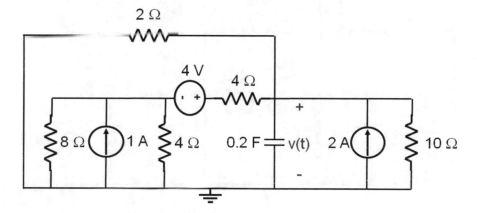

The Thevenin equivalent for this circuit connected to the capacitor is not known and therefore must be derived. We will once again begin by finding the open-circuit voltage for the capacitor terminals.

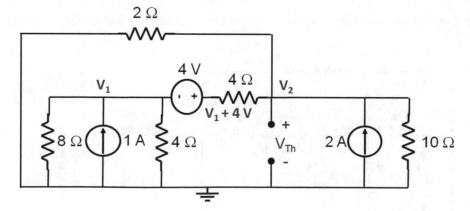

$$1 = \frac{V_1}{8} + \frac{V_1}{4} + \frac{V_1 + 4 - V_2}{4} \Rightarrow 8 = V_1 + 2V_1 + 2V_1 + 8 - 2V_2$$

$$\Rightarrow 0 = 5V_1 - 2V_2 \Rightarrow 5V_1 = 2V_2$$

$$2 = \frac{V_2 - V_1 - 4}{4} + \frac{V_2}{2} + \frac{V_2}{10} \Rightarrow 40 = 5V_2 - 5V_1 - 20 + 10V_2 + 2V_2$$

$$\Rightarrow 60 = 17V_2 - 5V_1$$

$$60 = 17V_2 - 2V_2 \Rightarrow V_2 = V_{Th} = 4 \text{ V}$$

We can then find the short-circuit current by resolving the circuit with the short in place.

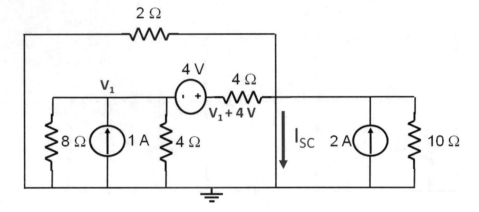

$$1 = \frac{V_1}{8} + \frac{V_1}{4} + \frac{V_1 + 4}{4} \Rightarrow 8 = V_1 + 2V_1 + 2V_1 + 8 \Rightarrow V_1 = 0$$

We can then use the node-voltage values to find the currents entering and leaving the nodes.

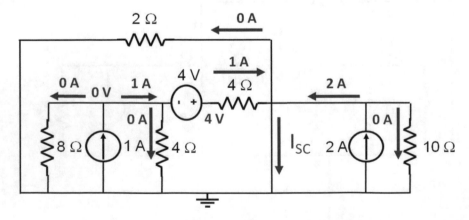

Therefore, $I_{SC} = 3$ A and $R_{Th} = V_{Th}/I_{SC} = 1.33333 \, \Omega$. Hence, for $0 \leq t < 0.3 \, s$ the equivalent circuit will be given by

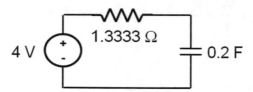

The voltage across the capacitor will thus be given by

$$v(t) = \begin{cases} 12 \text{ V} & t < 0 \\ \left(4 + 8e^{-3.75t}\right) \text{ V} & 0 \le t < 0.3 \text{ s} \end{cases}$$

Therefore, the voltage across the capacitor when the second switch is flipped, $t = 0.3$ s, is given by $v(0.3) = \left(4 + 8e^{-3.75 \cdot 0.3}\right)$ V $= 6.597$ V. This will serve as the initial voltage after the second switch is flipped. After flipping the second switch, the circuit topology is given by

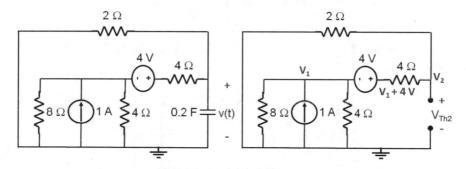

$$1 = \frac{V_1}{8} + \frac{V_1}{4} + \frac{V_1 + 4 - V_2}{4} \Rightarrow 8 = V_1 + 2V_1 + 2V_1 + 8 - 2V_2$$
$$\Rightarrow 0 = 5V_1 - 2V_2 \Rightarrow 5V_1 = 2V_2$$
$$0 = \frac{V_2 - V_1 - 4}{4} + \frac{V_2}{2} \Rightarrow 0 = V_2 - V_1 - 4 + 2V_2 \Rightarrow 4 = 3V_2 - V_1$$
$$4 = 3V_2 - \frac{2}{5}V_2 \Rightarrow V_2 = V_{Th2} = 1.53846 \text{ V}$$

Similarly, the short-circuit current can be found as

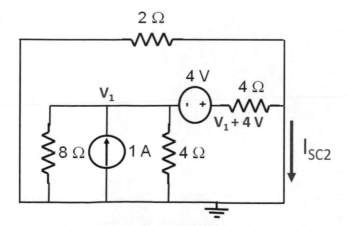

$$1 = \frac{V_1}{8} + \frac{V_1}{4} + \frac{V_1+4}{4} \Rightarrow 8 = V_1 + 2V_1 + 2V_1 + 8 \Rightarrow V_1 = 0$$

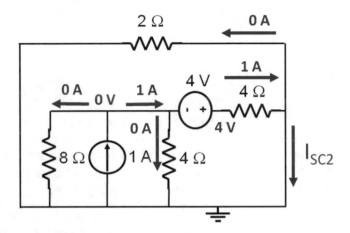

Therefore, $I_{SC} = 1$ A and $R_{Th2} = V_{Th2}/I_{SC2} = 1.53846\,\Omega$. Hence, for $t \geq 0.3$ s the equivalent circuit will be given by

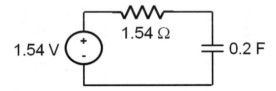

The voltage across the capacitor is thus given by

$$v(t) = \begin{cases} 12\,\text{V} & t < 0 \\ \left(4 + 8e^{-3.75t}\right)\text{V} & 0 \leq t < 0.3\,\text{s} \\ 1.54 + 5.06e^{-3.25(t-0.3)} & t \geq 0.3\,\text{s} \end{cases}$$

Notice that the switch changes both the time constant and the final voltage across the capacitor.

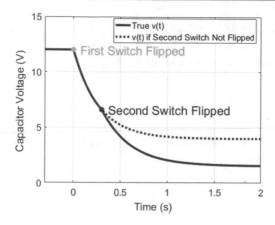

6.2 DC Step Response of Resistor–Inductor (RL) Circuits

The simplest inductor circuit consists of a current source, a resistor, and an inductor as shown in Fig. 6.5. Just as the RC circuit can be thought of as a Thevenin equivalent connected to a capacitor, this circuit can be considered a Norton equivalent connected to an inductor. Since the voltage across the inductor is dependent on the derivative of the current, the current cannot change instantaneously. Therefore, the current in the inductor before the switches are flipped serves as the initial condition for the differential equation. Likewise, since the inductor will eventually behave as a short circuit once the current stops changing, the final current in the inductor will be I_s just like the final voltage across the capacitor following a step response was V_s. If we sum the currents at the top node, we can derive the following differential equation for the current flow in the inductor, $i(t)$, after the switches have flipped.

$$I_s = \frac{v(t)}{R} + i(t) \Rightarrow I_s = \frac{L}{R} \cdot \frac{di(t)}{dt} + i(t) \Rightarrow \frac{di(t)}{dt} + \frac{i(t)}{(L/R)} = \frac{I_s}{(L/R)} \qquad (6.15)$$

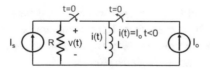

Fig. 6.5 Simple RL circuit

This equation has exactly the same form as Eq. (6.4) with the time constant RC replaced by a new time constant of L/R and the voltages replaced by currents. Therefore, the differential equation solution will have the same form and will be given by

$$i(t) = \begin{cases} I_o & t < 0 \\ I_s - (I_s - I_o)e^{-\frac{t}{(L/R)}} & t \geq 0 \end{cases} \qquad (6.16)$$

We can also generalize this expression to switches flipping at an arbitrary time, t_o, just as we did for the capacitor circuits to yield

$$i(t) = I_{\text{final}} - (I_{\text{final}} - I_{\text{initial}})e^{-\frac{t-t_o}{(L/R)}} \quad t \geq t_o \qquad (6.17)$$

Example 6.5 Find and plot the current in the inductor, $i(t)$, and the voltage across the inductor, $v(t)$, as a function of time for the circuit shown in Fig. 6.6. You may assume that circuit has been stable for a long time prior to flipping the switches.

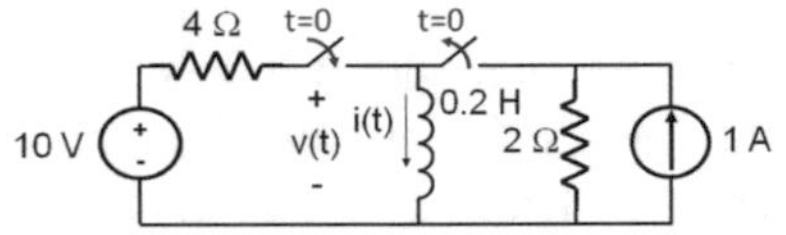

Fig. 6.6 RL circuit for Example 6.5

Solution:
Before the switches change position, $t < 0$, the circuit consists of an inductor, a resistor, and a current source in parallel.

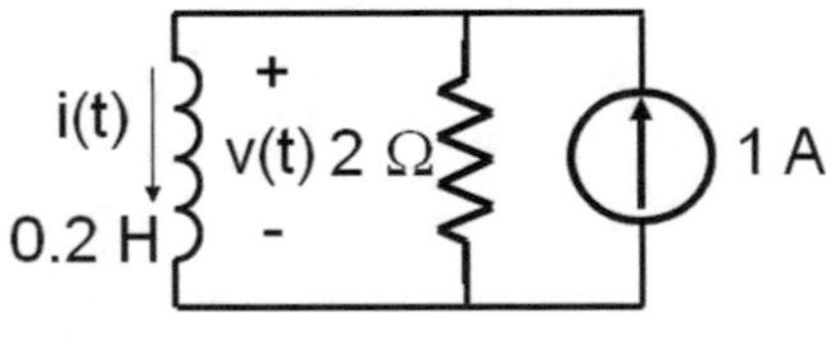

Since, the circuit has been connected in this fashion for a long time, the inductor will be acting as a short circuit. As a result, all of the current from the current source will be flowing in the inductor. If any of the current went through the resistor, $v(t)$ would not be zero which would require $i(t)$ to be changing over time according to Eq. (6.2). Therefore,

$$i(t) = 1\,\text{A} \quad t \leq 0\,\text{s}$$
$$v(t) = 0\,\text{ V} \quad t < 0\text{ s}$$

The current is defined for time less than or equal to zero because the current in the inductor cannot change instantaneously. However, the voltage across the inductor can change as soon as the switches change position, so we do not yet know its value. After the switches have changed position, the new circuit would look like

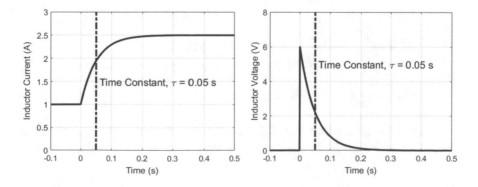

This circuit can be transformed into its Norton equivalent by a simple source transformation.

$$I_N = \frac{10 \ V}{4 \ \Omega} = 2.5 \ A$$

The current, $i(t)$, would then be given by

$$i(t) = I_{\text{final}} - (I_{\text{final}} - I_{\text{initial}})e^{-\frac{t}{(L/R)}} \quad t \geq 0$$
$$= 2.5 - (2.5 - 1)e^{-20t} \quad A \quad t \geq 0$$
$$= 2.5 - 1.5e^{-20t} \quad A \quad t \geq 0$$

Once the current in the inductor is known, the voltage can be found from

$$v(t) = L \cdot \frac{di(t)}{dt} = 0.2 \cdot \frac{d}{dt}\left(2.5 - 1.5e^{-20t}\right) = 6e^{-20t} \ V$$

Notice that the initial voltage across the inductor is 6 V. This is in agreement with the required 1 A of current flowing through the 4 Ω resistor causing a 4 V drop from the 10 V source before reaching the inductor.

Example 6.6 Find and plot the current in the inductor, $i(t)$, as a function of time for the circuit shown in Fig. 6.7. You may assume that the circuit has been stable for a long time prior to flipping the switch.

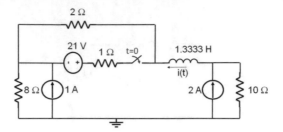

Fig. 6.7 RL circuit for Example 6.6

Solution:
The first step is to find the current flowing in the inductor before the switch changes position. To do this, we will redraw the equivalent circuit for the switch in the open position and the inductor replaced by a short circuit. Solving this circuit will allow us to find the value of $i(t)$ for $t < 0$.

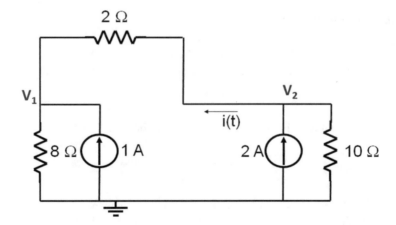

$$1\,A = \frac{V_1}{8\,\Omega} + \frac{V_1 - V_2}{2\,\Omega} \Rightarrow 8 = 5V_1 - 4V_2 \Rightarrow -5V_1 = -8 - 4V_2$$

$$2\,A = \frac{V_2}{10\,\Omega} + \frac{V_2 - V_1}{2\,\Omega} \Rightarrow 20 = 6V_2 - 5V_1$$

$$20 = 6V_2 - 8 - 4V_2 \Rightarrow 28 = 2V_2 \Rightarrow V_2 = 14\,V \Rightarrow V_1 = 12.8\,V$$

$$i(t) = \frac{V_2 - V_1}{2\,\Omega} = 0.6\,A \quad t \le 0$$

We can now redraw the circuit for $t > 0$. Once again, we will want to find the Norton equivalent circuit connected to the inductor. We will begin by finding the open-circuit voltage for the terminals.

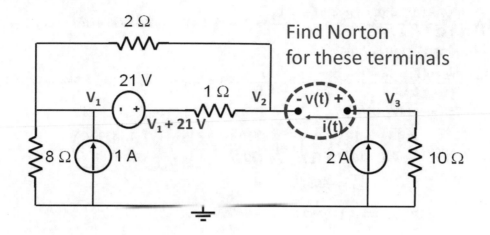

$$1\,A = \frac{V_1}{8\,\Omega} + \frac{V_1 - V_2}{2\,\Omega} + \frac{V_1 + 21\,V - V_2}{1\,\Omega} \Rightarrow 8 = V_1 + 4V_1 - 4V_2 + 8V_1 + 168 - 8V_2$$

$$-160 = 13V_1 - 12V_2$$

$$0 = \frac{V_2 - (V_1 + 21\,V)}{1\,\Omega} + \frac{V_2 - V_1}{2\,\Omega} \Rightarrow 0 = 2V_2 - 2V_1 - 42 + V_2 - V_1$$

$$42 = -3V_1 + 3V_2 \Rightarrow 168 + 12V_1 = 12V_2$$

$$-160 = 13V_1 - 12V_2 = 13V_1 - 168 - 12V_1 \Rightarrow V_1 = 8\,V \Rightarrow V_2 = 22\,V$$

$$2\,A = \frac{V_3}{10\,\Omega} \Rightarrow V_3 = 20\,V$$

$$V_{\text{Th}} = V_3 - V_2 = -2\,V$$

Now we need to find the short-circuit current.

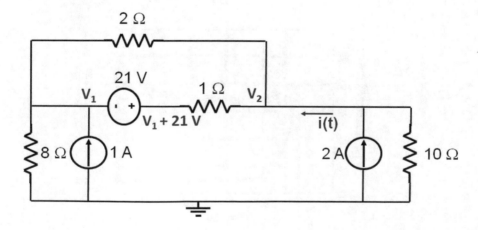

$$1\,A = \frac{V_1}{8\,\Omega} + \frac{V_1 - V_2}{2\,\Omega} + \frac{V_1 + 21\,V - V_2}{1\,\Omega} \Rightarrow 8 = V_1 + 4V_1 - 4V_2 + 8V_1 + 168 - 8V_2$$

$$-160 = 13V_1 - 12V_2$$

$$2 = \frac{V_2 - (V_1 + 21\,V)}{1\,\Omega} + \frac{V_2 - V_1}{2\,\Omega} + \frac{V_2}{10\,\Omega} \Rightarrow 20 = 10V_2 - 10V_1 - 210 + 5V_2 - 5V_1 + V_2$$

$$230 = -15V_1 + 16V_2$$

$$\begin{pmatrix} 13 & -12 \\ -15 & 16 \end{pmatrix} \begin{pmatrix} V_1 \\ V_2 \end{pmatrix} = \begin{pmatrix} -160 \\ 230 \end{pmatrix} \Rightarrow \begin{pmatrix} V_1 \\ V_2 \end{pmatrix} = \begin{pmatrix} 7.1429\,V \\ 21.0714\,V \end{pmatrix}$$

$$I_N = 2 - \frac{V_2}{10} = -0.1071\,A$$

$$R_{Th} = \frac{V_{Th}}{I_N} = 18.6667\,\Omega$$

Therefore, after the switch has flipped, the circuit could be redrawn as

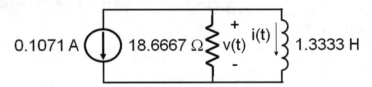

The current as a function of time is thus given by

$$i(t) = I_{final} - (I_{final} - I_{initial})e^{-\frac{t}{(L/R)}}\ t \geq 0$$
$$= -0.1071 - (-0.1071 - 0.6)e^{-14t}\ A\ t \geq 0$$
$$= -0.1071 - 0.7071e^{-14t}\ A\ t \geq 0$$

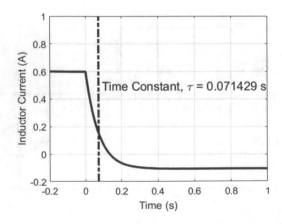

6.3   DC Step Response of Resistor–Inductor–Capacitor (RLC) Circuits

Thus far, we have only considered circuits that have a single inductor or a single capacitor. However, many circuits have a combination of multiple inductors and/or capacitors resulting in higher order differential equations. The best approach for solving these more complicated circuits is to use the Laplace Transform to solve the differential equations. This approach is known as s-domain circuit analysis. However, reviewing the Laplace transform and applying it to circuit analysis is beyond the scope of this chapter. Therefore, we will only focus on two circuit topologies each with one inductor, one capacitor, and one resistor. After deriving the differential equation, we will give the solution to the equation when a sudden change is applied to the input (i.e., switch flipped connecting a DC source).

6.3.1   Series RLC Circuit

Figure 6.8 shows a series RLC circuit consisting of a resistor, an inductor, and a capacitor connected in series with a DC voltage source. A switch connects the voltage source to the other circuit elements at time $t = 0$. Finding the current around the loop will allow us to find all of the voltages in the circuit. If we sum the voltages around the loop, we will have the following differential equation.

$$V_s = R \cdot i(t) + L \cdot \frac{\mathrm{d}i(t)}{\mathrm{d}t} + v_C(t) \tag{6.18}$$

Taking the derivative of (6.18) with respect to time gives

$$\frac{\mathrm{d}}{\mathrm{d}t}\left(V_s = R \cdot i(t) + L \cdot \frac{\mathrm{d}i(t)}{\mathrm{d}t} + v_C(t)\right) \Rightarrow 0 = R \cdot \frac{\mathrm{d}i(t)}{\mathrm{d}t} + L \cdot \frac{\mathrm{d}^2 i(t)}{\mathrm{d}t^2} + \frac{\mathrm{d}}{\mathrm{d}t}v_C(t)$$

$$\Rightarrow 0 = R \cdot \frac{\mathrm{d}i(t)}{\mathrm{d}t} + L \cdot \frac{\mathrm{d}^2 i(t)}{\mathrm{d}t^2} + \frac{i(t)}{C} \tag{6.19}$$

where we have substituted $i(t) = C \cdot \mathrm{d}v_C(t)/\mathrm{d}t$ for the time derivative of the capacitor voltage. If we multiply through by the capacitance, the final differential equation is given by

$$LC \cdot \frac{\mathrm{d}^2 i(t)}{\mathrm{d}t^2} + RC \cdot \frac{\mathrm{d}i(t)}{\mathrm{d}t} + i(t) = 0 \tag{6.20}$$

The solution to this differential equation has three possible forms depending on the values of R, L, and C. However, we know that all three forms should experience some form of exponential decay, similar to the RC and RL circuits analyzed previously, because eventually the transient response due to the

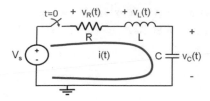

Fig. 6.8 Series RLC circuit

switch should decay away. Therefore, assuming the current has the form $i(t) = A \cdot e^{s \cdot t}$, we can substitute this solution into Eq. (6.20) and solve for the value of s.

$$LC \cdot \frac{d^2 i(t)}{dt^2} + RC \cdot \frac{di(t)}{dt} + i(t) = 0$$

$$LC \cdot \frac{d^2}{dt^2}(Ae^{st}) + RC \cdot \frac{d}{dt}(Ae^{st}) + (Ae^{st}) = 0 \qquad (6.21)$$

$$LC \cdot s^2 \cdot Ae^{st} + RC \cdot s \cdot Ae^{st} + Ae^{st} = 0$$

$$LC \cdot s^2 + RC \cdot s + 1 = 0$$

This quadratic equation should have two roots yielding two possible values for s given by

$$s = \frac{-RC \pm \sqrt{R^2 C^2 - 4LC}}{2LC} = -\frac{RC}{2LC} \pm \sqrt{\frac{R^2 C^2}{4L^2 C^2} - \frac{4LC}{4L^2 C^2}}$$

$$= -\frac{R}{2L} \pm \sqrt{\left(\frac{R}{2L}\right)^2 - \frac{1}{LC}} \qquad (6.22)$$

The simplest case would be when $\left(\frac{R}{2L}\right)^2 > \frac{1}{LC}$ as this would yield two distinct values of s and the current would follow a relatively simple exponential decay for t > 0 of the form

$$i(t) = A_1 \exp\left(-\left(\frac{R}{2L} + \sqrt{\left(\frac{R}{2L}\right)^2 - \frac{1}{LC}}\right)t\right) + A_2 \exp\left(-\left(\frac{R}{2L} - \sqrt{\left(\frac{R}{2L}\right)^2 - \frac{1}{LC}}\right)t\right) \quad (6.23)$$

where A_1 and A_2 are constants that depend on the initial conditions. This case is known as overdamped and would correspond to when the losses, i.e., R, dominate over the energy flow in and out of the capacitor and inductor.

The next simplest case would be when $\left(\frac{R}{2L}\right)^2 < \frac{1}{LC}$. For this condition, the circuit is called underdamped and the energy flow in and out of the capacitor and inductor would dominate. As a result, the circuit would ring or resonate. In addition, the roots of (6.22) would be imaginary. According to Euler's Identity, which will be reviewed in more detail in Chap. 7, an imaginary number in the exponent can be replaced by a sin and/or cos function. Therefore, the equation for $i(t)$ for the underdamped case for $t > 0$ is given by

$$i(t) = A_1 \exp\left(-\frac{R}{2L}t\right) \cos\left(t\sqrt{\frac{1}{LC} - \left(\frac{R}{2L}\right)^2}\right) + A_2 \exp\left(-\frac{R}{2L}t\right) \sin\left(t\sqrt{\frac{1}{LC} - \left(\frac{R}{2L}\right)^2}\right) \quad (6.24)$$

where once again A_1 and A_2 are found by the initial conditions.

For the final case, $\left(\frac{R}{2L}\right)^2 = \frac{1}{LC}$ and this would be known as the critically damped condition. For this case, the quadratic in Eq. (6.22) would have repeated roots. However, since we are solving a second-order differential equation, we still need an expression with two unknowns, A_1 and A_2, that will depend on the initial conditions. Therefore, the solution for $t > 0$ will have the form

$$i(t) = A_1 \exp\left(-\frac{R}{2L}t\right) + A_2 t \exp\left(-\frac{R}{2L}t\right) \tag{6.25}$$

An equation of the form $i(t) = A \cdot te^{s \cdot t}$ still satisfies the differential equation provided the system is critically damped as is shown below.

$$LC \cdot \frac{d^2 i(t)}{dt^2} + RC \cdot \frac{di(t)}{dt} + i(t) = 0$$

$$LC \cdot \frac{d^2}{dt^2}(Ate^{st}) + RC \cdot \frac{d}{dt}(Ate^{st}) + (Ate^{st}) = 0$$

$$\frac{d}{dt}(Ate^{st}) = Ae^{st} + At \cdot s \cdot e^{st} = Ae^{st}(1 + st) \tag{6.26}$$

$$\frac{d^2}{dt^2}(Ate^{st}) = \frac{d}{dt}(Ae^{st} + At \cdot s \cdot e^{st}) = Ase^{st} + A \cdot s \cdot e^{st} + At \cdot s^2 \cdot e^{st} = Ae^{st}(s^2 t + 2s)$$

$$[LC \cdot (s^2 t + 2s) + RC \cdot (1 + st) + t] \cdot Ae^{st} = 0$$

$$(LC \cdot s^2 + RC \cdot s + 1)t + LC \cdot (2s) + RC = 0$$

Now substitute in for $s = -\frac{R}{2L}$ which would be the only possible root when $\left(\frac{R}{2L}\right)^2 = \frac{1}{LC}$

$$\left(LC \cdot \left(\frac{R^2}{4L^2}\right) - RC \cdot \frac{R}{2L} + 1\right)t - LC \cdot \left(2\frac{R}{2L}\right) + RC = 0$$

$$\Rightarrow \left(LC \cdot \frac{1}{LC} - RC \cdot \frac{R}{2L} + 1\right)t - RC + RC = 0 \tag{6.27}$$

$$\Rightarrow \left(\frac{R^2 C}{2L} = 2\right) \Rightarrow \left(\frac{R^2 C}{4L} \cdot \frac{1}{LC} = 1 \cdot \frac{1}{LC}\right) \Rightarrow \left(\frac{R^2}{4L^2} = \frac{1}{LC}\right)$$

This is just the condition for the circuit when it is critically damped.

The expressions found for the overdamped, underdamped, and critically damped circuit can be simplified slightly by defining a natural frequency for the system, ω_n, and a damping ratio for the system, ζ, given by

$$\omega_n = \frac{1}{\sqrt{LC}} \quad \zeta = \frac{R}{2}\sqrt{\frac{C}{L}} \tag{6.28}$$

Based on these definitions, the overdamped circuit will have $\zeta > 1$, the underdamped circuit will have a $\zeta < 1$, and the critically damped circuit will have a $\zeta = 1$. The possible solutions to the differential equation for $t > 0$ also become

$$i(t) = \begin{cases} A_1 \exp\left(-\left(\zeta + \sqrt{\zeta^2 - 1}\right)\omega_n t\right) + A_2 \exp\left(-\left(\zeta - \sqrt{\zeta^2 - 1}\right)\omega_n t\right) & \zeta > 1 \\ A_1 \exp(-\omega_n t) + A_2 t \exp(-\omega_n t) & \zeta = 1 \\ \left(A_1 \cos\left(\left(\omega_n \sqrt{1 - \zeta^2}\right)t\right) + A_2 \sin\left(\left(\omega_n \sqrt{1 - \zeta^2}\right)t\right)\right) \exp(-\omega_n \zeta t) & \zeta < 1 \end{cases} \tag{6.29}$$

where it is helpful in the simplification to notice that $\frac{R}{2L} = \omega_n \zeta$. In addition, the differential equation itself can be written in the more general form as

$$\frac{1}{\omega_n^2} \cdot \frac{d^2 i(t)}{dt^2} + \frac{2\zeta}{\omega_n} \cdot \frac{di(t)}{dt} + i(t) = 0 \tag{6.30}$$

Now that we have derived the differential equation for the current and provided possible solutions depending on the damping, we need to find the constants A_1 and A_2 based on the initial conditions. To do so, we will assume that the circuit has been stable for a long time prior to flipping the switch. Therefore, the initial current flow must be zero. Likewise, the initial voltage across the resistor, $v_R(t) = R \cdot i(t)$, must also be zero since the current is zero. For the capacitor voltage, $v_C(t)$, we will assume that the capacitor is not storing any energy when the switch is flipped. Therefore, it will have an initial voltage of 0 V which cannot change instantaneously. Therefore, immediately after flipping the switch, all of the source voltage must appear across the inductor, $v_L(0) = V_s$. Since $v_L(t) = L \cdot di(t)/dt$, the initial slope of the current must be given by $(di(t)/dt)_{t=0} = V_s/L$.

We can now apply these initial conditions to the expressions in (6.29).

$$i(0) = 0 = \begin{cases} A_1 + A_2 & \zeta > 1 \\ A_1 & \zeta = 1 \\ A_1 & \zeta < 1 \end{cases} \tag{6.31}$$

Therefore, for the overdamped case, $A_1 + A_2$ is zero while for the underdamped and critically damped cases A_1 must be zero. Similarly, if we take the derivative of (6.29) and set it equal to V_s/L, we get

$$\left(\frac{di(t)}{dt}\right)_{t=0} = \frac{V_s}{L} = \begin{cases} -\left(\zeta + \sqrt{\zeta^2 - 1}\right)\omega_n A_1 - \left(\zeta - \sqrt{\zeta^2 - 1}\right)A_2 \omega_n & \zeta > 1 \\ -\omega_n A_1 + A_2 & \zeta = 1 \\ -\omega_n \zeta A_1 + A_2\left(\omega_n \sqrt{1 - \zeta^2}\right) & \zeta < 1 \end{cases} \tag{6.32}$$

Therefore, $i(t)$ for $t > 0$ is given by

$$i(t) = \begin{cases} \dfrac{\left(\frac{V_s}{L}\right)\left(\exp\left(-\left(\zeta - \sqrt{\zeta^2 - 1}\right)\omega_n t\right) - \exp\left(-\left(\zeta + \sqrt{\zeta^2 - 1}\right)\omega_n t\right)\right)}{2\omega_n \sqrt{\zeta^2 - 1}} & \zeta > 1 \\ \left(\frac{V_s}{L}\right) t \exp(-\omega_n t) & \zeta = 1 \\ \dfrac{\left(\frac{V_s}{L}\right)}{\left(\omega_n \sqrt{1 - \zeta^2}\right)} \exp(-\omega_n \zeta t) \sin\left(\left(\omega_n \sqrt{1 - \zeta^2}\right)t\right) & \zeta < 1 \end{cases} \tag{6.33}$$

With the current known, we can find the voltage across the resistor and the voltage across the inductor for $t > 0$ from $v_R(t) = R \cdot i(t)$ and $v_L(t) = L \cdot di(t)/dt$.

$$v_R(t) = \begin{cases} \dfrac{\zeta V_s\left(\exp\left(-\left(\zeta - \sqrt{\zeta^2 - 1}\right)\omega_n t\right) - \exp\left(-\left(\zeta + \sqrt{\zeta^2 - 1}\right)\omega_n t\right)\right)}{\sqrt{\zeta^2 - 1}} & \zeta > 1 \\ 2\omega_n V_s t \exp(-\omega_n t) & \zeta = 1 \\ \dfrac{2\zeta V_s}{\sqrt{1 - \zeta^2}} \exp(-\omega_n \zeta t) \sin\left(\left(\omega_n \sqrt{1 - \zeta^2}\right)t\right) & \zeta < 1 \end{cases} \tag{6.34}$$

$$v_L(t) = \begin{cases} \frac{V_s}{2}\left(\begin{array}{c} \left(1 - \frac{\zeta}{\sqrt{\zeta^2 - 1}}\right)\exp\left(-\left(\zeta - \sqrt{\zeta^2 - 1}\right)\omega_n t\right) \\ + \left(1 + \frac{\zeta}{\sqrt{\zeta^2 - 1}}\right)\exp\left(-\left(\zeta + \sqrt{\zeta^2 - 1}\right)\omega_n t\right) \end{array} \right) & \zeta > 1 \\ V_s(1 - \omega_n t)\exp(-\omega_n t) & \zeta = 1 \\ V_s \exp(-\omega_n \zeta t)\left(\begin{array}{c} \cos\left(\left(\omega_n\sqrt{1 - \zeta^2}\right)t\right) \\ -\frac{\zeta}{\sqrt{1 - \zeta^2}}\sin\left(\left(\omega_n\sqrt{1 - \zeta^2}\right)t\right) \end{array} \right) & \zeta < 1 \end{cases} \tag{6.35}$$

where we have once again used the simplification $\frac{R}{L} = 2\omega_n \zeta$. The value for the voltage across the capacitor will just be given by $v_C(t) = V_s - (v_R(t) + v_L(t))$ if we sum the voltages around the loop. Therefore, the capacitor voltage for $t > 0$ is

$$v_C(t) = \begin{cases} V_s - \frac{V_s}{2}\left(\begin{array}{c} \left(1 + \frac{\zeta}{\sqrt{\zeta^2 - 1}}\right)\exp\left(-\left(\zeta - \sqrt{\zeta^2 - 1}\right)\omega_n t\right) \\ + \left(1 - \frac{\zeta}{\sqrt{\zeta^2 - 1}}\right)\exp\left(-\left(\zeta + \sqrt{\zeta^2 - 1}\right)\omega_n t\right) \end{array} \right) & \zeta > 1 \\ V_s - V_s(1 + \omega_n t)\exp(-\omega_n t) & \zeta = 1 \\ V_s - V_s \exp(-\omega_n \zeta t)\left(\begin{array}{c} \cos\left(\left(\omega_n\sqrt{1 - \zeta^2}\right)t\right) \\ +\frac{\zeta}{\sqrt{1 - \zeta^2}}\sin\left(\left(\omega_n\sqrt{1 - \zeta^2}\right)t\right) \end{array} \right) & \zeta < 1 \end{cases} \tag{6.36}$$

Example 6.7 A Series RLC circuit has a V_s value of 10 V that is connected to the rest of the circuit at $t = 0$, an inductor of 2 mH, a capacitor of 80 μF, and a resistor of 40 Ω. Find and plot the current and the voltage across each circuit element as a function of time.

Solution:
The solution will depend on the level of damping. Therefore, we will first need to solve for ζ. For the given circuit parameters

$$\zeta = \frac{R}{2}\sqrt{\frac{C}{L}} = 4$$

Therefore, the system is overdamped, and the current, $i(t)$ will be given by

$$i(t) = \begin{cases} \frac{\left(\frac{V_s}{L}\right)\left(\exp\left(-\left(\zeta - \sqrt{\zeta^2 - 1}\right)\omega_n t\right) - \exp\left(-\left(\zeta + \sqrt{\zeta^2 - 1}\right)\omega_n t\right)\right)}{2\omega_n\sqrt{\zeta^2 - 1}} & t \geq 0 \\ 0 & t < 0 \end{cases}$$

where

$$\omega_n = \frac{1}{\sqrt{LC}} = 2500 \, \text{rad/s}$$

Substituting in for these values gives

$$i(t) = \begin{cases} 0.2582(\exp(-317.5t) - \exp(-19682t))\text{A} & t \geq 0 \\ 0 & t < 0 \end{cases}$$

Likewise, the voltages across each circuit element are given by

$$v_R(t) = \begin{cases} \dfrac{\zeta V_s\left(\exp\left(-\left(\zeta-\sqrt{\zeta^2-1}\right)\omega_n t\right) - \exp\left(-\left(\zeta+\sqrt{\zeta^2-1}\right)\omega_n t\right)\right)}{\sqrt{\zeta^2-1}} & t \geq 0 \\ 0 & t < 0 \end{cases}$$

$$= \begin{cases} 10.328(\exp(-317.5t) - \exp(-19682t))\text{V} & t \geq 0 \\ 0 & t < 0 \end{cases}$$

$$v_L(t) = \begin{cases} \dfrac{V_s}{2}\left(\begin{array}{l} \left(1 - \dfrac{\zeta}{\sqrt{\zeta^2-1}}\right)\exp\left(-\left(\zeta - \sqrt{\zeta^2 - 1}\right)\omega_n t\right) \\ + \left(1 + \dfrac{\zeta}{\sqrt{\zeta^2-1}}\right)\exp\left(-\left(\zeta + \sqrt{\zeta^2 - 1}\right)\omega_n t\right) \end{array} \right) & t \geq 0 \\ 0 & t < 0 \end{cases}$$

$$= \begin{cases} -0.164 \exp(-317.5t) + 10.164 \exp(-19682t)\text{V} & t \geq 0 \\ 0 & t < 0 \end{cases}$$

$$v_C(t) = \begin{cases} V_s - \dfrac{V_s}{2}\left(\begin{array}{l} \left(1 + \dfrac{\zeta}{\sqrt{\zeta^2-1}}\right)\exp\left(-\left(\zeta - \sqrt{\zeta^2 - 1}\right)\omega_n t\right) \\ + \left(1 - \dfrac{\zeta}{\sqrt{\zeta^2-1}}\right)\exp\left(-\left(\zeta + \sqrt{\zeta^2 - 1}\right)\omega_n t\right) \end{array} \right) & t \geq 0 \\ 0 & t < 0 \end{cases}$$

$$= \begin{cases} 10 - (10.164 \exp(-317.5t) - 0.164 \exp(-19682t))\text{V} & t \geq 0 \\ 0 & t < 0 \end{cases}$$

Plotting these voltages and the current gives

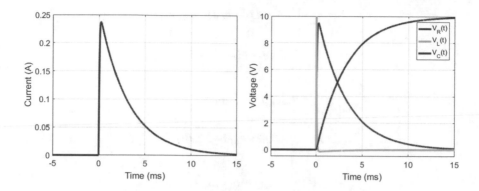

The current and corresponding voltage in the resistor increase very quickly before gradually decaying to zero. The voltage across the capacitor monotonically increases until it is fully charged at the source voltage. The voltage across the inductor changes instantaneously to the source voltage before quickly decaying to a negative voltage. This is due to the dominance of the $\exp\left(-\left(\zeta + \sqrt{\zeta^2 - 1}\right)\omega_n t\right)$ term at low time values. At later times, the $\exp\left(-\left(\zeta - \sqrt{\zeta^2 - 1}\right)\omega_n t\right)$ will dominate, and the voltage across the inductor will exponentially approach 0 V. The change in the inductor voltage is easier to visualize if we zoom in on our plot of voltages as shown below.

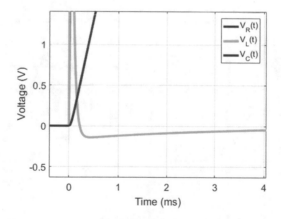

Example 6.8 Repeat Example 6.7 for a Series RLC circuit with a V_s value of 10 V that is connected to the rest of the circuit at $t = 0$, an inductor of 2 mH, a capacitor of 80 µF, and a resistor of 10 Ω.

Solution:
First, we will need to solve for ζ once again. For the given circuit parameters

$$\zeta = \frac{R}{2}\sqrt{\frac{C}{L}} = 1$$

Therefore, the system is critically damped. The current and voltages are thus given by

$$i(t) = \begin{cases} \left(\frac{V_s}{L}\right) t \exp(-\omega_n t) & t \geq 0 \\ 0 & t < 0 \end{cases}$$

$$= \begin{cases} 5000t \exp(-2500t)\text{A} & t \geq 0 \\ 0 & t < 0 \end{cases}$$

$$v_R(t) = \begin{cases} 2\omega_n V_s t \exp(-\omega_n t) & t \geq 0 \\ 0 & t < 0 \end{cases}$$

$$= \begin{cases} 50000t \exp(-2500t)\text{V} & t \geq 0 \\ 0 & t < 0 \end{cases}$$

$$v_L(t) = \begin{cases} V_s(1 - \omega_n t) \exp(-\omega_n t) & t \geq 0 \\ 0 & t < 0 \end{cases}$$

$$= \begin{cases} 10(1 - 2500t) \exp(-2500t)\text{V} & t \geq 0 \\ 0 & t < 0 \end{cases}$$

$$v_C(t) = \begin{cases} V_s - V_s(1 + \omega_n t) \exp(-\omega_n t) & t \geq 0 \\ 0 & t < 0 \end{cases}$$

$$= \begin{cases} 10 - 10(1 + 2500t) \exp(-2500t)\text{V} & t \geq 0 \\ 0 & t < 0 \end{cases}$$

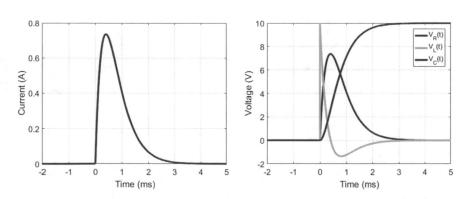

Example 6.9 Repeat Example 6.7 for a Series RLC circuit with a V_s value of 10 V that is connected to the rest of the circuit at $t = 0$, an inductor of 2 mH, a capacitor of 80 μF, and a resistor of 4 Ω.

Solution:
Finding the value for ζ this time yields

$$\zeta = \frac{R}{2}\sqrt{\frac{C}{L}} = 0.4$$

Therefore, the system is underdamped. The current and voltages are thus given by

$$i(t) = \begin{cases} \dfrac{\left(\frac{V_s}{L}\right)}{\left(\omega_n\sqrt{1-\zeta^2}\right)} \exp(-\omega_n\zeta t) \sin\left(\left(\omega_n\sqrt{1-\zeta^2}\right)t\right) & t \geq 0 \\ 0 & t < 0 \end{cases}$$

$$= \begin{cases} 2.1822 \exp(-1000t) \sin(2291.3t)\,\mathrm{A} & t \geq 0 \\ 0 & t < 0 \end{cases}$$

$$v_R(t) = \begin{cases} \dfrac{2\zeta V_s}{\sqrt{1-\zeta^2}} \exp(-\omega_n\zeta t) \sin\left(\left(\omega_n\sqrt{1-\zeta^2}\right)t\right) & t \geq 0 \\ 0 & t < 0 \end{cases}$$

$$= \begin{cases} 8.7287 \exp(-1000t) \sin(2291.3t)\,\mathrm{V} & t \geq 0 \\ 0 & t < 0 \end{cases}$$

$$v_L(t) = \begin{cases} V_s \exp(-\omega_n\zeta t) \left(\cos\left(\left(\omega_n\sqrt{1-\zeta^2}\right)t\right) - \dfrac{\zeta}{\sqrt{1-\zeta^2}} \sin\left(\left(\omega_n\sqrt{1-\zeta^2}\right)t\right) \right) & t \geq 0 \\ 0 & t < 0 \end{cases}$$

$$= \begin{cases} 10 \exp(-1000t)(\cos(2291.3t) - 0.4364 \sin(2291.3t))\,\mathrm{V} & t \geq 0 \\ 0 & t < 0 \end{cases}$$

$$v_C(t) = \begin{cases} V_s - V_s \exp(-\omega_n\zeta t) \left(\cos\left(\left(\omega_n\sqrt{1-\zeta^2}\right)t\right) + \dfrac{\zeta}{\sqrt{1-\zeta^2}} \sin\left(\left(\omega_n\sqrt{1-\zeta^2}\right)t\right) \right) & t \geq 0 \\ 0 & t < 0 \end{cases}$$

$$= \begin{cases} 10 - 10 \exp(-1000t)(\cos(2291.3t) + 0.4364 \sin(2291.3t))\,\mathrm{V} & t \geq 0 \\ 0 & t < 0 \end{cases}$$

Notice that the current as well as the voltage across the resistor and the voltage across the inductor oscillate between positive and negative values. In addition, the voltage across the capacitor oscillates about the source voltage before finally converging to the source voltage. This is due to the exchange of energy between the inductor and the capacitor.

6.3.2   Parallel RLC Circuit

Another basic circuit topology with a single resistor, inductor, and capacitor has all three circuit elements connected in parallel as shown in Fig. 6.9. For this circuit, the current from the current source is initially directed through a short circuit. However, at $t = 0$, the switch opens forcing the current to flow through the rest of the circuit. Finding the voltage, $v(t)$, will allow us to solve for all of the currents for this circuit. $v(t)$ can be found by summing all of the currents at the top node.

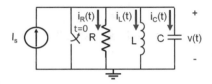

Fig. 6.9   Parallel RLC circuit

$$I_s = i_R(t) + i_L(t) + i_C(t) = \frac{v(t)}{R} + i_L(t) + C \cdot \frac{\mathrm{d}v(t)}{\mathrm{d}t} \tag{6.37}$$

Taking the derivative of (6.37) with respect to time yields

$$\frac{\mathrm{d}}{\mathrm{d}t}\left(I_s = \frac{v(t)}{R} + i_L(t) + C \cdot \frac{\mathrm{d}v(t)}{\mathrm{d}t}\right) \Rightarrow 0 = \frac{1}{R}\frac{\mathrm{d}v(t)}{\mathrm{d}t} + \frac{\mathrm{d}i_L(t)}{\mathrm{d}t} + C \cdot \frac{\mathrm{d}^2v(t)}{\mathrm{d}t^2}$$

$$\Rightarrow LC \cdot \frac{\mathrm{d}^2v(t)}{\mathrm{d}t^2} + \frac{L}{R}\frac{\mathrm{d}v(t)}{\mathrm{d}t} + v(t) = 0 \tag{6.38}$$

This second-order differential equation can be converted into its more general form by once again defining a natural frequency, ω_n, and a new damping ratio, ζ', given by

$$\omega_n = \frac{1}{\sqrt{LC}} \quad \zeta' = \frac{1}{2R}\sqrt{\frac{L}{C}} \tag{6.39}$$

With this substitution, Eq. (6.38) becomes

$$\frac{1}{\omega_n^2} \cdot \frac{\mathrm{d}^2v(t)}{\mathrm{d}t^2} + \frac{2\zeta'}{\omega_n} \cdot \frac{\mathrm{d}v(t)}{\mathrm{d}t} + v(t) = 0 \tag{6.40}$$

which is identical to Eq. (6.30) only with the voltage replacing the current. As a result, the solution to the differential equation for $t > 0$ will have the same form as was found previously. Therefore,

$$v(t) = \begin{cases} A_1 \exp\left(-\left(\zeta' + \sqrt{\zeta'^2 - 1}\right)\omega_n t\right) + A_2 \exp\left(-\left(\zeta' - \sqrt{\zeta'^2 - 1}\right)\omega_n t\right) & \zeta' > 1 \\ A_1 \exp(-\omega_n t) + A_2 t \exp(-\omega_n t) & \zeta' = 1 \\ \left(A_1 \cos\left(\left(\omega_n\sqrt{1 - \zeta'^2}\right)t\right) + A_2 \sin\left(\left(\omega_n\sqrt{1 - \zeta'^2}\right)t\right)\right)\exp(-\omega_n\zeta' t) & \zeta' < 1 \end{cases} \tag{6.41}$$

where the constants A_1 and A_2 are found from the initial conditions. Since the voltage on the capacitor cannot change instantaneously, the initial voltage, $v(0)$, must be zero. Also, since the current through the inductor cannot change instantaneously, and any current in the resistor would force a voltage across the capacitor, initially all of the current from the source must be flowing through the capacitor (i.e., $i_C(0) = I_s = C \cdot (dv(t)/dt)$). Therefore, immediately after the switch has flipped $(dv(t)/dt)_{t=0} = I_s/C$. Using these initial conditions, the final expression for the voltage, $v(t)$, can be found as

$$v(t) = \begin{cases} \left(\frac{I_s}{C}\right) \dfrac{\exp\left(-\left(\zeta' - \sqrt{\zeta'^2 - 1}\right)\omega_n t\right) - \exp\left(-\left(\zeta' + \sqrt{\zeta'^2 - 1}\right)\omega_n t\right)}{2\omega_n\sqrt{\zeta'^2 - 1}} & \zeta' > 1 \\[4mm] \left(\frac{I_s}{C}\right) t \exp(-\omega_n t) & \zeta' = 1 \\[4mm] \dfrac{\left(\frac{I_s}{C}\right)}{\left(\omega_n\sqrt{1 - \zeta'^2}\right)} \exp(-\omega_n\zeta' t) \sin\left(\left(\omega_n\sqrt{1 - \zeta'^2}\right)t\right) & \zeta' < 1 \end{cases} \tag{6.42}$$

With the voltage at $t > 0$ known, the currents in each of the branches can also be found.

$$i_R(t) = \begin{cases} \zeta' I_s \dfrac{\exp\left(-\left(\zeta' - \sqrt{\zeta'^2 - 1}\right)\omega_n t\right) - \exp\left(-\left(\zeta' + \sqrt{\zeta'^2 - 1}\right)\omega_n t\right)}{\sqrt{\zeta'^2 - 1}} & \zeta' > 1 \\[4mm] 2\omega_n I_s t \exp(-\omega_n t) & \zeta' = 1 \\[4mm] \dfrac{2\zeta' I_s}{\left(\sqrt{1 - \zeta'^2}\right)} \exp(-\omega_n\zeta' t) \sin\left(\left(\omega_n\sqrt{1 - \zeta'^2}\right)t\right) & \zeta' < 1 \end{cases} \tag{6.43}$$

$$i_C(t) = \begin{cases} \dfrac{I_s}{2}\left(\left(1 - \dfrac{\zeta'}{\sqrt{\zeta'^2 - 1}}\right)\exp\left(-\left(\zeta' - \sqrt{\zeta'^2 - 1}\right)\omega_n t\right) \\ + \left(1 + \dfrac{\zeta'}{\sqrt{\zeta'^2 - 1}}\right)\exp\left(-\left(\zeta' + \sqrt{\zeta'^2 - 1}\right)\omega_n t\right) \right) & \zeta' > 1 \\[6mm] I_s(1 - \omega_n t)\exp(-\omega_n t) & \zeta' = 1 \\[4mm] I_s \exp(-\omega_n\zeta' t)\left(\cos\left(\left(\omega_n\sqrt{1 - \zeta'^2}\right)t\right) \\ - \dfrac{\zeta'}{\sqrt{1 - \zeta'^2}}\sin\left(\left(\omega_n\sqrt{1 - \zeta'^2}\right)t\right) \right) & \zeta' < 1 \end{cases} \tag{6.44}$$

$$i_L(t) = \begin{cases} I_s - \dfrac{I_s}{2}\left(\left(1 + \dfrac{\zeta'}{\sqrt{\zeta'^2 - 1}}\right)\exp\left(-\left(\zeta' - \sqrt{\zeta'^2 - 1}\right)\omega_n t\right) \\ + \left(1 - \dfrac{\zeta'}{\sqrt{\zeta'^2 - 1}}\right)\exp\left(-\left(\zeta' + \sqrt{\zeta'^2 - 1}\right)\omega_n t\right) \right) & \zeta' > 1 \\[6mm] I_s - I_s(1 + \omega_n t)\exp(-\omega_n t) & \zeta' = 1 \\[4mm] I_s - I_s \exp(-\omega_n\zeta' t)\left(\cos\left(\left(\omega_n\sqrt{1 - \zeta'^2}\right)t\right) + \\ \dfrac{\zeta'}{\left(\sqrt{1 - \zeta'^2}\right)}\sin\left(\left(\omega_n\sqrt{1 - \zeta'^2}\right)t\right) \right) & \zeta' < 1 \end{cases} \tag{6.45}$$

6.4   Problems

Problem 6.1: The initial voltage across a 2 µF capacitor is 10 V. At time, t, equal to zero, the capacitor is discharged through a 400 Ω resistor as shown below.

What is the voltage across the resistor as a function of time? How much time does it take for the voltage across the capacitor to decrease by a factor of 2.

Problem 6.2: The initial voltage across a 4 µF capacitor is 5 V. At time, t, equal to zero, the capacitor is discharged through an 8 kΩ resistor as shown below.

What is the maximum voltage across the capacitor over all time? What is the voltage across the capacitor after 3 ms?

Problem 6.3: If an RC circuit should have a time constant of 2 ms and the capacitor has a value of 0.35 mF, what should be the value for the resistor?

Problem 6.4: If an RC circuit should have a time constant of 100 µs and the resistor has a value of 500 Ω, what should be the value for the capacitor?

Problem 6.5: The initial voltage across a 5 mF capacitor is 0 V. At time, t, equal to zero, the capacitor is connected to a 80 V source through a 4 kΩ resistor. What is the voltage across the capacitor and current flowing into the capacitor has a function of time?

Problem 6.6: Find the voltage across the capacitor, v(t) as a function of time for the circuit shown assuming that the switch has been opened for a very long period of time prior to closing the switch.

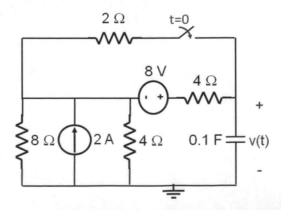

Problem 6.7: Find the voltage across the capacitor, v(t) as a function of time for the circuit shown assuming that the switch has been opened for a very long period of time prior to closing the switch.

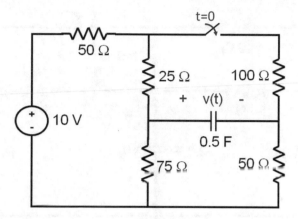

Problem 6.8: Find the voltage across the capacitor, v(t) as a function of time for the circuit shown assuming that the switch has been closed for a very long period of time prior to opening the switch.

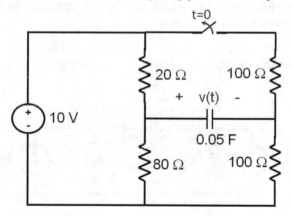

Problem 6.9: Find the voltage across the capacitor, v(t) as a function of time for the circuit shown assuming that the first switch closes at t = 0 and the second switch opens at t = 0.5 s. You may assume that the circuit has been stable for a long time prior to closing the first switch.

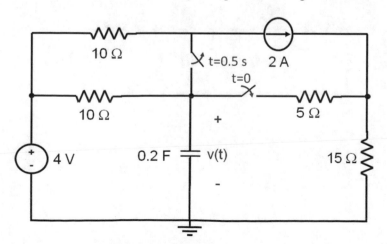

Problem 6.10: For the circuit shown for Problem 9, what is the voltage across the 5 Ω resistor, $v_R(t)$, as a function of time?

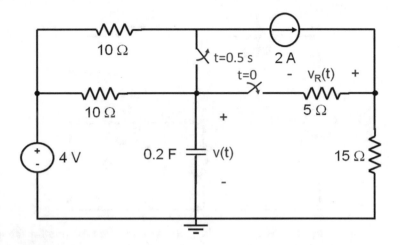

Problem 6.11: If an RL circuit should have a time constant of 2 ms and the inductor has a value of 0.5 H, what should be the value for the resistor?

Problem 6.12: Find the current in the inductor, $i(t)$, and the voltage across the inductor, $v(t)$, as a function of time for the circuit shown. You may assume that the circuit has been stable for a long time prior to flipping the switches.

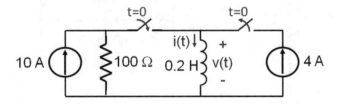

Problem 6.13: Find the current in the inductor, $i(t)$, and the voltage across the inductor, $v(t)$, as a function of time for the circuit shown. You may assume that the circuit has been stable for a long time prior to flipping the switches.

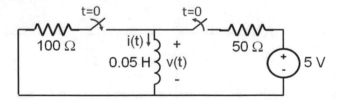

Problem 6.14: Find the current in the inductor, $i(t)$, as a function of time for the circuit shown. You may assume that the circuit has been stable for a long time prior to flipping the switch.

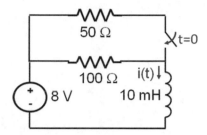

Problem 6.15: Find the current in the inductor, $i(t)$, as a function of time for the circuit shown. You may assume that the circuit has been stable for a long time prior to flipping the switch.

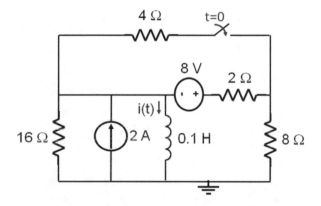

Problem 6.16: A Series RLC circuit has a V_s value of 8 V that is connected to the rest of the circuit at $t = 0$, an inductor of 40 mH, a capacitor of 100 μF, and a resistor of 15 Ω. Find the current as a function of time as well as the voltage across the capacitor for $t > 0$.

Problem 6.17: A Series RLC circuit has a V_s value of 8 V that is connected to the rest of the circuit at $t = 0$, an inductor of 40 mH, a capacitor of 100 μF, and a resistor of 150 Ω. Find the current as a function of time as well as the voltage across the inductor for $t > 0$.

Problem 6.18: A Series RLC circuit has a V_s value of 8 V that is connected to the rest of the circuit at $t = 0$, an inductor of 40 mH and a capacitor of 100 μF. What value of resistance would make the circuit critically damped and what is the current as a function of time for this case?

Problem 6.19: A Parallel RLC circuit has a I_s value of 4 A that is connected to the rest of the circuit at $t = 0$, an inductor of 10 mH, a capacitor of 25 μF, and a resistor of 100 Ω. Find the voltage as a function of time for $t > 0$.

Problem 6.20: A Parallel RLC circuit has a I_s value of 10 A that is connected to the rest of the circuit at $t = 0$, an inductor of 10 mH, a capacitor of 25 μF, and a resistor of 5 Ω. Find the voltage as a function of time for $t > 0$.

Problem 6.21: A Parallel RLC circuit has a I_s value of 2 A that is connected to the rest of the circuit at $t = 0$, an inductor of 10 mH and a capacitor of 25 μF. What value of resistance would make the circuit critically damped and what is the voltage as a function of time for this case?

Phasor-Domain Circuit Analysis 7

When analyzing and designing circuits for specific functions, solving and optimizing the circuit parameters would be very tedious if we were restricted to solving the complete differential equation for each case. Instead, we need a way to simplify the differential equations that arise due to inductors and capacitors into algebraic expressions. Phasor-domain circuit analysis allows us to make this simplification by assuming sinusoidal steady-state excitation and replacing the time derivatives with complex numbers.

7.1 Review of Complex Numbers

7.1.1 Rectangular Versus Polar Form

Before discussing the use of complex numbers to solve circuits with time-varying voltages and currents, it is helpful to review the basics of complex numbers. Many students have not seen complex numbers since high school, and the rigor may have varied drastically based on your teacher. Complex numbers are usually introduced in the rectangular form given by

$$x = a + i \cdot b \tag{7.1}$$

where x is the complex number, a is the real part of the complex number, and b is the imaginary part of the complex number. The imaginary number, i, is $\sqrt{-1}$. In electrical engineering, especially when applying complex numbers to circuit analysis, the use of the letter i for the imaginary number can lead to confusion as $i(t)$ is the current flow in the circuit. Therefore, to avoid this confusion, the letter j is normally used for the imaginary number. There is no difference between i and j. They are both symbolic representations for $\sqrt{-1}$. Therefore, the rectangular form for a complex number in electrical engineering is normally written as

$$x = a + j \cdot b \tag{7.2}$$

and this will be the representation used throughout the remainder of the text.

© Springer Nature Switzerland AG 2020
T. A. Bigelow, *Electric Circuits, Systems, and Motors*,
https://doi.org/10.1007/978-3-030-31355-5_7

In addition to being written in rectangular form, complex numbers can also be written in polar form as given by

$$x = ce^{j\theta} \tag{7.3}$$

where c is the magnitude of the complex number, often denoted $|x|$, and θ is the phase. Also, e is the Euler number where $\ln(e) = 1$. Writing the number in rectangular form is the same as expressing a point in space in Cartesian coordinates while writing the number in polar form is the same as specifying the point in polar coordinates as is illustrated in Fig. 7.1. In this graphical representation, the horizontal axis is the real part of the complex number while the vertical axis is the imaginary part of the complex number. The value of the complex number can then be uniquely specified by either the values of a and b or by the values of c and θ.

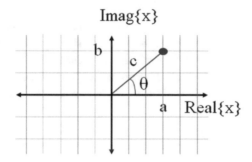

Fig. 7.1 Graphical representation of rectangular versus polar form of a complex number shown on the complex plane

Complex Number Notation: The polar form of a complex number can be written in multiple different ways depending on the context. Some of the more common notations are

$$x = ce^{j\theta} = c \cdot \exp(j\theta) = c\angle\theta$$

All of these forms are equivalent. Also, the phase angle can be given in either radians or degrees. When no degree symbol is present, you may assume the units are in radians.

From the graphical representation as well as basic trigonometry, it is clear that

$$\cos(\theta) = \frac{a}{c} \Rightarrow a = c \cdot \cos(\theta)$$
$$\sin(\theta) = \frac{b}{c} \Rightarrow b = c \cdot \sin(\theta) \tag{7.4}$$

which allows us to translate a complex number in polar form into a complex number in rectangular form. Likewise, from the Pythagorean Theorem and more basic trigonometry,

$$c^2 = a^2 + b^2 \Rightarrow c = \sqrt{a^2 + b^2}$$
$$\frac{b}{a} = \frac{c \cdot \sin(\theta)}{c \cdot \cos(\theta)} = \tan(\theta) \Rightarrow \theta = \operatorname{atan}\left(\frac{b}{a}\right) \tag{7.5}$$

which allows us to convert from the rectangular form to the polar form. From these expressions, it is clear that the value of θ can change by a factor of 2π while still representing the same complex number.

Example 7.1 Plot the location of each of the following complex numbers on the complex plane and convert them to the polar form with a phase between $-\pi$ and π radians.

$$x_1 = 3 + j6 \qquad x_2 = 3 - j6$$
$$x_3 = -3 + j6 \qquad x_4 = -3 - j6$$

Solution:
The plot of the complex numbers in the complex plane is shown below.

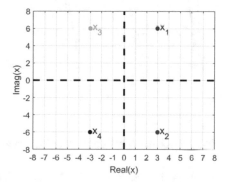

For this example, x_1 is in the first quadrant, x_3 is in the second quadrant, x_4 is in the third quadrant, and x_2 is in the fourth quadrant. Therefore, even before solving for the values of c and θ, we know

$$x_1 \Rightarrow 0 < \theta_1 < \pi/2$$
$$x_2 \Rightarrow -\pi/2 < \theta_2 < 0$$
$$x_3 \Rightarrow \pi/2 < \theta_3 < \pi$$
$$x_4 \Rightarrow -\pi < \theta_4 < -\pi/2$$

when the values are expressed in radians. If the values found for the phase lie outside of these ranges, then you know something is wrong. We can now use Eq. (7.5) to find the values of c and θ.

$$c = \sqrt{a^2 + b^2} \qquad\qquad \theta = \operatorname{atan}\left(\tfrac{b}{a}\right)$$
$$x_1: \qquad \sqrt{3^2 + 6^2} = 6.7082 \qquad \operatorname{atan}\left(\tfrac{6}{3}\right) = 1.1071 \text{ rad}$$
$$x_2: \qquad \sqrt{3^2 + (-6)^2} = 6.7082 \qquad \operatorname{atan}\left(\tfrac{-6}{3}\right) = -1.1071 \text{ rad}$$
$$x_3: \qquad \sqrt{(-3)^2 + 6^2} = 6.7082 \qquad \operatorname{atan}\left(\tfrac{-6}{3}\right) = -1.1071 \text{ rad}$$
$$x_4: \qquad \sqrt{(-3)^2 + (-6)^2} = 6.7082 \qquad \operatorname{atan}\left(\tfrac{-6}{-3}\right) = 1.1071 \text{ rad}$$

If we directly apply the equation with the arctangent, the phase found for x_3 and x_4 place the points in the wrong quadrant. This is because most calculators cannot distinguish between $\operatorname{atan}\left(\tfrac{b}{a}\right)$ and $\operatorname{atan}\left(\tfrac{-b}{-a}\right)$ or between $\operatorname{atan}\left(\tfrac{b}{a}\right)$ and $\operatorname{atan}\left(\tfrac{-b}{-a}\right)$. Therefore, we need to manually add or subtract a value of π to get the point into the correct quadrant.

$$\theta_3 = -1.1071 + \pi = 2.0344 \text{ rad}$$

$$\theta_4 = 1.1071 - \pi = -2.0344 \text{ rad}$$

The factor of π comes from the tangent function being periodic with a period of π (i.e., $\tan(\theta \pm \pi) = \tan(\theta)$) as shown below.

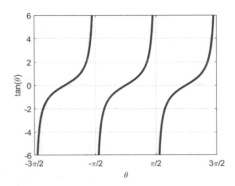

7.1.2 Addition, Subtraction, and Multiplication of Complex Numbers

When working with complex numbers, it is easy to make basic mistakes when manipulating them mathematically. The imaginary number, j, must be considered when adding, subtracting, or multiplying complex numbers. For example, if we have two complex numbers $x_1 = a_1 + jb_1 = c_1 e^{j\theta_1}$ and $x_2 = a_2 + jb_2 = c_2 e^{j\theta_2}$, then $x_1 + x_2 = (a_1 + a_2) + j(b_1 + b_2)$ as we can directly add the real and imaginary parts. Likewise, $x_1 - x_2 = (a_1 - a_2) + j(b_1 - b_2)$. However, if the number is in polar form, adding the complex numbers DOES NOT allow us to just add the magnitude and phase terms (i.e., $x_1 \pm x_2 \neq (c_1 \pm c_2)e^{j(\theta_1 \pm \theta_2)}$). If your calculator does not automatically add complex numbers, then you would need to convert both of the complex numbers to rectangular form prior to performing the addition or subtraction.

Also, when multiplying complex numbers in the rectangular form, the FOIL method should be utilized just like when multiplying binomials.

$$\begin{aligned} x_1 \cdot x_2 &= (a_1 + jb_1) \cdot (a_2 + jb_2) = a_1 a_2 + ja_1 b_2 + jb_1 a_2 + jb_1 jb_2 \\ &= a_1 a_2 + (j)^2 b_1 b_2 + j(a_1 b_2 + b_1 a_2) \end{aligned} \tag{7.6}$$

However, since $j = \sqrt{-1}$, $j^2 = -1$ simplifying Eq. (7.6) to

$$\begin{aligned} x_1 \cdot x_2 &= (a_1 + jb_1) \cdot (a_2 + jb_2) = a_1 a_2 + ja_1 b_2 + jb_1 a_2 + jb_1 jb_2 \\ &= (a_1 a_2 - b_1 b_2) + j(a_1 b_2 + b_1 a_2) \end{aligned} \tag{7.7}$$

Multiplying the numbers when they are in the polar form is a little simpler.

$$x_1 \cdot x_2 = c_1 e^{j\theta_1} \cdot c_2 e^{j\theta_2} = c_1 c_2 e^{j(\theta_1 + \theta_2)} \tag{7.8}$$

Example 7.2 Find x in polar and rectangular form if $x = 4e^{j\pi/3} - 2e^{-j\pi/6}$.

Solution:
Assuming you cannot just enter the numbers into your calculator, the first step is to convert both of the complex numbers in the expression into rectangular form.

$$4e^{j\pi/3} = 4\cos\left(\frac{\pi}{3}\right) + j4\sin\left(\frac{\pi}{3}\right) = 2 + j2\sqrt{3}$$
$$2e^{-j\pi/6} = 2\cos\left(\frac{-\pi}{6}\right) + j2\sin\left(-\frac{\pi}{6}\right) = \sqrt{3} - j1$$

Subtracting the corresponding real and imaginary parts gives

$$x = 4e^{j\pi/3}\ 2e^{-j\pi/6} - \left(2 + j2\sqrt{3}\right) - \left(\sqrt{3} - j1\right)$$
$$= \left(2 - \sqrt{3}\right) + j\left(2\sqrt{3} + 1\right) = 0.2679 + j4.4641$$

We can now convert this number into polar form.

$$c = \sqrt{a^2 + b^2} = 4.4721 \quad \theta = \text{atan}\left(\frac{b}{a}\right) = 1.5108$$
$$x = 4.4721 e^{j1.5108}$$

Example 7.3 Find x in polar and rectangular form if $x = (2 - j6) \cdot (3 + j3)$.

Solution:
We can use FOIL to first find x in rectangular form.

$$x = (2 - j6) \cdot (3 + j3) = 6 + j6 - j18 + 18 = 24 - j12$$

We can now convert this number into polar form.

$$c = \sqrt{a^2 + b^2} = 26.8328 \quad \theta = \text{atan}\left(\frac{b}{a}\right) = -0.4636$$
$$x = 26.8328 e^{-j0.4636}$$

Conversely, we could have first converted both numbers to polar form prior to the multiplication.

$$(2 - j6) = 6.3246 e^{-j1.2490}$$
$$(3 + j3) = 4.2426 e^{j0.7854}$$
$$(2 - j6) \cdot (3 + j3) = \left(6.3246 e^{-j1.2490}\right) \cdot \left(4.2426 e^{j0.7854}\right)$$
$$= 6.3246 \cdot 4.2426 e^{j(-1.2490 + 0.7854)} = 26.8328 e^{-j0.4636}$$

7.1.3 Division/Inversion of Complex Numbers

Division by a complex number is accomplished by first finding the inverse of the complex number and then multiplying by the inverse. The inverse is normally found by utilizing the complex conjugate of the complex number to make the number purely real. The complex conjugate of a complex number is found by changing the sign of the imaginary term. Therefore, if the complex number is given by $x = a + jb$, then the complex conjugate is given by $x^* = a - jb$. Likewise, if the complex number is given by $x = a - jb$, then the complex conjugate is given by $x^* = a + jb$. In polar form, this translates to the complex number $x = ce^{\pm j\theta}$ having the complex conjugate $x^* = ce^{\mp j\theta}$. Notice that

$$x \cdot x^* = |x|^2 = c^2 = a^2 + b^2 \tag{7.9}$$

Example 7.4 Find the complex conjugate of the following complex numbers.

$$x_1 = 3 + j6 \quad x_2 = 1 - j2$$
$$x_3 = -7 + j4 \quad x_4 = 5e^{-j0.2}$$

Solution:
For the rectangular form, if $x = a \pm jb$, $x^* = a \mp jb$. Therefore,

$$x_1^* = 3 - j6 \quad x_2^* = 1 + j2$$
$$x_3^* = -7 - j4$$

For the polar form, if $x = ce^{\pm j\theta}$, $x^* = ce^{\mp j\theta}$. Therefore,

$$x_4^* = 5e^{+j0.2}$$

Using the complex conjugate, we can invert a complex number by just multiplying the numerator and the denominator by the complex conjugate.

$$\frac{1}{x} = \frac{1}{a \pm jb} = \frac{1}{a \pm jb} \cdot \frac{(a \mp jb)}{(a \mp jb)} = \frac{a \mp jb}{a^2 \mp jab \pm jab - (jb)^2} = \frac{a \mp jb}{a^2 + b^2 \mp jab \pm jab} \tag{7.10}$$

The imaginary terms in the denominator then cancel leaving

$$\frac{1}{a \pm jb} = \frac{a \mp jb}{a^2 + b^2} = \frac{x^*}{|x|^2} = \frac{a}{a^2 + b^2} \mp \frac{jb}{a^2 + b^2} \tag{7.11}$$

When the number is in the polar form, the process is somewhat simpler but the result is the same.

$$\frac{1}{x} = \frac{1}{ce^{\pm j\theta}} = \frac{1}{ce^{\pm j\theta}} \cdot \frac{ce^{\mp j\theta}}{ce^{\mp j\theta}} = \frac{ce^{\mp j\theta}}{c^2} = \frac{x^*}{|x|^2} = \frac{1}{c}e^{\mp j\theta} \tag{7.12}$$

Example 7.5 Simplify the following complex number expressions and express your answer in polar form with a phase between $-\pi$ and π.

$$x_1 = \frac{1}{3 + j4} \qquad\qquad x_2 = \frac{1}{-11 + j9}$$
$$x_3 = \frac{1}{0.15 + 0.05j} + 14 + j5 - 12e^{j\frac{\pi}{7}} \quad x_4 = \frac{4 + j6}{1 - j3} + \frac{13e^{j\frac{\pi}{5}}}{2 + j7}$$

Solution:

When simplifying these expressions, we will show all of the steps. Many calculators can work directly with complex numbers eliminating the need for some of the manipulations shown.

$$x_1 = \frac{1}{3+j4} = \frac{1}{3+j4}\frac{(3-j4)}{(3-j4)} = \frac{3-j4}{25} = 0.12 - j0.16 = 0.2e^{-j0.9273}$$

x_2 can be simplified in the same way as x_1. However, this time we will first convert the denominator to polar form before finding the inverse.

$$x_2 = \frac{1}{-11+j9}$$

$$|-11+j9| = \sqrt{(-11)^2 + (9)^2} = 14.2127$$

$$\theta = \text{atan}\left(\frac{9}{-11}\right) = -0.6857 \text{ rad} \Rightarrow \textit{Not 2nd Quadrant}$$
$$\Rightarrow \theta = -0.6857 + \pi = 2.4559 \text{ rad}$$

$$x_2 = \frac{1}{14.2127e^{j2.4559}} = \frac{1}{14.2127}e^{-j2.4559} = 0.0704e^{-j2.4559}$$

For x_3 and x_4, we will first convert everything to rectangular form to facilitate adding the expressions together.

$$x_3 = \frac{1}{0.15+0.05j} + 14 + j5 - 12e^{j\frac{\pi}{7}} = \frac{0.15-0.75j}{(0.15)^2+(0.05)^2} + 14 + j5 - \left(12\cos\left(\frac{\pi}{7}\right) + j12\sin\left(\frac{\pi}{7}\right)\right)$$

$$= 40(0.15 - 0.05j) + 14 + j5 - (10.8116 + j5.2066)$$

$$= 6 - j2 + 14 + j5 - 10.8116 - j5.2066 = 9.1884 - j2.2066 = 9.4496e^{-j0.2357}$$

$$x_4 = \frac{4+j6}{1-j3} + \frac{13e^{j\frac{\pi}{5}}}{2+j7} = \frac{4+j6}{1-j3}\cdot\frac{(1+j3)}{(1+j3)} + \frac{\left(13\cos\left(\frac{\pi}{5}\right) + j13\sin\left(\frac{\pi}{5}\right)\right)}{2+j7}\frac{(2-j7)}{(2-j7)}$$

$$= \frac{4+j12+j6-18}{1+9} + \frac{21.0344 - j73.6205 + j15.2824 + 53.4885}{4+49}$$

$$= -1.4 + j1.8 + 1.4061 + -j1.1007 = 0.0061 + j0.6993 = 0.6993e^{j1.5621}$$

7.1.4   Useful Identities for Complex Numbers

In addition to the complex conjugate, there are other identities that can be useful when simplifying complex number expressions. The most important by far is the Euler's Identity given by

$$e^{j\theta} = \cos(\theta) + j\sin(\theta) \tag{7.13}$$

We have actually already used this identity to convert complex numbers from polar form to the rectangular form. Euler's Identity can also be written as

$$\cos(\theta) = \frac{e^{j\theta} + e^{-j\theta}}{2}$$

$$\sin(\theta) = \frac{e^{j\theta} - e^{-j\theta}}{2j}$$

(7.14)

The expressions given in Eq. (7.14) can be very useful when evaluating certain integrals.

Example 7.6 Use Euler's Identity to simplify and solve the following integral assuming $a > 0$.

$$\int_0^\infty e^{-ax} \cos(x) dx$$

Solution:
The first step is to replace the cosine expression with the complex exponentials from Euler's Identity.

$$\int_0^\infty e^{-ax} \cos(x) dx = \int_0^\infty e^{-ax} \left(\frac{e^{jx} + e^{-jx}}{2} \right) dx = \frac{1}{2} \int_0^\infty \left(e^{-ax} e^{jx} + e^{-ax} e^{-jx} \right) dx$$

$$= \frac{1}{2} \left[\int_0^\infty e^{-(a-j)x} dx + \int_0^\infty e^{-(a+j)x} dx \right] = \frac{1}{2} \left[\frac{e^{-(a-j)x}}{-(a-j)} + \frac{e^{-(a+j)x}}{-(a+j)} \right]_0^\infty$$

$$= \frac{1}{2} \left[\frac{1}{(a-j)} + \frac{1}{(a+j)} \right] = \frac{1}{2} \left[\frac{(a+j) + (a-j)}{a^2 + 1} \right] = \frac{a}{a^2 + 1}$$

Some addition identities for complex numbers, many of which are derived from Euler's Identity are given below.

$$e^{\pm j\pi n} = (-1)^n \qquad n = \text{any integer}$$

$$e^{\pm j\frac{\pi}{2}(2n+1)} = \pm(-1)^n j \qquad n = \text{any integer}$$

(7.15)

7.1.5 Solving Equations with Complex Numbers

Having reviewed basic operations with complex numbers, we can now use these operations to solve systems of equations. To begin, we will focus on a single equation with a single unknown.

$$(a_1 + jb_1)x + (a_2 + jb_2) = (a_3 + jb_3)$$

(7.16)

To solve this equation, we need to subtract $(a_2 + jb_2)$ from both sides and then divide by $(a_1 + jb_1)$ just like any other simple algebraic expression. This process will be illustrated in more detail in the following example.

Example 7.7 Solve the following equations involving complex numbers and express your answer in polar form with a phase between $-\pi$ and π.

(a) $(3+j)x = 11+j7$
(b) $(3+j4)x = -13-j14$
(c) $(5+j2)x + (3+j4) = (4-j1)$

Solution:

(a)

$$(3+j)x = 11+j7 \Rightarrow x = \frac{11+j7}{(3+j)} = \frac{11+j7}{(3+j)} \cdot \frac{(3-j)}{(3-j)} = \frac{33-j11+j21+7}{3^2+1^2}$$

$$x = \frac{40+j10}{10} = 4+j = 4.1231 e^{j0.245}$$

(b)

$$(3+j4)x = -13-j14 \Rightarrow x = \frac{-13-j14}{(3+j4)} = \frac{-13-j14}{(3+j4)} \cdot \frac{(3-j4)}{(3-j4)} = \frac{-95+j10}{25} = -3.8+j0.4$$

$$|x| = 3.821$$

Phase of $x = \angle x = \text{atan}\left(\frac{0.4}{-3.8}\right) = -0.1049 \text{ rad} \Rightarrow \textit{Not 2nd Quadrant}$

$$\Rightarrow \angle x = -0.1049 \text{ rad} + \pi = 3.0367 \text{ rad}$$

$$x = 3.821 e^{j3.0367}$$

(c)

$$(5+j2)x + (3+j4) = (4-j1) \Rightarrow x = \frac{(4-j1)-(3+j4)}{(5+j2)} = \frac{1-j5}{(5+j2)} \cdot \frac{(5-j2)}{(5-j2)} = \frac{-5-j27}{29}$$

$$x = -0.1724 - j0.931$$

$$|x| = 0.9469$$

Phase of $x = \angle x = \text{atan}\left(\frac{-0.931}{-0.1724}\right) = 1.3877 \text{ rad} \Rightarrow \textit{Not 3rd Quadrant}$

$$\Rightarrow \angle x = 1.3877 \text{ rad} - \pi = -1.7539 \text{ rad}$$

$$x = 0.9469 e^{-j1.7539}$$

When the number of equations and unknowns increases, the basic tools of algebra continue to apply. The complex numbers complicate the mathematics, but conceptually the process remains the same. We will once again use examples to illustrate the necessary steps.

Example 7.8 Solve for x and y in the following system of equations involving complex numbers and express your answer in polar form with a phase between $-\pi$ and π.

$$(4+j2)x + (1-j1)y = 7+j8$$
$$(1+j2)x + (2+j3)y = 9$$

Solution:
There are multiple approaches to solve this system of equations. One approach is to use complex conjugates to make the numbers multiplying one of the variables (either x or y) purely real. Finding a common term and subtracting the equations would then leave an expression with only one unknown. Multiplying the first equation by $(4 - j2)$ gives

$$(4 - j2) \cdot (4 + j2)x + (4 - j2) \cdot (1 - j1)y = (4 - j2) \cdot (7 + j8)$$
$$20x + (2 - j6)y = (44 + j18)$$

Likewise, multiplying the second equation by $(1 - j2)$ gives

$$(1 - j2) \cdot (1 + j2)x + (1 - j2) \cdot (2 + j3)y = (1 - j2) \cdot 9$$
$$5x + (8 - j1)y = (9 - j18)$$

In both cases, we are left with equations where the number multiplying x is purely real. If we multiply the new second equation by -4, we get

$$- 4 \cdot (5x + (8 - j1)y = (9 - j18))$$
$$- 20x + (-32 + j4)y = (-36 + j72)$$

This equation can now be added to $20x + (2 - j6)y = (44 + j18)$ to eliminate the x variable.

$$- 20x + (-32 + j4)y + 20x + (2 - j6)y = (-36 + j72) + (44 + j18)$$
$$(-30 - j2)y = 8 + j90 \Rightarrow y = \frac{8 + j90}{(-30 - j2)} = -0.4646 - j2.969 = 3.0052e^{-j1.726}$$

With y know, we can now solve for x.

$$5x + (8 - j1)y = (9 - j18) \Rightarrow$$
$$x = \frac{(9 - j18) - (8 - j1)y}{5}$$
$$x = 3.1372 + j1.0575 = 3.3106 \exp(j0.3251)$$

Another possible approach to solve the problem (depending on the tools available) is to translate the system of equations into a matrix equation. The problem can then be solved by inverting the complex matrix if one has access to MATLAB or a calculator capable of manipulating complex matrix equations. For this approach, the first step is to write the system of equations as a matrix.

$$\begin{pmatrix} 4 + j2 & 1 - j1 \\ 1 + j2 & 2 + j3 \end{pmatrix} \begin{pmatrix} x \\ y \end{pmatrix} = \begin{pmatrix} 7 + j8 \\ 9 \end{pmatrix}$$

Inverting the matrix and then multiplying it by the vector on the right hand side then gives

$$\begin{pmatrix} x \\ y \end{pmatrix} = \begin{pmatrix} 4 + j2 & 1 - j1 \\ 1 + j2 & 2 + j3 \end{pmatrix}^{-1} \begin{pmatrix} 7 + j8 \\ 9 \end{pmatrix} = \begin{pmatrix} 3.1372 + j1.0575 \\ -0.4646 - j2.9690 \end{pmatrix} = \begin{pmatrix} 3.3106e^{j0.3251} \\ 3.0052e^{-j1.726} \end{pmatrix}$$

Example 7.9 Solve for x and y in the following system of equations involving complex numbers and express your answer in polar form with a phase between $-\pi$ and π.

$$(3+j4)x+(4-j5)y = 18+j13$$
$$(4-j2)x+(-4-j5)y = -5+j10$$

Solution:

We will once again first solve the equations by using complex conjugate multiplication to combine the two equations into a single equation eliminating one of the variables.

$$(3-j4)\cdot(3+j4)x+(3-j4)\cdot(4-j5)y = (3-j4)\cdot(18+j13)$$
$$25x+(-8-j31)y = (106-j33)$$

$$(4+j2)\cdot(4-j2)x+(4+j2)\cdot(-4-j5)y = (4+j2)\cdot(-5+j10)$$
$$20x+(-6-j28)y = (-40+j30)$$

$$-4\cdot(25x+(-8-j31)y = (106-j33)) \Rightarrow -100x+(32+j124)y = (-424+j132)$$
$$5\cdot(20x+(-6-j28)y = (-40+j30)) \Rightarrow 100x+(-30-j140)y = (-200+j150)$$
$$-100x+(32+j124)y+100x+(-30-j140)y = (-424+j132)+(-200+j150)$$
$$(2-j16)y = (-624+j282) \Rightarrow y = -22.1538-j32.2308 = 42.4672e^{-j2.1196}$$

We can now solve for the other variable.

$$x = \frac{(18+j13)-(4-j5)(-22.1538-j32.2308)}{(3+j4)} = \frac{(18+j13)-(-269.22-j34.154)}{(3+j4)}$$

$$= \frac{287.77+j47.154}{(3+j4)} = 42.0769-j40.3846 = 58.3214e^{-j0.7649}$$

We can also use matrices to solve for the system of equations.

$$\begin{pmatrix} 3+j4 & 4-j5 \\ 4-j2 & -4-j5 \end{pmatrix}\begin{pmatrix} x \\ y \end{pmatrix} = \begin{pmatrix} 18+j13 \\ -5+j10 \end{pmatrix}$$

$$\Rightarrow \begin{pmatrix} x \\ y \end{pmatrix} = \begin{pmatrix} 3+j4 & 4-j5 \\ 4-j2 & -4-j5 \end{pmatrix}^{-1}\begin{pmatrix} 18+j13 \\ -5+j10 \end{pmatrix} = \begin{pmatrix} 58.3214e^{-j0.7649} \\ 42.4672e^{-j2.1196} \end{pmatrix}$$

7.2   Harmonic/Sinusoidal Signals

7.2.1   Definition of Harmonic Signals

As was stated previously, phasor domain circuit analysis is used to analyze linear circuits undergoing sinusoidal steady-state excitation. This means that the voltages and currents in the circuit have the form

$$v(t) = V_m \cos(\omega t + \theta_v)$$
$$i(t) = I_m \cos(\omega t + \theta_i)$$

(7.17)

In these equations, V_m and I_m are the maximum value of the voltage and current, respectively, ω is the angular frequency, and θ is the phase of the sinusoidal signal. From the trig identities

$$\cos(A + B) = \cos(A) \cos(B) - \sin(A) \sin(B)$$
$$\cos(x) = \sin\left(x + \frac{\pi}{2}\right)$$

(7.18)

The expressions in Eq. (7.17) are equivalent to

$$v(t) = V_m \cos(\omega t + \theta_v) = \overbrace{V_m \cos(\theta_v)}^{V_{m1}} \cos(\omega t) - \overbrace{V_m \sin(\theta_v)}^{V_{m2}} \sin(\omega t)$$
$$= V_{m1} \cos(\omega t) - V_{m2} \sin(\omega t)$$
$$i(t) = I_m \cos(\omega t + \theta_i) = I_{m1} \cos(\omega t) - I_{m2} \sin(\omega t)$$

(7.19)

and

$$v(t) = V_m \cos(\omega t + \theta_v) = V_m \sin\left(\omega t + \overbrace{\theta_v + \frac{\pi}{2}}^{\theta'_v}\right) = V_m \sin\left(\omega t + \theta'_v\right)$$

(7.20)

$$i(t) = I_m \cos(\omega t + \theta_i) = V_m \sin\left(\omega t + \theta'_i\right)$$

However, convention dictates that all harmonic signals will be expressed as cosines as shown in Eq. (7.17). Also, the phase, θ, is normally expressed in radians although one can always convert from radians to degrees using

$$(\# \text{degrees}) = \frac{180°}{\pi} (\# \text{radians})$$

(7.21)

A harmonic signal is shown graphically in Fig. 7.2. The signal is periodic which means it repeats for all time with no beginning or ending. The smallest time interval between corresponding regions on the waveform is called the period and is denoted T. The period and angular frequency are related by

$$\omega = \frac{2\pi}{T}$$

(7.22)

Often when describing harmonic signals, the frequency is expressed in Hz. The frequency in Hz is given by

$$f = \frac{1}{T}$$

(7.23)

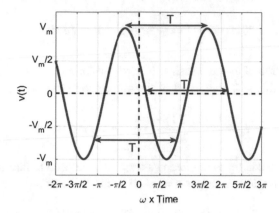

Fig. 7.2 Example harmonic signal illustrating signal amplitude and period

And, therefore,

$$\omega = 2\pi f \tag{7.24}$$

The maximum and minimum values of the signal are given by $+V_m$ and $-V_m$, respectively, with V_m defined as the amplitude of the waveform. The phase is related to the delay of the signal relative to $t = 0$. A negative value for the phase means that the peaks are shifted later in time while a positive value for the phase means that the peaks are occurring earlier in time as illustrated by Fig. 7.3.

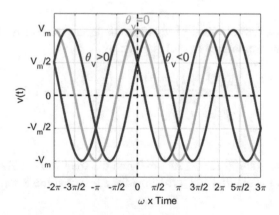

Fig. 7.3 Example harmonic signal illustrating impact of phase

Example 7.10 If $i(t) = 3\cos\left(20\pi t + \frac{2\pi}{3}\right)$ A find the period, the frequency in Hz, the frequency in rad/s, the amplitude, and the location for all zero crossings for $t > 0$.

Solution:
The angular frequency, ω, is the number multiplying the time in the expression for the cosine. Therefore, $\omega = 20\pi$ rad/s. The frequency in Hz for the signal is thus given by $f = \omega/2\pi = 10$ Hz, and the period is given by $T = 1/f = 0.1$ s. The amplitude of the signal is 3 A corresponding to a maximum current of $+3$ A and a minimum current of -3 A. In order to find the zero crossings, we need to find all of the values of time, t, for which the cosine is zero. The cosine is zero for all odd multiples of $\pi/2$. This can be expressed mathematically as

$$\left(20\pi t_o + \frac{2\pi}{3}\right) = \frac{\pi}{2}(2n+1) \quad n = \text{any integer}$$

where t_o is the time location of a zero crossing. Notice that for any integer, n, the term $(2n + 1)$ is an odd number. Therefore, this will give us all of the odd multiples of $\pi/2$. Solving for t_o gives

$$t_o = \frac{\left(\frac{\pi}{2}(2n+1) - \frac{2\pi}{3}\right)}{20\pi} \quad n = \text{any integer}$$

Since we are only interested in the zero crossings for $t > 0$, we need to find the values for n that give positive values for t_o. First, let's consider $n = -1$. For this value of n,

$$t_o = \frac{\left(\frac{\pi}{2}(-2+1) - \frac{2\pi}{3}\right)}{20\pi} = \frac{\left(-\frac{\pi}{2} - \frac{2\pi}{3}\right)}{20\pi}$$

which is not greater than zero. Now, consider $n = 0$.

$$t_o = \frac{\left(\frac{\pi}{2}(0+1) - \frac{2\pi}{3}\right)}{20\pi} = \frac{\left(\frac{\pi}{2} - \frac{2\pi}{3}\right)}{20\pi}$$

This value is also less than zero. Increasing to $n = 1$ gives

$$t_o = \frac{\left(\frac{\pi}{2}(2+1) - \frac{2\pi}{3}\right)}{20\pi} = \frac{\left(\frac{3\pi}{2} - \frac{2\pi}{3}\right)}{20\pi} = 41.6667 \text{ ms}$$

which is greater than zero. Larger values of n will continue to give positive numbers. Therefore, the location of all of the zero crossings for t > 0 is given by

$$t_o = \frac{\left(\frac{\pi}{2}(2n+1) - \frac{2\pi}{3}\right)}{20\pi} \text{ s} = \frac{\left(n - \frac{1}{6}\right)}{20} \text{ s} \quad n \geq 1$$
$$t_o = 50n - 8.3333 \text{ ms} \quad n \geq 1$$

Example 7.11 If $v(t) = 19\sin(2111t + 170°)$ mV find the frequency in Hz, the maximum value of the voltage, the location for the first zero crossings for $t > 0$, and the location for the first peak for $t > 0$.

Solution:
The first step in solving this problem is to translate the sin function into a cosine function as the convention is to use the cosine when defining the phase in circuit analysis. Therefore, if we use the trig identity $\cos(x \pm \pi/2) = \mp\sin(x)$, the voltage can be written as

$$v(t) = 19\sin(2111t + 170°) \text{ mV} = 19\cos(2111t + 170° - 90°) \text{ mV}$$
$$= 19\cos(2111t + 80°) \text{ mV}$$

Where we are making use of $\pi/2 = 90°$. Now, we need to translate the phase in degrees into a phase in radians using $(\# \text{ degrees}) = \frac{180°}{\pi}(\# \text{ radians})$ giving us

$$v(t) = 19\cos(2111t + 1.396263)\ \text{mV}$$

Now that it is in the standard form, the maximum voltage is given by 19 mV as the maximum value for the cosine function is 1. The frequency in Hz can also be found from

$$f = \omega/2\pi = 2111/2\pi = 335.9761\ \text{Hz}$$

The value for the first zero crossing is given by the smallest time value that is still greater than zero that satisfies

$$(2111t_o + 1.396263) = \frac{\pi}{2}(2n+1) \quad n = \text{any integer}$$
$$t_o = \frac{\frac{\pi}{2}(2n+1) - 1.396263}{2111}$$

Testing $n = -1$ gives

$$t_o = \frac{\frac{\pi}{2}(-2+1) - 1.396263}{2111} = \frac{-\frac{\pi}{2} - 1.396263}{2111}$$

which is less than zero. Increasing to $n = 0$ gives

$$t_o = \frac{\frac{\pi}{2}(0+1) - 1.396263}{2111} = 82.678\ \mu\text{s}$$

This value is greater than zero and thus corresponds to the first zero crossing.

The value for the first peak is given by the smallest time value that is still greater than zero that satisfies

$$(2111t_o + 1.396263) = 2n\pi \quad n = \text{any integer}$$
$$t_o = \frac{2n\pi - 1.396263}{2111}$$

The value of $2n\pi$ gives all of the even multiples of π which is when the cosine function has its maximum values. Obviously, n = −1 and n = 0 would give negative numbers, and therefore, we will begin with n = 1.

$$t_o = \frac{2\pi - 1.396263}{2111} = 2.315\ \text{ms}$$

This number is greater than zero and would thus correspond to the location of the first peak.

Example 7.12 Write the expression for $v(t)$ for the harmonic signal shown in Fig. 7.4.

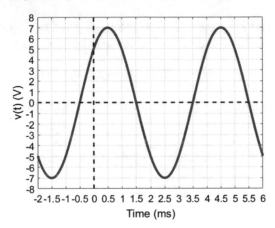

Fig. 7.4 Harmonic signal for Example 7.12

Solution:

The amplitude of the harmonic signal is 7 V as the maximum value of the voltage is 7 V and the minimum value of the voltage is -7 V. Also, the first peak value occurs at 0.5 ms while the second peak value occurs at 4.5 ms. Therefore, the period of the signal is T = 4.5 ms -0.5 ms = 4 ms. This corresponds to an angular frequency of

$$\omega = \frac{2\pi}{T} = 1570.8 \ \text{rad/s}$$

To find the phase of the signal, we can work with a peak value, a trough value, a zero crossing, or the amplitude of the signal at t = 0. If we use the first peak value shown at $t_{peak} = 0.5$ ms, then

$$\omega \cdot t_{peak} + \theta_v = 2\pi n \quad n = \text{any integer}$$

as this is when the cosine function has its maximums. Therefore,

$$\theta_v = 2\pi n - \omega \cdot t_{peak} = 2\pi n - 0.7854$$

If we want a value of the phase between $-\pi$ and π, then $n = 0$ is the correct choice and $\theta_v = -0.7854$ rad. This means the voltage would be given by

$$v(t) = 7\cos(1570.8t - 0.7854) \ V$$

If instead we focus on the first trough/minimum value shown as $t_{\text{trough}} = -1.5$ ms, then

$$\omega \cdot t_{\text{trough}} + \theta_v = \pi(2n + 1) \quad n = \text{any integer}$$

as this is when the cosine function has its minimums. Therefore,

$$\theta_v = \pi(2n + 1) - \omega \cdot t_{\text{trough}} = \pi(2n + 1) + 2.3562$$

If we want a value of the phase between $-\pi$ and π, then $n = -1$ is the correct choice, and we still get $\theta_v = -0.7854$ rad. If we use the first zero crossing shown at $t_o = -0.5$ ms, then

$$\omega \cdot t_o + \theta_v = \frac{\pi}{2}(2n+1) \quad n = \text{any integer}$$

as this is when the cosine function goes to zero. Therefore,

$$\theta_v = \frac{\pi}{2}(2n+1) - \omega \cdot t_o = \frac{\pi}{2}(2n+1) + 0.7854$$

If we want a value of the phase between $-\pi$ and π, then we have two possible values of n, $n = -1$ and $n = 0$. These values of n correspond to $-\pi/2$ and $\pi/2$ in the argument for the cosine. The $-\pi/2$ value corresponding to n = -1 would be for a zero crossing for when the cosine is going from a negative value to a positive value. Likewise, the $\pi/2$ value corresponding to $n = 0$ would be for when the cosine is going from a positive to a negative value. The time value of -0.5 ms is for a positive-going zero crossing. Therefore, an n value of -1 is the correct choice giving a phase of -0.7854 rad once again. Finally, we can solve for the phase using the value at t = 0 (or another known time location). If we zoom in on the graph at t = 0, the value of v(t) is given by 4.95 at t = 0.

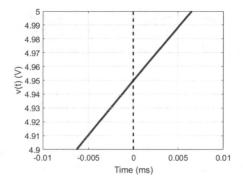

Therefore,

$$v(t) = 7\cos(\omega t + \theta_v) = 7\cos(\theta_v) = 4.95$$

$$\theta_v = \pm a\cos\left(\frac{4.95}{7}\right) + 2\pi n = \pm 0.7853 + 2\pi n$$

The n results from the cosine being periodic with a period of 2π. Since we want the phase to be between $-\pi$ and π, $n = 0$ is the correct choice. Therefore, $\theta_v = \pm 0.7853$. The $\pm$ in the phase expression is due to the cosine being an even function with $\cos(\theta_v) = \cos(-\theta_v)$. Graphically this would correspond with the slope of the cosine being negative or positive at $t = 0$. Taking the derivative of $v(t)$ with respect to time gives

$$\left(\frac{d}{dt}v(t)\right)_{t=0} = (-7\omega\sin(\omega t + \theta_v))_{t=0} = -7\omega\sin(\theta_v) = \begin{cases} -7774.6 & \theta_v = 0.7853 \\ 7774.6 & \theta_v = -0.7853 \end{cases}$$

Since from the graph, the voltage, $v(t)$, is increasing at $t = 0$, $\theta_v = -0.7853$ is the correct value for the phase. The other value for the phase would correspond to the case when the other side of the peak is at $t = 0$ as shown below.

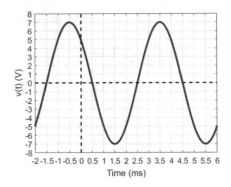

7.2.2   RMS Values

When working with harmonic signals, especially in power flow/generation applications, it is common to utilize the root-mean squared (RMS) values for the signals instead of the amplitude. The amplitude is the maximum voltage or current of the signal whereas the RMS values are given by

$$V_{\text{rms}} = \sqrt{\frac{1}{T} \int_{t_o}^{t_o+T} (v(t))^2 dt} \quad I_{\text{rms}} = \sqrt{\frac{1}{T} \int_{t_o}^{t_o+T} (i(t))^2 dt} \tag{7.25}$$

where $v(t)$ and/or $i(t)$ are periodic with a period T and t_o is any arbitrary starting time value for the signal. For sinusoidal signals of the form $A\cos(\omega t + \theta)$, the RMS quantity is given by

$$\sqrt{\frac{1}{T} \int_{t_o}^{t_o+T} (A\cos(\omega t + \theta))^2 dt} = \sqrt{\frac{A^2}{T} \int_{t_o}^{t_o+T} \cos^2\left(\omega\left(t + \frac{\theta}{\omega}\right)\right) dt} \tag{7.26}$$

However, if we let $t' = (t + \theta/\omega)$ and $t_o = -\theta/\omega$ and make use of the trigonometric identity $\cos^2(x) = 0.5(1 + \cos(2x))$, then Eq. (7.26) becomes

$$\sqrt{\frac{A^2}{T} \int_0^T \cos^2(\omega t')dt'} = \sqrt{\frac{A^2}{2T} \int_0^T (1 + \cos(2\omega t'))dt'}$$

$$= \sqrt{\frac{A^2}{2T}\left(T + \int_0^T \cos(2\omega t')dt'\right)} = \sqrt{\frac{A^2}{2T}\left(T + \frac{\sin(2\omega t')}{2\omega}\Big|_0^T\right)} \tag{7.27}$$

$$= \sqrt{\frac{A^2}{2T}\left(T + \frac{\sin\left(2\frac{2\pi}{T}T\right)}{2\omega}\right)} = \sqrt{\frac{A^2}{2T}\left(T + \frac{\sin(4\pi)}{2\omega}\right)} = \sqrt{\frac{A^2}{2}} = \frac{A}{\sqrt{2}}$$

Therefore, the RMS value for a harmonic signal is just the peak/amplitude value divided by the square root of 2.

$$V_{\text{rms}} = \frac{V_m}{\sqrt{2}} \quad I_{\text{rms}} = \frac{I_m}{\sqrt{2}} \tag{7.28}$$

Of course, the expressions given in Eq. (7.28) are only true for sinusoidal signals. Other periodic signals will have different relationships between the peak and RMS values as is illustrated in the next example.

Example 7.13 Find the RMS voltage for the waveform shown in Fig. 7.5.

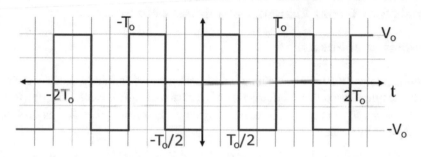

Fig. 7.5 Square wave for Example 7.13

Solution:
The RMS value can be found by integrating the square of the signal over any period. Let's select the period from $-T_o/2$ to $+T_o/2$.

$$V_{\text{rms}} = \sqrt{\frac{1}{T} \int_{t_o}^{t_o+T} (v(t))^2 dt} = \sqrt{\frac{1}{T_o} \int_{-T_o/2}^{+T_o/2} (v(t))^2 dt} = \sqrt{\frac{1}{T_o} \left(\int_{-T_o/2}^{0} (-V_o)^2 dt + \int_{0}^{+T_o/2} (V_o)^2 dt \right)}$$

$$= \sqrt{\frac{1}{T_o} \left(V_o^2 \cdot \frac{T_o}{2} + V_o^2 \cdot \frac{T_o}{2} \right)} = \sqrt{V_o^2} = V_o$$

Therefore, the peak value and the RMS value is the same for the square wave.

Example 7.13 Find the RMS voltage for $v(t) = 170 \cos\left(120\pi t + \frac{\pi}{6}\right)$ V.

Solution: The peak value of the voltage is 170 V and the signal is sinusoidal. Therefore, the RMS voltage is given by

$$V_{\text{rms}} = \frac{V_m}{\sqrt{2}} = \frac{170}{\sqrt{2}} = 120.21 \ V_{\text{rms}}$$

Sometimes, the entire time-domain expression is written as an RMS quantity as

$$v(t) = 120.21 \cos\left(120\pi t + \frac{\pi}{6}\right) V_{\text{rms}}$$

where the units V_{rms} indicate that number multiplying the cosine has already been divided by $\sqrt{2}$ to convert it into an RMS quantity.

Outlet Standards: The standard outlet in the United States provides 120 V_{rms} at 60 Hz. This voltage can be written as $v(t) = 120\cos(120\pi t + \theta_v)\ V_{rms} = 169.71\cos(120\pi t + \theta_v)\ V$. However, other countries have different standards for their wall outlets. Therefore, when traveling abroad, you need to bring adaptors to power your electronic devices.

7.3 Signals and Circuit Elements as Complex Numbers

7.3.1 Definition of Phasors

A phasor, while it may sound like something out of a science fiction movie, is just a complex number that carries the amplitude and phase information of a sinusoidal signal. Recall that one form of Euler's identity is $e^{\pm j\theta} = \cos(\theta) \pm j\sin(\theta)$. Therefore, a signal of the form $v(t) = V_m\cos(\omega t + \theta_v)$ or $i(t) = I_m\cos(\omega t + \theta_i)$ can be written as

$$v(t) = \text{Real}\{V_m\cos(\omega t + \theta_v) + jV_m\sin(\omega t + \theta_v)\} = \text{Real}\left\{V_m e^{j(\omega t + \theta_v)}\right\} = \text{Real}\left\{V_m e^{j\theta_v}e^{j\omega t}\right\}$$

$$i(t) = \text{Real}\{I_m\cos(\omega t + \theta_i) + jI_m\sin(\omega t + \theta_i)\} = \text{Real}\left\{I_m e^{j(\omega t + \theta_i)}\right\} = \text{Real}\left\{I_m e^{j\theta_i}e^{j\omega t}\right\}$$

$$(7.29)$$

where the phasor voltage and phasor current are given by

$$\tilde{V} = V_m e^{j\theta_v}$$
$$\tilde{I} = I_m e^{j\theta_i} \tag{7.30}$$

Since for most circuits, a sinusoidal excitation will result in sinusoidal voltages and currents throughout at the same frequency, the value of ω will not vary during the circuit analysis and can be ignored. Therefore, we only need to solve for the complex phasor voltages and phasor currents when solving circuits undergoing sinusoidal excitation.

Example 7.14 Convert the following sinusoidal signals into phasors.

$$v(t) = 6\cos\left(2t + \frac{\pi}{3}\right)\ V$$
$$i(t) = 8\sin\left(3t + \frac{\pi}{6}\right)\ mA$$

Solution: For the voltage expression, the amplitude of the voltage is 6 V while the phase is $\pi/3$ radians. Therefore, the phasor voltage is given by

$$\tilde{V} = 6e^{j\frac{\pi}{3}}\ V$$

For the current, the first step is to convert the signal into a cosine as the phasor representation is always defined with respect to the cosine. It would be possible to define a phasor type of representation with respect to the sin function, but this is not standard practice. The current can be translated into a cosine using the trig identity $\sin(x) = \cos(x - \pi/2)$. Therefore,

$$i(t) = 8\sin\left(3t + \frac{\pi}{6}\right)\ mA = 8\cos\left(3t + \frac{\pi}{6} - \frac{\pi}{2}\right)\ mA = 8\cos\left(3t - \frac{\pi}{3}\right)\ mA$$

The phasor current is then given by

$$\tilde{I} = 8e^{-j\frac{\pi}{3}} \text{ mA}$$

Example 7.15 Convert the following sinusoidal signals into phasors.

$$i(t) = 4\cos(2t) + 3\sin(2t) \text{ A}$$

Solution: There are two possible approaches to this problem. First, we can use the trig identity $\cos(x \pm y) = \cos(x)\cos(y) \mp \sin(x)\sin(y)$ to write the expression as a single cosine function. From this identity, we know that $I_m\cos(\theta_i) = 4$ and $-I_m\sin(\theta_i) = 3$. Therefore,

$$(I_m\cos(\theta_i))^2 + (-I_m\sin(\theta_i))^2 = 4^2 + 3^2 = 25$$

$$I_m^2\left(\overbrace{\cos^2(\theta_i) + \sin^2(\theta_i)}^{=1}\right) = 25 \Rightarrow I_m = 5 \text{ A}$$

Also,

$$\frac{I_m\sin(\theta_i)}{I_m\cos(\theta_i)} = \frac{-3}{4} \Rightarrow \tan(\theta_i) = -0.75 \Rightarrow \theta_i = -0.6435 \text{ rad}$$

Thus, the phasor representation is given by $\tilde{I} = 5e^{-j0.6435}$ A. The second approach is to first translate both the cosine and sine expressions into phasors and then add the phasors.

$$4\cos(2t) \text{ A} \Rightarrow 4 \text{ A}$$

$$3\sin(2t) \text{ A} = 3\cos\left(2t - \frac{\pi}{2}\right) = 3e^{-j\frac{\pi}{2}} \text{ A} = -j3 \text{ A}$$

$$\tilde{I} = 4 - j3 \text{ A} = 5e^{-j0.6435} \text{ A}$$

Example 7.16 Can you convert the following sinusoidal signal into a phasor?

$$i(t) = 4\cos(2t) + 3\cos(t) \text{ A}$$

Solution: No, the signal cannot be converted into a phasor because it is a function of two different frequencies. The phasor domain assumes that only one frequency is present and that it is not changed anywhere in the circuit. The term $4\cos(2t)$ can be converted into a phasor, and the term $3\cos(t)$ can be converted into a phasor. The summation cannot because of the two different frequencies. To solve a circuit at different frequencies using phasor domain circuit analysis, one must use superposition where each frequency component is solved separately.

7.3.2   Circuit Elements in the Phasor Domain

The primary advantage of using phasors in circuit analysis is that it allows us to replace the differential equations with algebraic expressions involving complex numbers. If you recall, the original differential equations result from the current/voltage relationships for inductors and capacitors. For example, the current through an inductor and voltage across an inductor are related by $v(t) = L \cdot \frac{di(t)}{dt}$. Therefore, if $i(t)$ is a sinusoidal signal given by $i(t) = I_m\cos(\omega t + \theta_i)$, the voltage across the inductor must be given by

$$v(t) = L\frac{d}{dt}\left(I_m \cos(\omega t + \theta_i)\right) = -\omega L I_m \sin(\omega t + \theta_i) = \omega L I_m \cos\left(\omega t + \theta_i + \frac{\pi}{2}\right) \qquad (7.31)$$

Changing the voltage and current into phasors gives

$$\tilde{I} = I_m e^{j\theta_i}$$

$$\tilde{V} = -\omega L I_m e^{j\left(\theta_i - \frac{\pi}{2}\right)} = j\omega L \cdot I_m e^{j\theta_i} = j\omega L \cdot \tilde{I} \qquad (7.32)$$

$$\Rightarrow \frac{\tilde{V}}{\tilde{I}} = j\omega L$$

Therefore, the time derivative is replaced by multiplication by $j\omega L$. The ωL changes the amplitude, while the j term shifts the phase by a factor of $+\pi/2$. This positive phase shift means that the voltage will lead the current (or current will lag the voltage), and the voltage will peak before the current for all time values as shown in Fig. 7.6. Every circuit for which the inductance dominates will have the voltage leading the current.

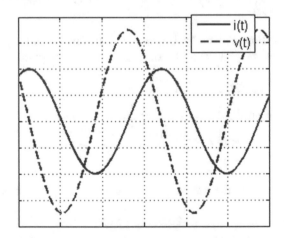

Fig. 7.6 Voltage leading the current (current lagging the voltage) for an inductor

In a similar manner, if the voltage across a capacitor is given by $v(t) = V_m \cos(\omega t + \theta_v)$, then the current through the capacitor must be given by

$$i(t) = C\frac{d}{dt}\left(V_m \cos(\omega t + \theta_v)\right) = -\omega C V_m \sin(\omega t + \theta_v) = \omega C V_m \cos\left(\omega t + \theta_v + \frac{\pi}{2}\right) \qquad (7.33)$$

Translating all of the expressions to phasors gives

$$\tilde{V} = V_m e^{j\theta_v}$$

$$\tilde{I} = -\omega C V_m e^{j\left(\theta_v - \frac{\pi}{2}\right)} = j\omega C \cdot V_m e^{j\theta_v} = j\omega C \cdot \tilde{V} \qquad (7.34)$$

$$\Rightarrow \frac{\tilde{V}}{\tilde{I}} = \frac{1}{j\omega C}$$

The voltage and current amplitude for the capacitor will thus differ by a factor of $1/\omega C$ while the phase of the voltage will be less than the phase of the current by a factor of $\pi/2$. This means that the current will peak before the voltage (current leads the voltage, voltage lags current) as shown in Fig. 7.7. The current will always lead the voltage when the circuit is capacitive.

Lastly, we can also find the relationship between current and voltage for a resistor in the phasor domain. Assuming, $i(t) = I_m \cos(\omega t + \theta_i)$, we have

$$v(t) = RI_m \cos(\omega t + \theta_i) \Rightarrow \tilde{V} = R\tilde{I} \Rightarrow \frac{\tilde{V}}{\tilde{I}} = R \tag{7.35}$$

Therefore, the amplitude of the voltage and current differ by a factor of R for the resistor, but the two are in phase as shown in Fig. 7.8.

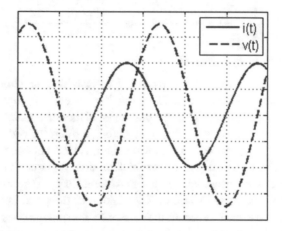

Fig. 7.7 Voltage lagging the current (current leading the voltage) for a capacitor

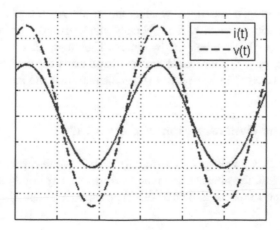

Fig. 7.8 Voltage and current in phase for a resistor

Example 7.17 For the waveforms shown in Fig. 7.9, determine if the voltage leads the current (inductive), the voltage lags the current (capacitive), or the voltage and current are in phase (resistive).

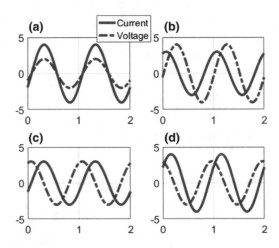

Fig. 7.9 Voltage/current values for Example 7.17

Solution: For the signals shown in A, the current and voltage have maximum, minimum, and zero crossings at exactly the same points in time. Therefore, the current and voltage are in phase, and the circuit is resistive. For the signals shown in B, the maximum of the current occurs earlier in time (shifted to the left) relative to the maximum of the voltage. Therefore, the current leads the voltage (voltage lags the current), and the circuit is capacitive. For C, the maximum of the voltage precedes the maximum of the current. Therefore, the voltage leads the current, and the circuit is inductive. The signals shown in D can be a little confusing as the current has its first maximum for t > 0 before the voltage. However, leading and lagging is not determined based on the position of the peaks relative to zero, but on the position of the peaks relative to each other. In the case of D, the voltage does peak before the current. Therefore, the voltage leads the current, and the circuit is inductive. Remember, it is important to compare the relative position of the nearest peaks when determining which quantity leads and lags.

Memory Tool: When trying to remember whether the current or voltage leads for an inductor or a capacitor, the mnemonic device "ELI the ICE man" can be useful. The voltage, also called the electromotive force, can be symbolized by an E. Therefore, ELI has E (voltage) before I (current) in an inductor (L). This would mean voltage leads the current. Similarly, ICE has I (current) before E (voltage) in a capacitor (C) indicating that the current leads the voltage.

7.3.3 Impedance and Generalization of Ohm's Law

In 7.3.2, we saw that the phasor voltage and current for inductors and capacitors could be related by a complex number. In general, the phasor current and voltage for any combination of circuit elements can be related by a complex number known as the impedance in a generalized form of Ohm's Law.

$$\tilde{V} = Z \cdot \tilde{I} \tag{7.36}$$

The impedance, Z, in this equation has units of Ω just like the resistance in the original Ohm's Law. The impedance for resistors, inductors, and capacitors is given by

$$
\begin{array}{lll}
\text{Resistors} & \tilde{V} = R \cdot \tilde{I} \Rightarrow Z_R = R \\
\text{Inductors} & \tilde{V} = j\omega L \cdot \tilde{I} \Rightarrow Z_L = j\omega L & (7.37) \\
\text{Capacitors} & \tilde{V} = \frac{1}{j\omega C} \cdot \tilde{I} \Rightarrow Z_C = \frac{1}{j\omega C}
\end{array}
$$

These impedances can be combined in series and in parallel just like we combined resistors in Chap. 1. The equivalent impedance for a set of impedances in series and in parallel are given by

$$
\begin{aligned}
Z_{eq} &= Z_1 + Z_2 + Z_3 + \cdots + Z_{N-1} + Z_N \\
&= \sum_{k=1}^{N} Z_k
\end{aligned} \tag{7.38}
$$

and

$$
Z_{eq} = \frac{1}{\frac{1}{Z_1} + \frac{1}{Z_2} + \frac{1}{Z_3} + \cdots + \frac{1}{Z_{N-1}} + \frac{1}{Z_N}} = \frac{1}{\sum\limits_{k=1}^{N} \frac{1}{Z_k}}, \tag{7.39}
$$

respectively.

Example 7.18 Find the equivalent impedance seen by the current source, Z_{eq}, and the voltage, $v(t)$, in the time domain for the circuit shown in Fig. 7.10 if

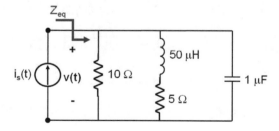

Fig. 7.10 Circuit for Example 7.18

$$
i_s(t) = 10\cos\left(2 \cdot 10^5 t\right) \, A.
$$

Solution: The first step is to convert the circuit into the phasor domain. This means expressing all of the independent sources as phasors and translating the values for the inductors and capacitors to their respective impedances.

$$
Z_L = j\omega L = j\left(2 \cdot 10^5\right)\left(50 \cdot 10^{-6}\right) = j10 \, \Omega
$$

$$
Z_C = \frac{1}{j\omega C} = \frac{1}{j(2 \cdot 10^5)(10^{-6})} = -j5 \, \Omega
$$

$$
i_s(t) = 10\cos\left(2 \cdot 10^5 t\right) A \Rightarrow 10 \, A
$$

Replacing the circuit parameters with these parameters gives

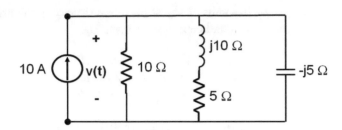

Notice that the impedance of the capacitor is negative while the impedance of the inductor is positive. This is true for all capacitors and inductors. Also, the inductor and the resistor are in series, so their impedances can be added giving

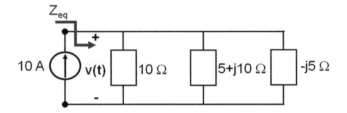

However, these three impedances are in parallel, so Z_{eq} is given by

$$Z_{eq} = \frac{1}{\frac{1}{Z_1} + \frac{1}{Z_2} + \frac{1}{Z_3}} = \frac{1}{\frac{1}{10} + \frac{1}{5+j10} + \frac{1}{-j5}} = \frac{1}{0.1 + \frac{5-j10}{(5+j10)(5-j10)} + j0.2}$$

$$= \frac{1}{0.1 + \frac{5-j10}{125} + j0.2} = \frac{1}{0.1 + 0.04 - j0.08 + j0.2} = \frac{1}{0.14 + j0.12} = 4.1176 - j3.5294 \ \Omega$$

With Z_{eq} known, we can use Ohm's Law to find the voltage.

$$\tilde{V} = Z \cdot \tilde{I} = 10 \cdot Z_{eq} = 41.1765 - j35.2941 = 54.2326e^{-j0.7086} \ V$$

However, $54.2326e^{-j0.7086} \ V$ is the voltage in the phasor domain. We need to find the voltage in the time domain. Knowing the difference between the phasor domain and the time domain is very important when solving circuits.

$$v(t) = 54.2326 \cos\left(2 \cdot 10^5 t - 0.7086\right) \ V$$

Example 7.19 Repeat Example 7.18 with $i_s(t) = 10\cos(6 \cdot 10^4 t) \ A$

Solution: The only difference between this example and the previous example is that the frequency of the source has changed. However, since the frequency is different the impedances are different. The new impedances are given by

$$Z_L = j\omega L = j\left(6 \cdot 10^4\right)\left(50 \cdot 10^{-6}\right) = j3 \ \Omega$$

$$Z_C = \frac{1}{j\omega C} = \frac{1}{j(6 \cdot 10^4)(10^{-6})} = -j16.667 \ \Omega$$

The circuit in the phasor domain is thus given by

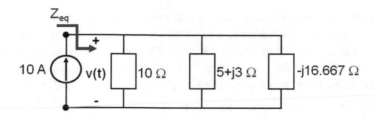

$$Z_{eq} = \frac{1}{\frac{1}{Z_1} + \frac{1}{Z_2} + \frac{1}{Z_3}} = \frac{1}{\frac{1}{10} + \frac{1}{5+j3} + \frac{1}{-j16.6667}} = \frac{1}{0.1 + 0.1471 - j0.0882 + j0.06}$$

$$= \frac{1}{0.2471 - j0.0282} = 3.9954 + j0.4566 \ \Omega$$

$$\tilde{V} = Z \cdot \tilde{I} = 10 \cdot Z_{eq} = 39.9543 + j4.5662 = 40.2144 e^{+j0.1138} \ V$$

$$\Rightarrow v(t) = 40.2144 \cos\left(6 \cdot 10^4 t + 0.1138\right) \ V$$

Example 7.20 For the circuit shown in Fig. 7.10, find the excitation frequency for which the current, $i_s(t)$, and the voltage, $v(t)$, will be in phase.

Solution: Notice that when the frequency was $6 \cdot 10^4$ rad/s in Example 7.19, the impedance had a positive imaginary part whereas in Example 7.18 the imaginary portion of the impedance was negative. Likewise, the phase of the voltage was positive for Example 7.19 while it was negative previously. Therefore, there should be a frequency between $6 \cdot 10^4$ rad/s and $2 \cdot 10^5$ rad/s where the imaginary portion of the impedance should go to zero, and, consequently, the voltage and current should be in phase. To find this frequency, we need to determine the value of ω that would cause the imaginary portion of Z_{eq} to go to zero. However, if the imaginary part of Z_{eq} is zero, then the imaginary part of $1/Z_{eq}$ must also be zero. Therefore,

$$Z_{eq} = \frac{1}{\frac{1}{Z_1} + \frac{1}{Z_2} + \frac{1}{Z_3}}$$

$$\Rightarrow \frac{1}{Z_{eq}} = \frac{1}{10} + \frac{1}{5 + j\omega L} + j\omega C = \frac{1}{10} + \frac{5 - j\omega L}{25 + (\omega L)^2} + j\omega C$$

$$= \frac{1}{10} + \frac{5}{25 + (\omega L)^2} - j\frac{\omega L}{25 + (\omega L)^2} + j\omega C$$

$$\mathrm{Imag}\left\{\frac{1}{Z_{eq}}\right\} = \frac{-\omega L}{25 + (\omega L)^2} + \omega C = 0$$

$$\Rightarrow \frac{\omega L}{25 + (\omega L)^2} = \omega C \Rightarrow \frac{L}{25 + (\omega L)^2} = C \Rightarrow \frac{L}{C} = 25 + (\omega L)^2$$

$$\Rightarrow \omega = \frac{1}{L}\sqrt{\frac{L}{C} - 25} = \frac{1}{50 \cdot 10^{-6}}\sqrt{\frac{50 \cdot 10^{-6}}{10^{-6}} - 25} = 1 \cdot 10^5 \ \text{rad/s}$$

7.4 Phasor Domain Circuit Analysis

By transforming the independent voltage and current sources into phasors and expressing the values of inductors and capacitors as impedances, we can now use all of the circuit analysis techniques that were introduced in Chap. 3 to solve circuits. Kirchhoff's Current and Voltage Laws are still applicable allowing us to still use the node-voltage method and the mesh-current method to solve the circuit. The only difference is that the generated equations are now complex as will be illustrated through the following examples.

Example 7.20 Find the node-voltages in the time domain and in the phasor domain for the circuit shown in Fig. 7.11 if the values for the independent sources are given by

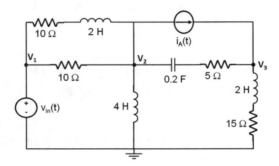

Fig. 7.11 Circuit for Example 7.20

$$v_{in}(t) = 5\cos(2t - 0.2)$$
$$i_A(t) = 2\cos(2t + 0.4)$$

Solution: The first step is to translate the circuit elements into the phasor domain. The impedances for the inductors are given by

$$j\omega L :$$
$$4\,H \rightarrow j8\ \Omega$$
$$2\,H \rightarrow j4\ \Omega$$

Similarly, the impedance for the capacitor is given by

$$\frac{1}{j\omega C} : 0.2\,F \rightarrow -j2.5\ \Omega$$

Also,

$$v_{in}(t) \rightarrow 5e^{-j0.2}\ V \text{ and } i_A(t) \rightarrow 2e^{+j0.4}\ A$$

Therefore, the circuit can be redrawn as

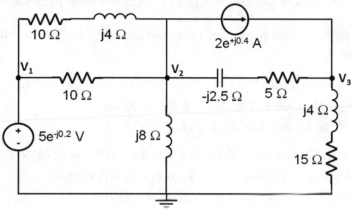

Now, let's solve this circuit using the node-voltage method. The voltage at V_1 is already known to be $5e^{-j0.2}$ V due to the presence of the independent voltage source. Therefore, we only need to sum currents at nodes V_2 and V_3. We actually cannot sum currents at V_1 as we cannot know the current flowing in the independent voltage source until after the values of V_1 and V_2 are both known.

$$Node\ V_2: \frac{\tilde{V}_2 - \tilde{V}_1}{10\ \Omega} + \frac{\tilde{V}_2 - \tilde{V}_1}{10 + j4\ \Omega} + \frac{\tilde{V}_2}{j8\ \Omega} + \frac{\tilde{V}_2 - \tilde{V}_3}{5 - j2.5\ \Omega} + \tilde{I}_A = 0$$

$$\frac{\tilde{V}_2}{10\ \Omega} + \frac{\tilde{V}_2}{10 + j4\ \Omega} + \frac{\tilde{V}_2}{j8\ \Omega} + \frac{\tilde{V}_2}{5 - j2.5\ \Omega} - \frac{\tilde{V}_3}{5 - j2.5\ \Omega}$$

$$= \frac{\tilde{V}_1}{10\ \Omega} + \frac{\tilde{V}_1}{10 + j4\ \Omega} - \tilde{I}_A$$

$$\tilde{V}_2 \left(\begin{array}{c} \frac{1}{10\ \Omega} + \frac{1}{10 + j4\ \Omega} \\ + \frac{1}{j8\ \Omega} + \frac{1}{5 - j2.5\ \Omega} \end{array} \right) - \tilde{V}_3 \left(\frac{1}{5 - j2.5\ \Omega} \right) = 5e^{-j0.2} \left(\frac{1}{10\ \Omega} + \frac{1}{10 + j4\ \Omega} \right) - 2e^{+j0.4}$$

$$\tilde{V}_2 \left(\begin{array}{c} 0.1 + 0.0862 - j0.0345 \\ -j0.125 + 0.16 + j0.08 \end{array} \right) - \tilde{V}_3(0.16 + j0.08)$$

$$= 5e^{-j0.2}(0.1 + 0.0862 - j0.0345) - 2e^{+j0.4}$$

$$\tilde{V}_2(0.3462 - j0.0795) - \tilde{V}_3(0.16 + j0.08) = 5e^{-j0.2}(0.1862 - j0.0345) - 2e^{+j0.4}$$

$$\tilde{V}_2(0.3462 - j0.0795) - \tilde{V}_3(0.16 + j0.08) = -0.9639 - j1.1328$$

$$Node\ V_3: \frac{\tilde{V}_3 - \tilde{V}_2}{5 - j2.5\ \Omega} + \frac{\tilde{V}_3}{15 + j4\ \Omega} - \tilde{I}_A = 0$$

$$\frac{\tilde{V}_3}{5 - j2.5\ \Omega} + \frac{\tilde{V}_3}{15 + j4\ \Omega} + \frac{-\tilde{V}_2}{5 - j2.5\ \Omega} = \tilde{I}_A$$

$$\tilde{V}_3 \left(\frac{1}{5 - j2.5\ \Omega} + \frac{1}{15 + j4\ \Omega} \right) - \tilde{V}_2 \left(\frac{1}{5 - j2.5\ \Omega} \right) = 2e^{+j0.4}\ A$$

$$\tilde{V}_3(0.222 + j0.0634) - \tilde{V}_2(0.16 + j0.08) = 1.8421 + j0.7788$$

Now, let's make the coefficient for $\tilde{V}_2$ real in both equations by multiplying with the appropriate complex conjugate.

$$(0.3462 + j0.0795)\left[\tilde{V}_2(0.3462 - j0.0795) - \tilde{V}_3(0.16 + j0.08) = -0.9639 - j1.1328\right]$$
$$0.1262\tilde{V}_2 - \tilde{V}_3(0.049 + j0.0404) = -0.2437 - j0.4688$$
$$\tilde{V}_2 - \tilde{V}_3(0.3886 + j0.3203) = -1.9312 - j3.7153$$

$$(0.16 - j0.08)\left[\tilde{V}_3(0.222 + j0.0634) - \tilde{V}_2(0.16 + j0.08) = 1.8421 + j0.7788\right]$$
$$-0.032\tilde{V}_2 + \tilde{V}_3(0.0406 - j0.0076) = 0.3570 - j0.0228$$
$$-\tilde{V}_2 + \tilde{V}_3(1.2697 - j0.2386) = 11.1577 - j0.7111$$

Now add the two equations

$$\tilde{V}_3(0.8811 - j0.5589) = 9.2265 - j4.4265$$
$$\Rightarrow \tilde{V}_3 = 9.7396 + j1.1541 \ V$$

Therefore, $\tilde{V}_3$ in the phasor domain is given by $9.8078e^{+j0.1179} \ V$, and $\tilde{V}_3$ in the time domain is given by $9.8078\cos(2t + 0.1179) \ V$. With $\tilde{V}_3$ known, we can solve for $\tilde{V}_2$.

$$\tilde{V}_2 = \tilde{V}_3(0.3886 + j0.3203) + -1.9312 - j3.7153$$
$$= 1.4841 - j0.1473 \ V$$

Hence, $\tilde{V}_2$ in the phasor domain is given by $1.4914e^{-j0.0989} \ V$, and $\tilde{V}_2$ in the time domain is given by $1.4914\cos(2t - 0.0989) \ V$.

When solving systems of equations, especially complex equations, it is best to confirm the accuracy of the solution by determining if the final answers satisfy the initial equations. In phasor domain circuit analysis, this means substituting the values into the original node-voltage or mesh current equations. If we do that for our solution, we get

Node V_2 :
$$\frac{1.4914e^{-j0.0989} - 5e^{-j0.2}}{10} + \frac{1.4914e^{-j0.0989} - 5e^{-j0.2}}{10 + j4}$$
$$+ \frac{1.4914e^{-j0.0989}}{j8} + \frac{1.4914e^{-j0.0989} - 9.8078e^{+j0.1179}}{5 - j2.5} + 2e^{+j0.4} = 0$$
$$\frac{-3.4162 + j0.8461}{10} + \frac{-3.4162 + j0.8461}{10 + j4} + \frac{1.4841 - j0.1473}{j8} + \frac{-8.2555 - j1.3014}{5 - j2.5} = -1.8421 - j0.7788$$
$$-0.3416 + j0.0846 - 0.2653 + j0.1907 - 0.0184 - j0.1855 - 1.2168 - j0.8687$$
$$= -1.8421 - j0.7788$$
$$-1.8421 - j0.7788 = -1.8421 - j0.7788 \quad Q.E.D.$$

Node V_3 :

$$\frac{9.8078e^{+j0.1179} - 1.4914e^{-j0.0989}}{5 - j2.5} + \frac{9.8078e^{+j0.1179}}{15 + j4} - 2e^{+j0.4} = 0$$

$$\frac{8.255 + j1.3014}{5 - j2.5} + \frac{9.7396 + j1.1541}{15 + j4} = 1.8421 + j0.7788$$

$$1.2168 + j0.8687 + 0.6254 - j0.0898 = 1.8421 + j0.7788$$

$$1.8421 + j0.7788 = 1.8421 + j0.7788 \quad Q.E.D.$$

It is important to check all of the equations because if an algebra mistake was made when simplifying only one of the equations, then the solutions would agree with all of the equations except for the equation where the mistake was made.

Example 7.21 Find the node-voltage, V_1, in the time domain and in the phasor domain for the circuit shown in Fig. 7.12 using both the node-voltage method and the mesh-current method if $R_1 = 45\ \Omega$, $R_2 = 75\ \Omega$, $L_1 = 5\ H$, $C_1 = 5\ mF$, $v_A(t) = 20\cos\left(10t - \frac{\pi}{2}\right)\ V$, and $v_B(t) = 10\cos(10t)\ V$.

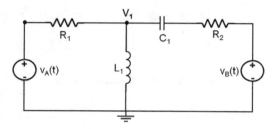

Fig. 7.12 Circuit for Example 7.21

Solution: Once again, the first step is to translate the circuit elements into the phasor domain.

$$j\omega L_1 : 5\ H \rightarrow j50\ \Omega$$

$$\frac{1}{j\omega C} : 5\ mF \rightarrow -j20\ \Omega$$

$$v_A(t) \rightarrow 20e^{-\frac{j\pi}{2}}\ V = -j20\ V \quad v_B(t) \rightarrow 10\ V$$

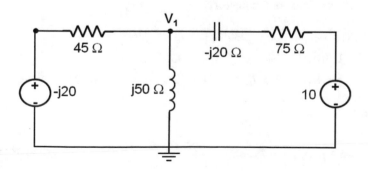

Node-Voltage Method:

We can now write an equation for the currents at node V_1.

$$\frac{\tilde{V}_1 + j20}{45\ \Omega} + \frac{\tilde{V}_1}{j50\ \Omega} + \frac{\tilde{V}_1 - 10}{75 - j20\ \Omega} = 0$$

$$\tilde{V}_1\left(\frac{1}{45} + \frac{1}{j50} + \frac{1}{75 - j20}\right) = \frac{10}{75 - j20} - \frac{j20}{45}$$

$$\tilde{V}_1(0.0222 - j0.02 + 0.0124 + j0.0033) = 0.1245 + j0.0332 - j0.4444$$

$$\tilde{V}_1(0.0347 - j0.0167) = 0.1245 - j0.4112$$

$$\tilde{V}_1 = \frac{0.1245 - j0.4112}{0.0347 - j0.0167} = 7.5497 - j8.2294$$

Therefore, the voltage at V_1 is given by $11.1679e^{-j0.8284}$ V and $11.1679\cos(10t - 0.8284)$ V in the phasor domain and time domain, respectively.

Mesh-Current Method:

Summing the voltages around the loop associated with voltage source v_A gives

$$- - j20 + 45\tilde{I}_A + j50\left(\tilde{I}_A - \tilde{I}_B\right) = 0 \Rightarrow (45 + j50)\tilde{I}_A - j50\tilde{I}_B = -j20$$

Similarly, summing the voltages around the loop associated with voltage source v_B gives

$$(75 - j20)\tilde{I}_B + 10 + j50\left(\tilde{I}_B - \tilde{I}_A\right) = 0 \Rightarrow -j50\tilde{I}_A + (75 + j30)\tilde{I}_B = -10$$

Now, multiply by the appropriate complex conjugates to make the coefficient for $\tilde{I}_A$ purely real in both equations.

$$(45 - j50) \cdot \left[(45 + j50)\tilde{I}_A - j50\tilde{I}_B = -j20\right]$$
$$\Rightarrow 4525\tilde{I}_A + (-2500 - j2250)\tilde{I}_B = -1000 - j900 \Rightarrow 181\tilde{I}_A + (-100 - j90)\tilde{I}_B = -40 - j36$$
$$(j50) \cdot \left[-j50\tilde{I}_A + (75 + j30)\tilde{I}_B = -10\right]$$
$$\Rightarrow 2500\tilde{I}_A + (-1500 + j3750)\tilde{I}_B = -j500 \Rightarrow 10\tilde{I}_A + (-6 + j15)\tilde{I}_B = -j2$$

Now combine these equations to eliminate $\tilde{I}_A$

$$1810\tilde{I}_A + (-1000 - j900)\tilde{I}_B = -400 - j360$$
$$- 1810\tilde{I}_A + (1086 - j2715)\tilde{I}_B = j362$$
$$\Rightarrow (86 - j3615)\tilde{I}_B = -400 + j2 \Rightarrow \tilde{I}_B = 110.6202e^{-j1.5996}\ \text{mA}$$

Now solve for $\tilde{I}_A$

$$10\tilde{I}_A + (-6 + j15)\tilde{I}_B = -j2 \Rightarrow \tilde{I}_A = \frac{-j2 - (-6 + j15)\tilde{I}_B}{10} = 310.7502e^{-j2.1411}\ \text{mA}$$

Now we can use $\tilde{I}_A$ and $\tilde{I}_A$ to get the voltage at V_1.

$$\tilde{V}_1 = j50(\tilde{I}_A - \tilde{I}_B) = 7.5497 - j8.2294 = 11.1679e^{-j0.8284}\ V \Rightarrow 11.1679\cos(10t - 0.8284)\ V$$

Example 7.22 Find $v_{out}(t)$ in the time domain for the circuit shown in Fig. 7.13 if $v_{in}(t) = 8\cos(5t) + 4\cos(10t)\ V$.

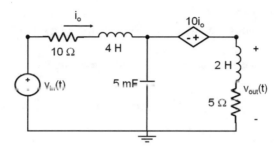

Fig. 7.13 Circuit for Example 7.22

Solution: The input voltage consists of two different frequencies. Therefore, the circuit will need to be solved for each frequency independently using superposition. This is done by first finding the phasor domain solution for $\omega = 5$ rad/s and converting it into the time domain. Then, we solve for the phasor domain solution for $\omega = 10$ rad/s and convert it into the time domain. The final solution is then the summation of the two time domain solutions.

Solution for $\omega = 5$ rad/s:
First, redraw the circuit in the phasor domain for the $\omega = 5$ rad/s case.

$$j\omega L : \begin{cases} 4\ H \to j20\ \Omega \\ 2\ H \to j10\ \Omega \end{cases}$$

$$\frac{1}{j\omega C} : 5\ mF \to -j40\ \Omega$$

The dependent source forms a super node. Therefore, if the voltage on the right is $\tilde{V}_{out1}$, then the voltage on the left side of the super node is $\tilde{V}_{out1} - 10\tilde{I}_{o1}$.

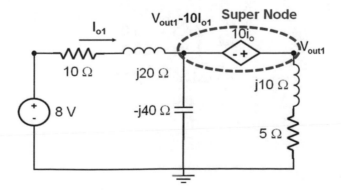

Summing the currents at the super node gives

$$\frac{\tilde{V}_{out1} - 10\tilde{I}_{o1} - 8}{10 + j20} + \frac{\tilde{V}_{out1} - 10\tilde{I}_{o1}}{-j40} + \frac{\tilde{V}_{out1}}{5 + j10} = 0$$

$$\tilde{V}_{out1}\left(\frac{1}{10 + j20} + \frac{1}{-j40} + \frac{1}{5 + j10}\right) - 10\tilde{I}_{o1}\left(\frac{1}{10 + j20} + \frac{1}{-j40}\right) = \frac{8}{10 + j20}$$

$$\tilde{V}_{out1}(0.02 - j0.04 + j0.025 + 0.04 - j0.08) - 10\tilde{I}_{o1}(0.02 - j0.04 + j0.025)$$
$$= 0.16 - j0.32$$

$$\tilde{V}_{out1}(0.06 - j0.095) - \tilde{I}_{o1}(0.2 - j0.15) = 0.16 - j0.32$$

where the dependent source introduces another equation given by

$$\tilde{I}_{o1} = \frac{8 - (\tilde{V}_{out1} - 10\tilde{I}_{o1})}{10 + j20} \Rightarrow (10 + j20)\tilde{I}_{o1} = 8 - \tilde{V}_{out1} + 10\tilde{I}_{o1}$$

$$\tilde{V}_{out1} + j20\tilde{I}_{o1} = 8$$

Now eliminate $\tilde{I}_{o1}$ from the expressions.

$$(0.2 + j0.15) \cdot \left[\tilde{V}_{out1}(0.06 - j0.095) - \tilde{I}_{o1}(0.2 - j0.15) = 0.16 - j0.32\right]$$
$$\Rightarrow \tilde{V}_{out1}(0.0262 - j0.01) - 0.0625\tilde{I}_{o1} = 0.08 - j0.04$$
$$\Rightarrow \tilde{V}_{out1}(0.42 - j0.16) - \tilde{I}_{o1} = 1.28 - j0.64$$
$$-j20 \cdot \left[\tilde{V}_{out1} + j20\tilde{I}_{o1} = 8\right]$$
$$\Rightarrow -j20\tilde{V}_{out1} + 400\tilde{I}_{o1} = -160j$$
$$\Rightarrow -j0.05\tilde{V}_{out1} + \tilde{I}_{o1} = -0.4j$$
$$\tilde{V}_{out1}(0.42 - j0.21) = 1.28 - j1.04 \Rightarrow \tilde{V}_{out1} = 3.5122e^{-j0.2187} \text{ V}$$

Therefore, the contribution from the first frequency, $\omega = 5$ rad/s, is given by

$$v_{out1}(t) = 3.5122\cos(5t - 0.2187) \text{ V}$$

Solution for $\omega = 10$ rad/s:
Once again, the circuit should be redrawn with the new frequency used to calculate the new impedance values.

$$j\omega L : \begin{cases} 4\ H \to j40\ \Omega \\ 2\ H \to j20\ \Omega \end{cases}$$

$$\frac{1}{j\omega C} : 5\ mF \to -j20\ \Omega$$

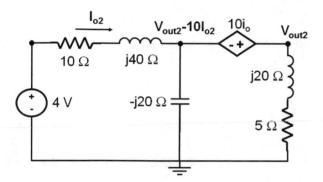

Therefore, the equations to be solved are

$$\frac{\tilde{V}_{out2} - 10\tilde{I}_{o2} - 4}{10 + j40} + \frac{\tilde{V}_{out2} - 10\tilde{I}_{o2}}{-j20} + \frac{\tilde{V}_{out2}}{5 + j20} = 0$$

$$\tilde{V}_{out2}\left(\frac{1}{10+j40} + \frac{1}{-j20} + \frac{1}{5+j20}\right) - 10\tilde{I}_{o2}\left(\frac{1}{10+j40} + \frac{1}{-j20}\right) = \frac{4}{10+j40}$$

$$\tilde{V}_{out2}(0.0059 - j0.0235 + j0.05 + 0.0118 - j0.0471)$$
$$- 10\tilde{I}_{o2}(0.0059 - j0.0235 + j0.05) = 0.0235 - j0.0941$$

$$\tilde{V}_{out2}(0.0177 - j0.0206) - \tilde{I}_{o2}(0.0588 + j0.2647) = 0.0235 - j0.0941$$

$$\tilde{I}_{o2} = \frac{4 - \left(\tilde{V}_{out2} - 10\tilde{I}_{o2}\right)}{10 + j40} \Rightarrow (10 + j40)\tilde{I}_{o2} = 4 - \tilde{V}_{out2} + 10\tilde{I}_{o2}$$

$$\tilde{V}_{out2} + j40\tilde{I}_{o2} = 4$$

As was mentioned previously, these equations can also be solved using matrices.

$$\begin{pmatrix} 0.0177 - j0.0206 & -0.0588 - j0.2647 \\ 1 & j40 \end{pmatrix} \begin{pmatrix} \tilde{V}_{out2} \\ \tilde{I}_{o2} \end{pmatrix} = \begin{pmatrix} 0.0235 - j0.0941 \\ 4 \end{pmatrix}$$

$$\Rightarrow \begin{pmatrix} \tilde{V}_{out2} \\ \tilde{I}_{o2} \end{pmatrix} = \begin{pmatrix} 0.0177 - j0.0206 & -0.0588 - j0.2647 \\ 1 & j40 \end{pmatrix}^{-1} \begin{pmatrix} 0.0235 - j0.0941 \\ 4 \end{pmatrix}$$

$$= \begin{pmatrix} 3.1795 - j1.2308 \\ 0.0308 - j0.0205 \end{pmatrix} = \begin{pmatrix} 3.4094 \exp(-j0.3693)\ V \\ 0.0370 \exp(-j0.588)\ A \end{pmatrix}$$

Therefore, the contribution from the second frequency, $\omega = 10$ rad/s, is given by

$$v_{out2}(t) = 3.4094 \cos(10t - 0.3693)\ V$$

The total response is then given by the sum of the contribution at each frequency.

$$v_{out}(t) = v_{out1}(t) + v_{out2}(t) = 3.5122 \cos(5t - 0.2187)\ V + 3.4094 \cos(10t - 0.3693)\ V$$

7.5 Problems

Problem 7.1: Write each of the following complex numbers in polar form ($Ae^{j\theta}$) with phase between $-\pi$ and π.

(a) $3+j7$
(b) $3+j7$
(c) $-3+j7$
(d) $-3+j7$

Problem 7.2: Find x in rectangular and polar form with a phase between $-\pi$ and π if $x = 2e^{j\pi/4} - 2e^{-j\pi/8} + 4 - j$.

Problem 7.3: Find x in rectangular and polar form with a phase between $-\pi$ and π if $x = 2e^{j\pi/4}(4+j3)$.

Problem 7.4: Find x in rectangular and polar form with a phase between $-\pi$ and π if $x = (8-j5)(1+j2)$.

Problem 7.5: Find x in rectangular and polar form with a phase between $-\pi$ and π if $x = j5(3+j2)$.

Problem 7.6: Find x in rectangular and polar form with a phase between $-\pi$ and π if $x = 2e^{j0.2} - 3e^{-j0.4} - 4 - 2j$.

Problem 7.7: Find the complex conjugate of the following complex numbers and express the answer in both polar and rectangular form.

$$x_1 = 2+j8 \quad x_2 = 5e^{j0.5}$$
$$x_3 = -7-j2 \quad x_4 = 4e^{-j2}$$

Problem 7.8: Simplify the following complex number expressions and express your answer in rectangular and polar form with a phase between $-\pi$ and π.

$$x_1 = \frac{1}{2+j2} \quad\quad x_2 = \frac{1}{-12-j8}$$
$$x_3 = \frac{1}{0.25-0.5j} + 4 + j2 \quad x_4 = \frac{3+j6}{1-j2}$$

Problem 7.9: Use Euler's Identity to simplify and solve the following integral.

$$\int_0^\infty e^{-x}\sin(x)dx$$

Problem 7.10: Solve the following equations involving complex numbers and express your answer in polar form with a phase between $-\pi$ and π.

(a) $(1+j2)x = (4-j5)$
(b) $(-2+j)x = 9$
(c) $3 - (2+j3)x = (4-j12)$
(d) $(4-j5)x + (3+j6) = (25+j10)$

Problem 7.11: Solve for x and y in the following system of equations involving complex numbers and express your answer in polar form with a phase between $-\pi$ and π.

$$(1+j2)x + 2y = 8$$
$$4x + (2+j3)y = 16$$

Problem 7.12: Solve for x and y in the following system of equations involving complex numbers and express your answer in polar form with a phase between $-\pi$ and π.

$$(5+j2)x + (3-j7)y = 9$$
$$(-3+j15)x + (6+j2)y = 15$$

Problem 7.13: Solve for x and y in the following system of equations involving complex numbers and express your answer in polar form with a phase between $-\pi$ and π.

$$(-3+j9)x + (2-j2)y = (7-j3)$$
$$(4+j5)x + (2+j1)y = (2+j6)$$

Problem 7.14: Solve for x and y in the following system of equations involving complex numbers and express your answer in polar form with a phase between $-\pi$ and π.

$$(6+j9)x + (4-j5)y + (1+j2)z = (8-j2)$$
$$(2-j)x + (3+j6)y + (0.7+j4)z = (9+j12)$$
$$(4+j2)x + (8+j2)y + (2+j2)z = (1-j2)$$

Problem 7.15: Find the voltage for the waveform shown below in the time domain. Your phase must be between $-\pi$ and π radians.

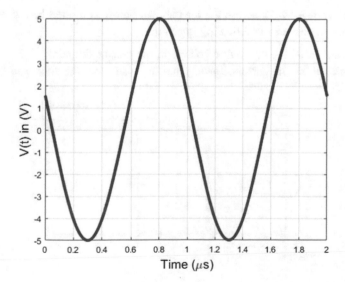

Problem 7.16: A voltage is given by $v(t) = 13\cos(1200t - 110°)$ mV. Find

(a) The frequency in Hz.
(b) The RMS value for the voltage.
(c) The phasor representation of v(t) with the angle expressed in radians.
(d) The first time in ms for $t > 0$ for which v(t) has a zero crossing.
(e) The first time in ms for $t > 0$ for which v(t) reaches its peak positive value.

Problem 7.17: Convert the following sinusoidal signals into phasors with phases between $-\pi$ and π radians.

$$v_1(t) = 9\cos(80t - 2.34) \ V$$
$$v_2(t) = 9\cos(40t - 2.34) \ V$$
$$v_3(t) = 9\sin(40t - 2.34) \ V$$

Problem 7.18: Convert the following sinusoidal signals into phasors with phases between $-\pi$ and π radians.

$$i_1(t) = 3\cos(2t) \ A$$
$$i_2(t) = 10\cos(5t + 0.5) \ A$$
$$i_3(t) = 3\cos(2t) + 10\cos(5t + 0.5) \ A$$

Problem 7.19: Convert the following sinusoidal signals into phasors with phases between $-\pi$ and π radians.

$$v_1(t) = 5\cos(4t) + 10\sin(4t) \ V$$
$$v_2(t) = 5\cos(4t) + 10\cos(4t - 0.5) \ V$$

Problem 7.20: The voltage, $v_s(t)$, across the load shown below is plotted below where $C_o = 33 \ \mu F$. Use this information to answer the following questions.

(a) Write an expression for $\tilde{V}_s$, the source voltage in the phasor domain.
(b) Write an expression for the current flowing into the load from the source in the time domain.

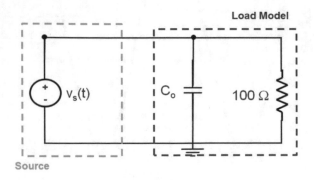

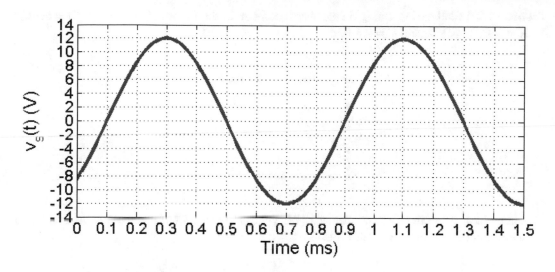

Problem 7.21: For the following circuit, assuming $v_{in}(t) = 5\cos(100t)$ V_{rms}, what physically realizable value of C_o that will give you an impedance seen by the source that is purely real.

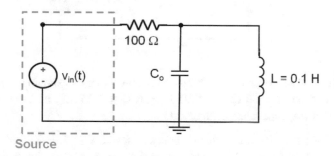

Problem 7.22: Find the current in the phasor domain and the time domain for the following circuit if $R = 500\ \Omega$, $C = 1\ \mu F$, and $v(t) = 10\cos(1000t)$ V.

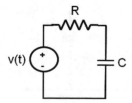

Problem 7.23: Find the current, i_{in}, in the phasor domain for the following circuit if $v_{in}(t) = 5\cos(2t)$ V, R = 5 Ω, L = 2 H, and C = 0.5 F.

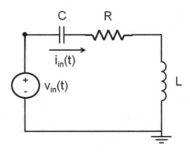

Problem 7.24: The circuit shown below has a voltage source, $v_s(t)$, connected to a resistor, an inductor, a capacitor, and an unknown load.

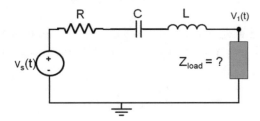

If we know that $v_s(t) = 8\cos(100t + 0.45)$ V and an oscilloscope measurement indicates that $v_1(t) = 3\cos(100t + 0.33)$ V when $R = 50$ Ω, $L = 0.25$ H, and $C = 0.25$ mF, find the current flowing into the unknown load and the impedance of the load.

Problem 7.25: For the circuit shown below, $R = 5$ Ω, $L = 0.25$ H, $C = 0.1$ F, and $v_{in}(t) = 20\cos(4t - 0.375)$ V. Find V_1 in the phasor domain and V_2 in the time domain.

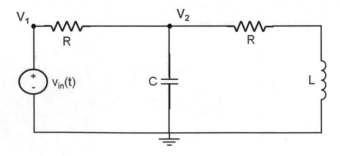

Problem 7.26: Find the node voltage V_1 in the phasor domain for the circuit shown below if $R_1 = 45\ \Omega$, $R_2 = 150\ \Omega$, $L_1 = 0.2\ H$, $C_1 = 333\ mF$, $v_A(t) = 20\cos\left(t - \frac{\pi}{2}\right)\ V$, and $v_B(t) = 10\cos(t)\ V$. the following circuit.

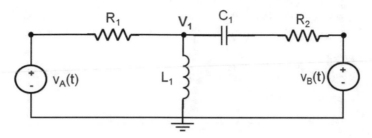

Problem 7.27: Write the node-voltage equations that would need to be solved to perform the phasor domain circuit analysis for the following circuit.

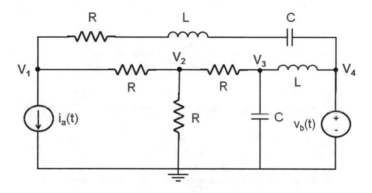

Problem 7.28: Find the magnitude and phase of $v_A(t)$, $v_B(t)$, and $v_C(t)$ if

$$v_1(t) = 3\cos(10t)\ V \quad v_2(t) = 2\sin(10t)\ V \quad i_1(t) = 2\cos\left(10t + \frac{\pi}{4}\right)\ A$$

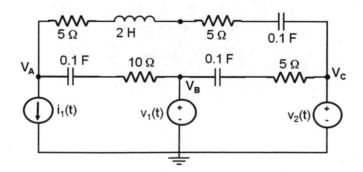

Problem 7.29: Find the magnitude and phase of $v_1(t)$, $v_2(t)$, and $v_3(t)$ if $v_{in}(t) = 9\cos(650t)$ V.

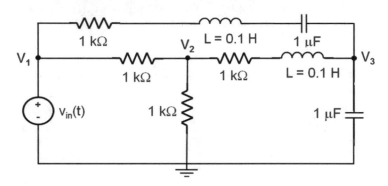

Problem 7.30: Find $v_1(t)$, $v_2(t)$, and $v_3(t)$ as functions of time for the following circuit if

$$v_a(t) = 5\cos\left(0.2t - \frac{\pi}{2}\right) V \quad i(t) = 2\cos(0.2t) A.$$

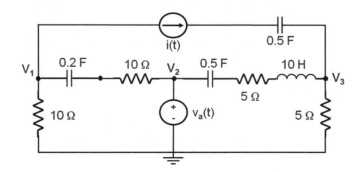

Problem 7.31: For the circuit shown below, let $R_1 = 200\ \Omega$, $R_2 = 50\ \Omega$, $R_3 = 50\ \Omega$, and $C_1 = 15$ mF. Find $v_o(t)$ if $v_{in}(t) = 10\cos\left(60t - \frac{\pi}{3}\right) + 2\cos\left(15t - \frac{\pi}{3}\right)$ V.

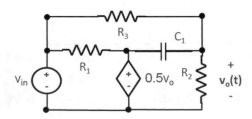

Phasor-Domain Power Analysis

8

Power flow for circuits undergoing harmonic excitation is very similar to the DC power analysis that was covered in Chap. 4. However, there are some key differences. First, since the voltages and currents vary with respect to time, the power flow will also vary over time. Therefore, the time-average power flow is of greatest interest when assessing efficiency. Second, rather than having just one power term, the flow of energy associated with the reactive components (i.e., the inductors and capacitors) must also be quantified. Therefore, multiple power terms are needed to define the flow of energy in the circuit. Lastly, when maximizing power transfer, the imaginary portion of the impedance must also be considered. Discussing these different aspects of power flow is the focus of this chapter.

8.1 Power Quantities for Harmonic Signals

8.1.1 Instantaneous Power, Average Power, and Reactive Power

As we saw in Chap. 4, power flow at any instant in time is given by

$$p(t) = i(t)v(t) \tag{8.1}$$

where the current is assumed to be flowing into the positive voltage terminal as shown in Fig. 8.1. Since both the current and the voltage have a sinusoidal dependence on time, Eq. (8.1) can be written as

$$
\begin{aligned}
p(t) = v(t)i(t) &= V_{\mathrm{m}} \cos(\omega t + \theta_v) I_{\mathrm{m}} \cos(\omega t + \theta_i) \\
&= V_{\mathrm{m}} I_{\mathrm{m}} \cos(\omega t + \theta_v) \cos(\omega t + \theta_i) \\
&= \frac{V_{\mathrm{m}} I_{\mathrm{m}}}{2} (\cos(\omega t + \theta_v - \omega t - \theta_i) + \cos(2\omega t + \theta_v + \theta_i)) \\
&= \frac{V_{\mathrm{m}} I_{\mathrm{m}}}{2} (\cos(\theta_v - \theta_i) + \cos(2\omega t + \theta_v + \theta_i)) \\
&= \frac{V_{\mathrm{m}} I_{\mathrm{m}}}{2} (\cos(\theta_v - \theta_i) + \cos(2\omega t + 2\theta_i + (\theta_v - \theta_i)))
\end{aligned}
\tag{8.2}
$$

where we have used the trigonometric identity

© Springer Nature Switzerland AG 2020
T. A. Bigelow, *Electric Circuits, Systems, and Motors*,
https://doi.org/10.1007/978-3-030-31355-5_8

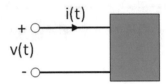

Fig. 8.1 Sign convention for power calculations

$$\cos(A)\cos(B) = \frac{\cos(A-B) + \cos(A+B)}{2} \tag{8.3}$$

The second cosine term in Eq. (8.2) can be simplified by

$$\cos(A \pm B) = \cos(A)\cos(B) \mp \sin(A)\sin(B) \tag{8.4}$$

to yield

$$p(t) = v(t)i(t) = \frac{V_m I_m}{2}\left(\begin{array}{l}\cos(\theta_v - \theta_i) + \cos(2\omega t + 2\theta_i)\cos(\theta_v - \theta_i) \\ -\sin(2\omega t + 2\theta_i)\sin(\theta_v - \theta_i)\end{array}\right) \tag{8.5}$$

If we change our reference time by setting $t' = (t + \theta_i/\omega)$, Eq. (8.5) becomes

$$\begin{aligned} p(t') &= \frac{V_m I_m}{2}\left(\cos(\theta_v - \theta_i) + \cos(2\omega t')\cos(\theta_v - \theta_i) - \sin(2\omega t')\sin(\theta_v - \theta_i)\right) \\ &= P + P\cos(2\omega t') - Q\sin(2\omega t') \end{aligned} \tag{8.6}$$

where

$$P = \frac{V_m I_m}{2}\cos(\theta_v - \theta_i) = V_{\text{rms}}I_{\text{rms}}\cos(\theta_v - \theta_i) \tag{8.7}$$

and

$$Q = \frac{V_m I_m}{2}\sin(\theta_v - \theta_i) = V_{\text{rms}}I_{\text{rms}}\sin(\theta_v - \theta_i) \tag{8.8}$$

The quantity P in Eq. (8.7) is the time-average power that is actually used or supplied by the circuit element and it has units of watts (W). This can be seen by integrating Eq. (8.6) with respect to time using

$$P_{\text{avg}} = P = \frac{1}{T}\int_T p(t)dt \tag{8.9}$$

The terms involving $\sin(2\omega t')$ and $\cos(2\omega t')$ will go to zero upon integrating over a single period leaving only the time-independent quantity P. Once again, a positive value for P would indicate an energy sink while a negative value for P denotes an energy source in the circuit.

The quantity Q in Eq. (8.8) is the reactive power. In order to distinguish the reactive power from the time-average power, the units of volt-amp reactive (VAR) is used instead of watts. Technically,

VAR and W are the same units. However, the difference in terminology emphasizes the difference in the quantity being discussed. Reactive power is the power that is associated with the energy flow into and out of energy storage elements such as inductors and capacitors. The sign of the reactive power cannot be used to distinguish between energy sources or sinks.

Example 8.1 Find the time-average power and reactive power associated with an inductor, a capacitor, and a resistor.

Solution:
Inductor: For an inductor, we have

$$\tilde{V} = Z \cdot \tilde{I} = j\omega L \cdot \tilde{I} = \omega L \cdot \tilde{I} \cdot e^{j\frac{\pi}{2}} \Rightarrow \theta_v = \theta_i + \frac{\pi}{2}$$

$$P = \frac{V_m I_m}{2} \cos(\theta_v - \theta_i) = \frac{\omega L I_m \cdot I_m}{2} \cos\left(\theta_i + \frac{\pi}{2} - \theta_i\right) = \frac{\omega L I_m^2}{2} \cos\left(\frac{\pi}{2}\right) = 0 \text{ W}$$

$$Q = \frac{V_m I_m}{2} \sin(\theta_v - \theta_i) = \frac{\omega L I_m \cdot I_m}{2} \sin\left(\theta_i + \frac{\pi}{2} - \theta_i\right) = \frac{\omega L I_m^2}{2} \sin\left(\frac{\pi}{2}\right) = \frac{\omega L I_m^2}{2}$$

Therefore, there is no time-average power delivered to an ideal inductor. Energy can be stored in the inductor, but no energy is lost. Also, the value of Q for an inductor is always positive. Therefore, an inductive load will always have a positive Q value.

Capacitor: For a capacitor, we have

$$\tilde{V} = Z \cdot \tilde{I} = \frac{1}{j\omega C} \cdot \tilde{I} = \frac{1}{\omega C} \cdot \tilde{I} \cdot e^{-j\frac{\pi}{2}} \Rightarrow \theta_v = \theta_i - \frac{\pi}{2}$$

$$P = \frac{V_m I_m}{2} \cos(\theta_v - \theta_i) = \frac{\left(\frac{I_m}{\omega C}\right) \cdot I_m}{2} \cos\left(\theta_i - \frac{\pi}{2} - \theta_i\right) = \frac{I_m^2}{2\omega C} \cos\left(-\frac{\pi}{2}\right) = 0 \text{ W}$$

$$Q = \frac{V_m I_m}{2} \sin(\theta_v - \theta_i) = \frac{\left(\frac{I_m}{\omega C}\right) \cdot I_m}{2} \sin\left(\theta_i - \frac{\pi}{2} - \theta_i\right) = \frac{I_m^2}{2\omega C} \sin\left(-\frac{\pi}{2}\right) = -\frac{I_m^2}{2\omega C}$$

Therefore, once again, we have no energy lost in the capacitor as the time-average power must be zero. This time, however, the value of Q must be negative. It is negative even though the capacitor is not a source of energy. All capacitive loads will have negative values of Q.

Resistor: For a resistor, we have

$$\tilde{V} = Z \cdot \tilde{I} = R \cdot \tilde{I} \Rightarrow \theta_v = \theta_i$$

$$P = \frac{V_m I_m}{2} \cos(\theta_v - \theta_i) = \frac{(R I_m) \cdot I_m}{2} \cos(\theta_i - \theta_i) = \frac{R I_m^2}{2} \cos(0) = \frac{R I_m^2}{2}$$

$$Q = \frac{V_m I_m}{2} \sin(\theta_v - \theta_i) = \frac{(R I_m) \cdot I_m}{2} \sin(\theta_i - \theta_i) = \frac{R I_m^2}{2} \sin(0) = 0 \text{ VAR}$$

For the resistor, the reactive power, Q, is always zero. Therefore, purely resistive loads will have no reactive power component. This does not mean that there are no inductors or capacitors in the load as the impedance of the inductors and capacitors may just be canceling at the operating frequency of the circuit. Resistive loads will also always have real values for the time-average power, P, indicating they are pulling energy out of the circuit.

Example 8.2 For each of the following voltages and currents, find the time-average power and reactive power associated with each circuit element and identify if the load is inductive, capacitive, or purely resistive.

(a) $v(t) = 10\cos(100t - 0.35)$ V $i(t) = 2\cos(100t + 0.42)$ A

(b) $v(t) = 10\cos(100t - 0.35)$ V $i(t) = 2\cos(100t - 0.35)$ A

(c) $v(t) = 10\cos(100t - 2.75)$ V $i(t) = 2\cos(100t + 3.1)$ A

(d) $v(t) = 10\cos(100t + 0.35)$ V $i(t) = 2\cos(100t - 0.42)$ A

Solution:

(a) :

$$P = \frac{V_m I_m}{2}\cos(\theta_v - \theta_i) = \frac{10 \cdot 2}{2}\cos(-0.35 - 0.42) = 7.1791\,\text{W}$$
$$Q = \frac{V_m I_m}{2}\sin(\theta_v - \theta_i) = \frac{10 \cdot 2}{2}\sin(-0.35 - 0.42) = -6.9614\,\text{VAR}$$

Load is capacitive.

(b) :

$$P = \frac{V_m I_m}{2}\cos(\theta_v - \theta_i) = \frac{10 \cdot 2}{2}\cos(-0.35 - 0.35) = 10\,\text{W}$$
$$Q = \frac{V_m I_m}{2}\sin(\theta_v - \theta_i) = \frac{10 \cdot 2}{2}\sin(-0.35 - 0.35) = 0\,\text{VAR}$$

Load is purely resistive.

(c) :

$$P = \frac{V_m I_m}{2}\cos(\theta_v - \theta_i) = \frac{10 \cdot 2}{2}\cos(-2.75 - 3.1) = 9.0763\,\text{W}$$
$$Q = \frac{V_m I_m}{2}\sin(\theta_v - \theta_i) = \frac{10 \cdot 2}{2}\sin(-2.75 - 3.1) = 4.1976\,\text{VAR}$$

Load is inductive.

(d) :

$$P = \frac{V_m I_m}{2}\cos(\theta_v - \theta_i) = \frac{10 \cdot 2}{2}\cos(0.35 - 0.42) = 7.1791\,\text{W}$$
$$Q = \frac{V_m I_m}{2}\sin(\theta_v - \theta_i) = \frac{10 \cdot 2}{2}\sin(0.35 - 0.42) = 6.9614\,\text{VAR}$$

Load is inductive.

8.1.2 Power Factor

When doing power calculations, it is often useful to compare the actual time-average power to the power flow that would be occurring if the load were purely resistive. As we will see later, purely resistive loads have the most efficient power delivery. One quantity for doing this is the power factor, pf, defined as

$$\text{pf} = \cos(\theta_v - \theta_i) \tag{8.10}$$

When the voltage and current are in phase (i.e., resistive load), the power factor is 1. Any deviation in the phase between the voltage and current due to reactive components will lower the power factor. The power factor is often expressed in percent. For example, a pf of 0.75 will often be written as 75.

Example 8.3 For the circuit shown in Fig. 8.2, assuming $v_{in}(t) = 5\cos(314t)V_{rms}$, what physically realizable value of C_o will give you a power factor (pf) at the source of one?

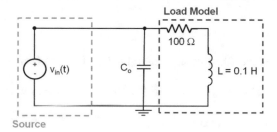

Fig. 8.2 Circuit for Example 8.3

Solution:

A pf of 1 means that the impedance seen by the source must be purely real. The impedance seen by the source is given by

$$Z_{eq} = \cfrac{1}{j\omega C_o + \cfrac{1}{R_{load} + j\omega L}} = \cfrac{1}{j\omega C_o + \cfrac{R_{load} - j\omega L}{R_{load}^2 + (\omega L)^2}}$$

$$= \frac{R_{load}^2 + (\omega L_{load})^2}{j\omega C_o \left(R_{load}^2 + (\omega L)^2 \right) + R_{load} - j\omega L}$$

Therefore,

$$j\omega C_o \left(R_{load}^2 + (\omega L)^2 \right) - j\omega L = 0$$

$$\Rightarrow C_o = \frac{L}{R_{load}^2 + (\omega L)^2} = 9.1025 \,\mu F$$

Relationship between pf and impedance: When finding the phase difference between the voltage and the current for any purpose including the pf, it can be helpful to notice that $(\theta_v - \theta_i) = \theta_z$ where θ_z is the phase angle for the impedance. Therefore, the pf $= \cos(\theta_z)$ and thus only depends on the load and not the source that is suppling power to the load.

8.1.3   Apparent Power

In addition to the time-average power, P, and reactive power, Q, the apparent power is another power quantity that is important when matching generators to loads. The apparent power is the volt-amp capacity required to supply the average power demanded by the load. The maximum apparent power that can be supplied from the source must exceed the apparent power demanded by the load. Apparent power is given by

$$|S| = \sqrt{P^2 + Q^2} = \frac{V_m I_m}{2} = V_{rms} I_{rms} \qquad (8.11)$$

Obviously, the fraction of time-average power to apparent power is the power factor. The units of apparent power are volt-amps (VA).

Example 8.4 A motor has a power factor of 67. The motor is connected to a 115 V_{rms} source and is drawing 4.4 A_{rms}. What is the time-average power, reactive power, and apparent power being supplied to the motor?

Solution:
The apparent power is just the RMS voltage multiplied by the RMS current.

$$|S| = V_{rms} I_{rms} = (115 \, V_{rms})(4.4 \, A_{rms}) = 506 \, VA$$

Also, the time-average power is given by

$$P = V_{rms} I_{rms} \cos(\theta_v - \theta_i) = V_{rms} I_{rms} pf = |S| pf = (506 \, VA) \cdot 0.67 = 339.02 \, W$$

The reactive power is a little more complicated as there are two possible solutions depending on whether the voltage leads or lags the current.

$$\cos(\theta_v - \theta_i) = 0.67 \Rightarrow (\theta_v - \theta_i) = \pm 0.83659 \, rad$$
$$\Rightarrow \sin(\theta_v - \theta_i) = \pm 0.74236$$
$$Q = V_{rms} I_{rms} \sin(\theta_v - \theta_i) = |S| \sin(\theta_v - \theta_i) = \pm 375.635 \, VAR$$

However, since motors are naturally inductive due to the coils needed to generate the required magnetic fields, +375.635 VAR is the best possible answer.

8.1.4   Complex Power

As we saw in Chap. 7, the analysis of AC circuits is normally accomplished in the phasor domain using complex numbers. Therefore, it is useful to be able to solve for the power quantities using the phasor domain voltages and currents rather than first translating the values into the time domain expressions. From Euler's Identity, we know that $e^{j\theta} = \cos(\theta) + j\sin(\theta)$. We can define a complex power quantity, S, given by

$$
\begin{aligned}
S &= \frac{1}{2} \tilde{V} \cdot \tilde{I}^* = \tilde{V}_{\text{rms}} \cdot \tilde{I}_{\text{rms}}^* \\
&= \frac{1}{2} V_m e^{j\theta_v} \cdot \left(I_m e^{j\theta_i} \right)^* = \frac{1}{2} V_m e^{j\theta_v} \cdot \left(I_m e^{-j\theta_i} \right) \\
&= \frac{1}{2} V_m I_m e^{j(\theta_v - \theta_i)} = \frac{1}{2} V_m I_m (\cos(\theta_v - \theta_i) + j\sin(\theta_v - \theta_i)) \\
&= P + jQ
\end{aligned}
\tag{8.12}
$$

The units for complex power are the same as the units for apparent power (i.e., VA).

8.2 Maximum Power Transfer

As was discussed in Chap. 4, maximizing the delivery of power from a source to a load is often desirable. Inefficiencies can lead to poor power delivery or a loss of sensitivity from sensors. For purely resistive circuits, maximum power delivery was shown to occur when the load resistance was made equal to the source resistance. When inductances and capacitances are present, the resistances become impedances and phasor domain circuit analysis is used to solve for the circuits. However, the goal is still to maximize the usable power delivered to the load. For AC circuits, the usable power is always the time-average power.

Consider the Thevenin equivalent circuit shown in Fig. 8.3 operating at frequency ω. The impedance of the source has real and imaginary parts given by R_{Th} and X_{Th}, respectively, while the impedance of the load has real and imaginary parts given by R_L and X_L. When maximizing power transfer for AC circuits, both the real and the imaginary parts of the impedances should be considered.

The complex power delivered to the load is given by

$$
S_{\text{load}} = \frac{1}{2} \tilde{V}_{\text{load}} \cdot (\tilde{I}_{\text{load}})^* = \frac{1}{2} \overbrace{\left(\tilde{V}_{\text{Th}} \frac{Z_L}{Z_{\text{Th}} + Z_L} \right)}^{\substack{\text{Voltage from} \\ \text{Voltage Divider}}} \overbrace{\left(\frac{\tilde{V}_{\text{Th}}}{Z_{\text{Th}} + Z_L} \right)^*}^{\substack{\text{Current from} \\ \text{Ohm's Law}}} = \frac{\left| \tilde{V}_{\text{Th}} \right|^2 Z_L}{2|Z_{\text{Th}} + Z_L|^2}
$$

$$
= \frac{\left| \tilde{V}_{\text{Th}} \right|^2 (R_L + jX_L)}{2|R_{\text{Th}} + jX_{\text{Th}} + R_L + jX_L|^2} = \frac{\left| \tilde{V}_{\text{Th}} \right|^2 (R_L + jX_L)}{2\left((R_{\text{Th}} + R_L)^2 + (X_{\text{Th}} + X_L)^2 \right)}
\tag{8.13}
$$

where we have made use of the identity $x \cdot x^* = |x|^2$ for both the voltages in the numerator and the impedances in the denominator. Since the real part of the complex power is the usable time-average power, P_{load} is given by

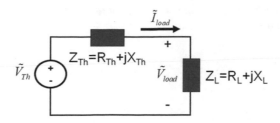

Fig. 8.3 Thevenin equivalent for AC circuit illustrating maximum power transfer

$$P_{\text{load}} = \frac{\left|\tilde{V}_{\text{Th}}\right|^2 R_{\text{L}}}{2\left((R_{\text{Th}} + R_{\text{L}})^2 + (X_{\text{Th}} + X_{\text{L}})^2\right)} \tag{8.14}$$

If we take the derivative of the power to the load with respect to X_{L} and setting it equal to zero gives

$$\frac{\partial P_{\text{load}}}{\partial X_{\text{L}}} = \frac{\left|\tilde{V}_{\text{Th}}\right|^2 R_{\text{L}}}{2} \frac{\partial}{\partial X_{\text{L}}} \left(\frac{1}{\left((R_{\text{Th}} + R_{\text{L}})^2 + (X_{\text{Th}} + X_{\text{L}})^2\right)}\right)$$

$$= -\frac{\left|\tilde{V}_{\text{Th}}\right|^2 R_{\text{L}}}{2} \left(\frac{2(X_{\text{Th}} + X_{\text{L}})}{\left((R_{\text{Th}} + R_{\text{L}})^2 + (X_{\text{Th}} + X_{\text{L}})^2\right)^2}\right) = 0 \Rightarrow X_{\text{L}} = -X_{\text{Th}} \tag{8.15}$$

Therefore, the maximum power transfer is achieved when the reactance of the load, X_{L}, cancels the reactance of the source, X_{Th}. This means that if the source has a capacitive component, the load should have an inductive component. Likewise, if the source has an inductive component, then the load should have a capacitive component. The key is to have the total impedance seen by the source be purely real (i.e., pf = 1). This can normally be accomplished relatively easily by adding inductors or capacitors to the load. Now that the optimal imaginary part of the load impedance is known, we can find the best real part of the load impedance by taking the derivative of P_{load} with respect to R_{L}.

$$\frac{\partial P_{\text{load}}}{\partial R_{\text{L}}} = \frac{\left|\tilde{V}_{\text{Th}}\right|^2}{2} \frac{\partial}{\partial R_{\text{L}}} \left(\frac{R_{\text{L}}}{\left((R_{\text{Th}} + R_{\text{L}})^2 + (X_{\text{Th}} + X_{\text{L}})^2\right)}\right)_{X_{\text{L}} = -X_{\text{Th}}} = \frac{\left|\tilde{V}_{\text{Th}}\right|^2}{2} \frac{\partial}{\partial R_{\text{L}}} \left(\frac{R_{\text{L}}}{(R_{\text{Th}} + R_{\text{L}})^2}\right)$$

$$= -\frac{\left|\tilde{V}_{\text{Th}}\right|^2}{2} \left(\frac{(R_{\text{Th}} + R_{\text{L}})^2 - 2R_{\text{L}}(R_{\text{Th}} + R_{\text{L}})}{(R_{\text{Th}} + R_{\text{L}})^2}\right) = 0$$

$$\Rightarrow (R_{\text{Th}} + R_{\text{L}}) - 2R_{\text{L}} = 0 \Rightarrow R_{\text{L}} = R_{\text{Th}}$$

$$\tag{8.16}$$

Therefore, just as we saw in Chap. 4, R_{L} should be matched to R_{Th} for maximum power transfer.

However, as stated in Chap. 4, one should never increase the source resistance to match the load resistance. Equation (8.16) was derived assuming R_{L} was the quantity that could be changed. The best power transfer is always achieved by keeping the source resistance as small as possible. Once the source resistance has been minimized, the next step is to match the load resistance to that source resistance.

Example 8.5 For the circuit shown in Fig. 8.4, find the values of C_o and R_o that result in the maximum power transfer to R_o when $v(t) = 10\cos(\omega t)$ V where $\omega = 20{,}000$ rad/s. Also, plot the time-average power delivered to R_o as a function of R_o at $\omega = 20{,}000$ rad/s as well as a function of ω at the optimal value found for R_o.

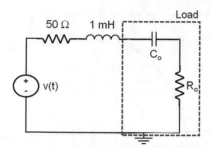

Fig. 8.4 Circuit for Example 8.5

Solution:

We first need to find expressions for Z_L and Z_{Th}. For this problem, Z_{Th} is the series combination of the 50 Ω resistor and the 1 mH inductor while Z_L is the series combination of R_o and C_o. Therefore,

$$Z_{Th} = 50 + j\omega(1\,\text{mH}) = 50 + j20\,\Omega$$

$$Z_L = R_o + \frac{1}{j\omega C_o} = R_o - j\frac{1}{\omega C_o}$$

Setting the real part of Z_L to the real part of Z_{Th} and the imaginary part of Z_L to the negative of the imaginary part of Z_{Th} gives

$$R_o = 50\,\Omega$$

$$\frac{1}{\omega C_o} = 20\,\Omega \Rightarrow C_o = \frac{1}{20\omega} = 2.5\,\mu\text{F}$$

The time-average power delivered to R_o is given by

$$P_{\text{load}} = \frac{\left|\tilde{V}_{Th}\right|^2 R_L}{2\left((R_{Th} + R_L)^2 + (X_{Th} + X_L)^2\right)} = \frac{100 R_o}{2\left((50 + R_o)^2 + \left(\omega 0.001 - \frac{1}{\omega C_o}\right)^2\right)}$$

$$= \frac{50 R_o}{\left((50 + R_o)^2 + \left(\frac{\omega}{1000} - \frac{400{,}000}{\omega}\right)^2\right)}$$

Therefore, if $\omega = 20{,}000$ rad/s, we get

$$P_{\text{load}} = \frac{50 R_o}{\left((50 + R_o)^2 + (20 - 20)^2\right)} = \frac{50 R_o}{(50 + R_o)^2}$$

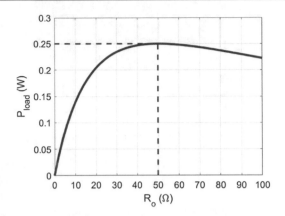

Likewise, if we vary the frequency while keeping R_o at 50 Ω, we get

$$P_{\text{load}} = \frac{2,500}{\left(10,000 + \left(\frac{\omega}{1000} - \frac{400,000}{\omega}\right)^2\right)}$$

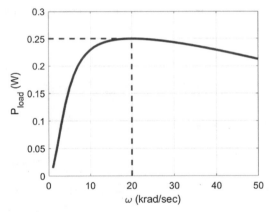

As the resistance of the load and frequency change, the power delivery to the load is also negatively impacted.

Practical Hint: When attempting to match the load impedance to the source impedance for maximum power transfer several factors can reduce the efficiency of the transfer. First, the impedance of the source as well as the load can change due to temperature especially in uncontrolled environments. Second, in many applications, the signals of interests contain multiple frequency components. Therefore, not every component will be transferred at its maximum efficiency. The overall impact on system performance will depend on the bandwidth or range of frequencies in the signal.

8.3   Load Matching

While we know that the load should be the complex conjugate of the source impedances to achieve maximum power transfer, matching the real part of the impedances is often challenging. The imaginary parts can be easily canceled by adding inductors and capacitors to the load without increasing the power lost. However, altering the load impedance to match the source impedance is not

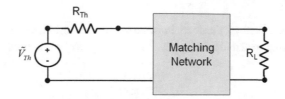

Fig. 8.5 Matching network placed between source and load impedance to maximize power transfer to the load

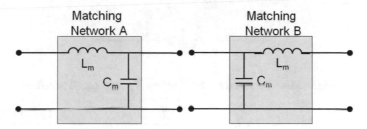

Fig. 8.6 Two basic circuit topologies for the matching network

immediately obvious. Fortunately, it is also possible to match the real part of the impedances by designing a matching network between the source and load as shown in Fig. 8.5. We will assume that the source and load impedances are purely real since it is trivial to add inductors and capacitors to remove any imaginary components.

Two basic topologies for the matching network are shown in Fig. 8.6. Both topologies consist of a single inductor and a single capacitor. Therefore, the matching network does not increase the losses in the system as would happen should a resistor be added. To match a load resistance R_L to a source resistance R_{Th} using the first matching network (Matching Network A), the inductor and capacitor should be

$$C_m = \frac{\sqrt{\frac{R_L}{R_{Th}}-1}}{\omega R_L} \quad L_m = \frac{R_{Th}\sqrt{\frac{R_L}{R_{Th}}-1}}{\omega} \tag{8.17}$$

Likewise, to match the load and source resistance using the second matching network (Matching Network B), the inductor and capacitor should be

$$C_m = \frac{\sqrt{\frac{R_{Th}}{R_L}-1}}{\omega R_{Th}} \quad L_m = \frac{R_L\sqrt{\frac{R_{Th}}{R_L}-1}}{\omega} \tag{8.18}$$

Notice that when $R_L > R_{Th}$, the first matching network (Matching Network A) should be used with the second matching network (Matching Network B) used when $R_L < R_{Th}$. Otherwise, the inductor and capacitor would have imaginary values which are not physically possible.

Example 8.6 Show that the values given for the inductor and capacitor in Eq. (8.17) will match the load resistance to the source resistance for Matching Network A.

Solution:

We know that when connected to the load impedance as shown below, the total impedance of the matching network and load resistance, Z_{in}, should be purely real and should be equal to R_{Th}.

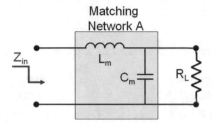

Z_{in} consists of L_m in series with the parallel combination R_L and C_m. Therefore, Z_{in} is given by

$$Z_{in} = j\omega L_m + \frac{1}{\frac{1}{R_L} + j\omega C_m} = j\omega L_m + \frac{R_L}{1 + j\omega C_m R_L} = j\omega L_m + \frac{R_L(1 - j\omega C_m R_L)}{1 + (\omega C_m R_L)^2}$$

$$= \frac{R_L}{1 + (\omega C_m R_L)^2} + j\left(\omega L_m - \frac{\omega C_m R_L^2}{1 + (\omega C_m R_L)^2}\right)$$

If we now use the values from Eq. (8.17), we get

$$Z_{in} = \frac{R_L}{1 + \left(\omega R_L \frac{\sqrt{\frac{R_L}{R_{Th}} - 1}}{\omega R_L}\right)^2} + j\left(\omega \frac{R_{Th}\sqrt{\frac{R_L}{R_{Th}} - 1}}{\omega} - \frac{\omega R_L^2 \frac{\sqrt{\frac{R_L}{R_{Th}} - 1}}{\omega R_L}}{1 + \left(\omega R_L \frac{\sqrt{\frac{R_L}{R_{Th}} - 1}}{\omega R_L}\right)^2}\right)$$

$$= \frac{R_L}{1 + \frac{R_L}{R_{Th}} - 1} + j\left(R_{Th}\sqrt{\frac{R_L}{R_{Th}} - 1} - \frac{R_L\sqrt{\frac{R_L}{R_{Th}} - 1}}{1 + \frac{R_L}{R_{Th}} - 1}\right) = R_{Th} + j\left(R_{Th}\sqrt{\frac{R_L}{R_{Th}} - 1} - R_{Th}\sqrt{\frac{R_L}{R_{Th}} - 1}\right) = R_{Th}$$

Example 8.7 Show that the values given for the inductor and capacitor in Eq. (8.18) will match the load resistance to the source resistance for Matching Network B.

Solution:

Once again, when connected to the load impedance as shown below, the total impedance of the matching network and load resistance, Z_{in}, should be purely real and should be equal to R_{Th}. Therefore, $1/Z_{in} = 1/R_{Th}$

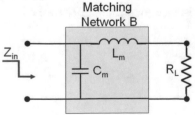

$1/Z_{in}$ is given by

$$\frac{1}{Z_{in}} = \frac{1}{R_L + j\omega L_m} + j\omega C_m = \frac{R_L - j\omega L_m}{R_L^2 + (\omega L_m)^2} + j\omega C_m = \frac{R_L}{R_L^2 + (\omega L_m)^2} + j\left(\omega C_m - \frac{\omega L_m}{R_L^2 + (\omega L_m)^2}\right)$$

$$\frac{1}{Z_{in}} = \frac{R_L}{R_L^2 + \left(\omega \frac{R_L\sqrt{\frac{R_{Th}}{R_L} - 1}}{\omega}\right)^2} + j\left(\omega \frac{\sqrt{\frac{R_{Th}}{R_L} - 1}}{\omega R_{Th}} - \frac{\omega \frac{R_L\sqrt{\frac{R_{Th}}{R_L} - 1}}{\omega}}{R_L^2 + \left(\omega \frac{R_L\sqrt{\frac{R_{Th}}{R_L} - 1}}{\omega}\right)^2}\right)$$

$$= \frac{R_L}{R_L^2 + R_L^2\left(\frac{R_{Th}}{R_L} - 1\right)} + j\left(\frac{\sqrt{\frac{R_{Th}}{R_L} - 1}}{R_{Th}} - \frac{R_L\sqrt{\frac{R_{Th}}{R_L} - 1}}{R_L^2 + R_L^2\left(\frac{R_{Th}}{R_L} - 1\right)}\right)$$

$$= \frac{R_L}{R_L R_{Th}} + j\left(\frac{\sqrt{\frac{R_{Th}}{R_L} - 1}}{R_{Th}} - \frac{R_L\sqrt{\frac{R_{Th}}{R_L} - 1}}{R_L R_{Th}}\right) = \frac{1}{R_{Th}} + j\left(\frac{\sqrt{\frac{R_{Th}}{R_L} - 1}}{R_{Th}} - \frac{\sqrt{\frac{R_{Th}}{R_L} - 1}}{R_{Th}}\right) = \frac{1}{R_{Th}}$$

Example 8.8 Design a matching network to match a 10 Ω load to a 50 Ω source at a frequency of 1 Mrad/s.

Solution:

Since the load impedance is less than the source impedance, Matching Network B should be used with

$$C_m = \frac{\sqrt{\frac{R_{Th}}{R_L} - 1}}{\omega R_{Th}} = \frac{\sqrt{\frac{50}{10} - 1}}{10^6 \cdot 50} = 40\,\text{nF}$$

$$L_m = \frac{R_L\sqrt{\frac{R_{Th}}{R_L} - 1}}{\omega} = \frac{10\sqrt{\frac{50}{10} - 1}}{10^6} = 20\,\mu\text{H}$$

Example 8.9 Design a matching network to match an 850 Ω load to a 50 Ω source at a frequency of 1 krad/s.

Solution:

Since the load impedance is greater than the source impedance, Matching Network A should be used with

$$C_m = \frac{\sqrt{\frac{R_L}{R_{Th}} - 1}}{\omega R_L} = \frac{\sqrt{\frac{850}{50} - 1}}{1000 \cdot 1000} = 16\,\mu\text{F}$$

$$L_m = \frac{R_{Th}\sqrt{\frac{R_L}{R_{Th}} - 1}}{\omega} = \frac{50\sqrt{\frac{850}{50} - 1}}{1000} = 0.2\,\text{H}$$

8.4 Problems

Problem 8.1 For each of the following voltages and currents, find the time-average power and reactive power associated with each circuit element and identify if the load is inductive, capacitive, or purely resistive.

(a) $v(t) = 5\cos(50t + 0.5)\,\text{V} \quad i(t) = 2\cos(50t + 0.75)\,\text{A}$

(b) $v(t) = 6\cos(50t - 0.3\pi)\,\text{V} \quad i(t) = 4\cos(50t + 1.4\pi)\,\text{A}$

(c) $v(t) = 2\cos(100t - 1.8)\,\text{V} \quad i(t) = 3\cos(100t - 2.5)\,\text{A}$

(d) $v(t) = 8\cos(25t + 0.3\pi)\,\text{V} \quad i(t) = 0.5\cos(25t - 1.7\pi)\,\text{A}$

Problem 8.2 What is the power factor for the voltage/current relationships given in Problem 1?

Problem 8.3 Can the power factor for a load ever be negative? Why or why not?

Problem 8.4 This problem uses the following circuit with $v_{in}(t) = 5\cos(2000t + 0.83)\,V_{rms}$.

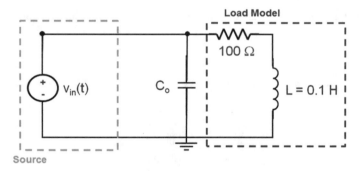

(a) What physically realizable value of C_o will give you a power factor (pf) at the source of one?
(b) Assuming C_o is the value you found in part a, what is the time-average power and reactive power delivered to the load as labeled in the above diagram? Hint: The capacitance is not part of the load.

Problem 8.5 For the circuit shown below, find the source frequency, ω, which would result in a power factor of 1.0 at the source if $L_o = 0.2$ H and $Co = 2.5$ μF.

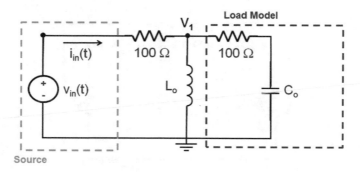

Problem 8.6 This problem uses the following circuit with $v_2(t) = 5\cos(1000\,t + 0.72)\,V$.

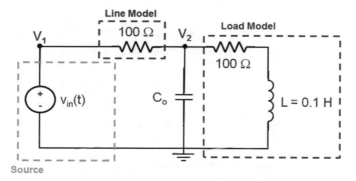

(a) Find the time-average power (P) and apparent power (|S|) going into the load.

(b) What physically realizable value of C_o will give you a power factor (pf) at the source of one?

Problem 8.7 A motor has a power factor of 75. The motor is connected to a 230 V_{rms} source and is drawing 10 A_{rms}. What is the time-average power, reactive power, and apparent power being supplied to the motor?

Problem 8.8 For each of the following phasor domain voltages and currents, find the complex power, time-average power, reactive power, and apparent power associated with the circuit element.

(a) $\tilde{V} = 5\,V$ $\tilde{I} = 0.4\exp(-j0.5)\,A$

(b) $\tilde{V} = 100\exp(j0.8)\,V_{rms}$ $\tilde{I} = 3\exp(j2)\,A_{rms}$

(c) $\tilde{V} = 9\exp(-j0.25)\,V_{rms}$ $\tilde{I} = j2\,A_{rms}$

(d) $\tilde{V} = 50\exp(-j0.75)\,V$ $\tilde{I} = 4\exp(j0.25)\,A_{rms}$

Problem 8.9 For the circuit shown below, find the values of C_o and R_o that result in the maximum power transfer to R_o when $v(t) = 7\cos(1000\,t)\,\text{V}$.

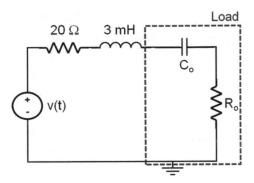

Problem 8.10 For the circuit shown below, find the values of C_o and R_{Th} that result in the maximum power transfer to R_o when $v(t) = 15\cos(8000\,t)\,\text{V}$.

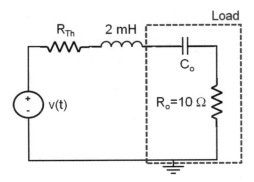

Problem 8.11 For the circuit shown below, find the values of C_o and R_o that result in the maximum power transfer to the load when $v_{in}(t) = 8\cos(25000\,t)\,\text{V}$.

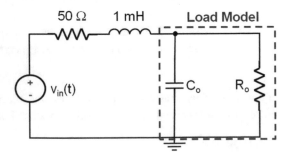

Problem 8.12 Design a matching network to match a $8\ \Omega$ load to a $20\ \Omega$ source at a frequency of 1 MHz.

Problem 8.13 Design a matching network to match a $20\ \Omega$ load to a $8\ \Omega$ source at a frequency of 1 MHz.

Problem 8.14 Design a matching network to match a $1000\ \Omega$ load to a $50\ \Omega$ source at a frequency of 5 MHz.

Operational Amplifiers

9

When designing electric circuits, frequently the goal is to prepare the signals for processing by a computer or microcontroller. This normally involves amplifying and filtering the signals before they can be accurately captured. One of the basic building blocks for many amplifiers and filters is the Operational Amplifier (Op-Amp). These circuit elements require external power in order to function making them active elements. Operational amplifiers consist of multiple transistors which is why they need an external power source. In this chapter, we will discuss operational amplifiers and discuss some of their uses and limitations.

9.1 Overview of Amplifier Types

In chapter one, we introduced the concept of dependent sources. For these sources, the voltage or current produced by the source was dependent on some other voltage or current in the circuit. These dependent sources are basically amplifiers. However, just as independent voltage and current sources also often have parasitic resistances as manifested by the Thevenin Resistance (R_{Th}), dependent sources (i.e., amplifiers) will also have these resistances. The models for the four basic amplifier types are shown in Fig. 9.1. The voltage amplifier has a voltage as both its input and its output with a voltage-dependent voltage source. The transresistance amplifier has a voltage as an output and a current as an input and, therefore, needs a current-dependent voltage source. The transconductance amplifier has a voltage as an input and a current as an output with a voltage-dependent current source. Lastly, the current amplifier has a current as both an input and an output with a current dependent current source.

It is important to design and utilize the correct type of amplifier when optimizing the performance of your system. In order to classify an amplifier into the proper category, one needs to look at the resistance values. For example, simply because the amplifier is described in terms of an input and output voltage does not make it a voltage amplifier. An input source can always be changed from its Norton to its Thevenin Equivalent. Hence, the input of any amplifier can always be expressed in terms of either an input voltage or an input current. The key is to consider the input resistance, R_{in}, and output resistance, R_{out}, of the amplifier. Every amplifier that is expecting an input current will have a low input resistance to maximize the current flow while every amplifier that is expecting an input voltage will have a high input resistance to maximize the voltage at the input of the amplifier. Likewise, amplifiers designed to output a current will have a high output resistance so that the majority of the current flows into the load connected to the amplifier. Likewise, amplifiers designed to

© Springer Nature Switzerland AG 2020
T. A. Bigelow, *Electric Circuits, Systems, and Motors*,
https://doi.org/10.1007/978-3-030-31355-5_9

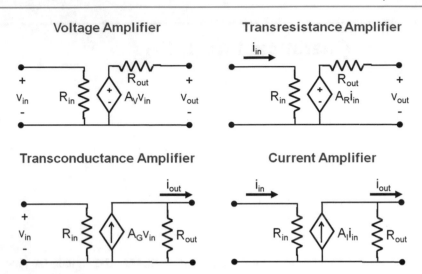

Fig. 9.1 Basic amplifier configurations using dependent sources

output a voltage will have a low output resistance so that most of the voltage is across the load connected to the amplifier. The relative sizes of the input and output resistances can always be used to identify the type of amplifier.

Example 9.1 Classify the following amplifiers as either a voltage amplifier, a current amplifier, a transresistance amplifier, or a transconductance amplifier.

(a) $R_{in} = 50\ \Omega$ $R_{out} = 50\ \Omega$ Gain $= 100$ V/V
(b) $R_{in} = 1\ M\Omega$ $R_{out} = 50\ \Omega$ Gain $= 100$ V/A
(c) $R_{in} = 1\ M\Omega$ $R_{out} = 50\ M\Omega$ Gain $= 100$ A/A
(d) $R_{in} = 1000\ \Omega$ $R_{out} = 5\ M\Omega$ Gain $= 100$ A/V

Solution When determining the type of amplifier, the units on the gain term do not matter. Any amplifier can be expressed with either a voltage or a current at the input or the output. There may even be some advantages to expressing the amplifier performance as some gain in V/V, for example, even though the structure of the amplifier should be V/A. The key is to focus on the impedance values when identifying the type of amplifier.

(a) For this amplifier, the input and the output resistances are low. Since R_{in} is low, the input of the amplifier is current. Likewise, since R_{out} is low, the output of the amplifier is voltage. Therefore, this is a **transresistance amplifier** (voltage out/current in).
(b) For this amplifier, the input resistance is high and the output resistance is low. Since R_{in} is high, the input of the amplifier is voltage. Likewise, since R_{out} is low, the output of the amplifier is voltage. Therefore, this is a **voltage amplifier** (voltage out/voltage in).
(c) For this amplifier, the input and output resistances are high. Since R_{in} is high, the input of the amplifier is voltage. Likewise, since R_{out} is high, the output of the amplifier is current. Therefore, this is a **transconductance amplifier** (current out/voltage in).
(d) For this amplifier, the output resistance is high. The input resistance is harder to classify. Resistances in the $k\Omega$ range could be considered either high or low depending on the other resistances in the circuit. Given no additional information, it is probably best to classify the

amplifier as a relatively poor **current amplifier** (current out/current in) and treat the input resistance as relatively low. The other possibility would be a poor **transconductance amplifier** (current out/voltage in). It largely depends on the output resistance of the circuit or sensor connected to the input of the amplifier. In either case, the amplifier should be redesigned with either a much lower or a much higher input impedance.

> **Lab Hint**: Of the different amplifier designs, the voltage amplifier (i.e., voltage-dependent voltage source) and the transresistance amplifier (i.e., current-dependent voltage source) are the most important as computers/microcontrollers take voltage signals at their input while sensors can output either voltage or currents depending on the physics of the sensor. The goal of the amplifier to increase the signal from the sensor and amplify it for analysis in the computer/microcontroller. Therefore, the output from the amplifier should normally be a voltage.

9.2 Ideal Op-Amp

9.2.1 Circuit Model for Op-Amp

The circuit diagram for the Op-Amp is shown in Fig. 9.2. The Op-Amp consists of two voltage inputs identified as the inverting and non-inverting inputs. In the circuit diagram, the inverting input is denoted by a negative sign and the non-inverting input is denoted by a positive sign. For the ideal Op-Amp the resistance between the inverting and non-inverting inputs is infinite. Therefore, no current can flow into these inputs. The output of the amplifier is also intended to provide a voltage output and has ideally no output resistance. Since the Op-Amp is an active device, it requires external power as indicated by the positive and negative power supply terminals in the diagram. The output voltage for the Op-Amp cannot exceed the voltages supplied to its terminals in either the positive or negative direction.

The current and voltage relationships for the terminals of the Op-Amp are described in more detail in Fig. 9.3. In this diagram, the Op-Amp is biased with $+V_{CC}$ and $-V_{CC}$ at the positive and negative power supply terminals. These terminals also can provide current, i_{c+} and i_{c-}, which can flow from the output terminal, i_o. As was stated previously,

$$i_p \cong i_n \cong 0 \tag{9.1}$$

Therefore,

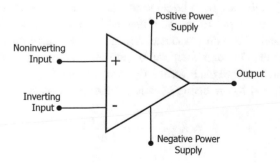

Fig. 9.2 Circuit diagram for Op-Amp

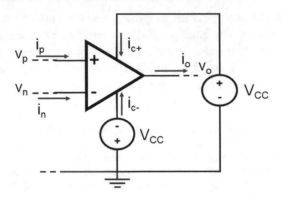

Fig. 9.3 Current and Voltage relationships for the Op-Amp

$$i_o = i_{c+} + i_{c-} \tag{9.2}$$

One common mistake when analyzing circuits is to assume that $i_o = 0$ because i_p and i_n must be zero. This mistake is more common when the power supply terminals used to bias the Op-Amp are not explicitly drawn. However, the output current, i_o, can and does come from the power supply terminals.

The output voltage, v_o, is technically given by

$$v_o \cong \begin{cases} -V_{CC} & A_v(v_p - v_n) < -V_{CC} \\ A_v(v_p - v_n) & |A_v(v_p - v_n)| \leq V_{CC} \\ +V_{CC} & A_v(v_p - v_n) > V_{CC} \end{cases} \tag{9.3}$$

where the gain term A_v is very large. Therefore, the output voltage will saturate very close to the bias/power supply voltages unless v_p is very close to v_n. When using the Op-Amp, negative feedback is used to make $v_p \cong v_n$ by taking some fraction of the output, v_o, and applying it to the inverting Op-Amp input (i.e., v_n). This effectively subtracts some of the output from the input forcing v_p and v_n to be close and thus preventing v_o from saturating at either $\pm V_{CC}$. The feedback is best illustrated when the Op-Amp is in the non-inverting amplifier configuration.

9.2.2 Non-inverting Amplifier Configuration

The circuit diagram for the non-inverting amplifier configuration is shown in Fig. 9.4. As is common when analyzing Op-Amp circuits, the power supply terminals used to bias the Op-Amp are not explicitly shown even though they must be present. The source or sensor producing the signal that needs to be amplified at the input to the amplifier is expressed as its Thevenin equivalent with Thevenin voltage, v_{Th}, and Thevenin resistance, R_{Th}. However, since no current can flow into the non-inverting terminal of the Op-Amp, there can be no voltage drop across R_{Th}. Therefore, $v_p = v_{Th}$. Also, since no current can flow into the inverting Op-Amp terminal, the voltage v_n is just a fraction of v_o where the resistors R_1 and R_2 are acting as voltage dividers. Hence,

$$v_n = v_o \frac{R_1}{R_1 + R_2} = \frac{v_o}{1 + R_2/R_1} \tag{9.4}$$

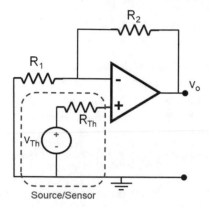

Fig. 9.4 Non-inverting amplifier configuration with an Op-Amp

Therefore,

$$v_o = A_v(v_p - v_n) = A_v(v_{Th} - v_n) = A_v\left(v_{Th} - \frac{v_o}{(1 + R_2/R_1)}\right)$$

$$\Rightarrow v_o + v_o\left(\frac{A_v}{(1 + R_2/R_1)}\right) = A_v v_{Th} \tag{9.5}$$

$$\Rightarrow v_o = \frac{A_v v_{Th}}{1 + \left(\frac{A_v}{(1 + R_2/R_1)}\right)}$$

However, since A_v is very large, $\left(\frac{A_v}{(1 + R_2/R_1)}\right) \gg 1$. Hence,

$$v_o \cong \frac{A_v v_{Th}}{\left(\frac{A_v}{(1 + R_2/R_1)}\right)} = (1 + R_2/R_1)v_{Th} \tag{9.6}$$

Therefore, the output of the amplifier is the same as the input only scaled by a factor of $(1 + R_2/R_1)$.

$$\text{Gain} = \left(1 + \frac{R_2}{R_1}\right) \text{V/V} \tag{9.7}$$

This same result can also be achieved even faster by recognizing that because of the large gain of the Op-Amp, $v_p \cong v_n$. Using this approximation, we have

$$v_p = v_{Th} = v_n = v_o \frac{R_1}{R_1 + R_2} = \frac{v_o}{1 + R_2/R_1}$$

$$\Rightarrow v_{Th} = \frac{v_o}{1 + R_2/R_1} \Rightarrow v_o = (1 + R_2/R_1)v_{Th} \tag{9.8}$$

Before concluding, we need to confirm that the non-inverting amplifier configuration is, in fact, a voltage amplifier. Simply because the input and output of the amplifier were expressed as voltages does not mean that the amplifier itself is a voltage amplifier. Instead, we must find the input and output resistance of the amplifier as was discussed previously. The input resistance to the amplifier would be the resistance seen by the Thevenin equivalent source connected at the input of the circuit. This would be the resistance seen looking into the non-inverting amplifier terminal as given by

$$R_{\text{in}} = \frac{v_\text{p}}{i_\text{p}} \tag{9.9}$$

This resistance is very high (ideally infinite) as the current, i_p, is very small (ideally zero). Therefore, the amplifier configuration is expecting a voltage input. Similarly, the output resistance can be found mathematically by connecting the output of the Op-Amp to ground and solving for the resulting short-circuit current, $(i_\text{o})_{\text{short circuit}}$. For the non-inverting Op-Amp configuration, this would be given by

$$(i_\text{o})_{\text{short circuit}} = \frac{v_\text{o}}{R_{\text{out\_Op-Amp}}} \Rightarrow R_{\text{out}} = R_{\text{out\_Op-Amp}} \tag{9.10}$$

Since the output resistance of the Op-Amp is ideally very small, the output resistance of the amplifier configuration is ideally very small. Therefore, the amplifier is outputting a voltage. Since the amplifier configuration has both a voltage input and a voltage output, the non-inverting amplifier configuration is a voltage amplifier.

Lab Hint: One should never directly connect the output of an Op-Amp to ground as this would cause excessive current flow that would damage the Op-Amp. Instead, in the lab, the Op-Amp should be connected to a known resistance that would not allow the Op-Amp to exceed its maximum allowable current. The internal resistance of the Op-Amp could then be calculated from the resulting voltage and current values. The maximum current for an Op-Amp should be given in the specifications for the Op-Amp provided by the manufacturer. For most Op-Amps, the output resistance will be in the range of 10–50 Ω. However, one should never design a circuit that is strongly dependent on the output resistance of the Op-Amp as this resistance can vary drastically for Op-Amps with the same part number and will typically have a strong dependence on temperature as well.

Example 9.2 Design a voltage amplifier with a gain of 100 V/V using an Op-Amp.

Solution The non-inverting amplifier configuration is a voltage amplifier. The gain of the amplifier is given by $(1 + R_2/R_1)$. Therefore, $R_2 = 99$ kΩ and $R_1 = 1$ kΩ will give us a gain of 100 V/V as shown below.

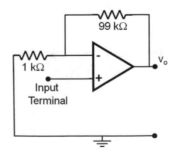

9.2.3 Inverting Amplifier Configuration

Another very common amplifier configuration using the Op-Amp is the inverting configuration shown in Fig. 9.5. Once again, the source connected to the input of the Op-Amp has been modeled by its Thevenin equivalent. Hence, R_{Th} is the resistance of the source or sensor producing the signal that needs to be amplified and not part of the amplifier configuration. However, it can still impact the gain of the circuit.

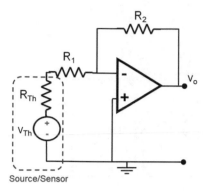

Fig. 9.5 Inverting amplifier configuration with an Op-Amp

The gain for the inverting amplifier configuration can be found by utilizing $v_p \cong v_n$ and summing the currents at the inverting terminal of the Op-Amp. The non-inverting terminal is connected directly to the ground node, so $v_p \cong v_n = 0$ V. Therefore, the currents flowing into the node at the inverting terminal are given by

$$
\begin{aligned}
\text{Current in } R_1 &= \frac{v_{Th} - 0}{R_{Th} + R_1} \\
\text{Current in } R_2 &= \frac{v_o - 0}{R_2} \\
i_n &= 0
\end{aligned}
\tag{9.11}
$$

Summing these currents gives

$$
\frac{v_{Th} - 0}{R_{Th} + R_1} + \frac{v_o - 0}{R_2} = 0 \Rightarrow v_o = -v_{Th} \frac{R_2}{R_{Th} + R_1}
\tag{9.12}
$$

Therefore, the gain can be expressed as

$$
\text{Gain} = \frac{v_o}{v_{Th}} = -\frac{R_2}{R_{Th} + R_1} \ \text{V/V}
\tag{9.13}
$$

Assuming $R_{Th} \ll R_1$, Eq. (9.13) becomes

$$
\text{Gain} = \frac{v_o}{v_{Th}} \cong -\frac{R_2}{R_1} \ \text{V/V}
\tag{9.14}
$$

Equation (9.14) is the most common expression for the gain of the inverting amplifier, but that does not make the inverting configuration a true voltage amplifier. The input impedance for the inverting amplifier is R_1 as this is the resistance seen by the source connected to the input of the amplifier. Minimizing R_1 will maximize the gain of the amplifier. Therefore, R_1, the input impedance, should be a relatively small value when the amplifier is designed properly. Hence, the inverting amplifier configuration is expecting a current input and is thus best described as a transresistance amplifier.

The inverting configuration redrawn as a transresistance amplifier is shown in Fig. 9.6. Since the Norton current, i_N, is related to the Thevenin voltage, v_{Th}, by $i_N = v_{Th}/R_{Th}$, the gain of the transresistance amplifier can be found from Eq. (9.12) and is given by

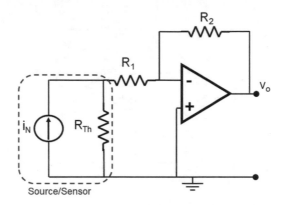

Fig. 9.6 Inverting amplifier configuration redrawn as a transresistance amplifier

$$v_o = -i_n \frac{R_{Th} R_2}{R_{Th} + R_1} \Rightarrow \text{Gain} = \frac{v_o}{i_N} = -\frac{R_{Th} R_2}{R_{Th} + R_1} \text{ V/A} \tag{9.15}$$

Assuming $R_{Th} \gg R_1$, as would be the case for a source/sensor designed to output a current gives

$$\text{Gain} = \frac{v_o}{i_N} = -R_2 \text{ V/A} \tag{9.16}$$

Example 9.3 A photodiode sensor has a Thevenin resistance of 10 MΩ and outputs a current of 25 μA when exposed to a light source. Design a transresistance amplifier to amplify this signal level so that $|v_o| = 0.5$ V using an Op-Amp.

Solution The desired amplifier must have a gain of

$$|\text{Gain}| = \left|\frac{v_o}{i_N}\right| = \frac{0.5 \text{ V}}{25 \text{ μA}} = 20000 \text{ V/A}$$

If we use the inverting configuration and select a value for R_1 that is much less than the 10 MΩ Thevenin resistance (i.e., $R_1 = 0$ Ω), then an R_2 value of 20 kΩ would give us the desired gain value. The final amplifier design would then be given by

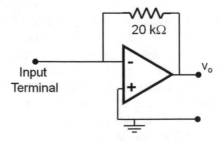

9.2.4 Instrumentation/Differential Amplifier

The instrumentation or differential amplifier is a very useful amplifier in many measurement systems as it amplifies the difference between two signals. This would allow any noise that is common to both signals to be eliminated. The circuit diagram for the instrumentation amplifier is shown in Fig. 9.7. Node voltages and branch currents have also been labeled to facilitate our analysis.

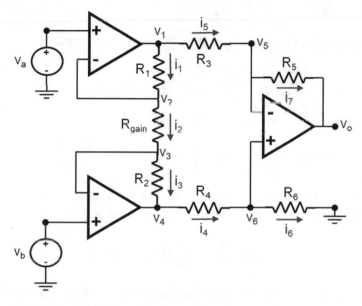

Fig. 9.7 Instrumentation/differential amplifier constructed using three Op-Amps

For this amplifier configuration, voltage v_a and v_b have been applied to the non-inverting terminals of the first two Op-Amps. Since the gain of the Op-Amps is assumed to be large, the voltages at the inverting and non-inverting terminals must be approximately the same due to the feedback. Hence, $v_2 = v_a$ and $v_3 = v_b$. With v_2 and v_3 known, we can find the current i_2 as $i_2 = (v_2 - v_3)/R_{gain} = (v_a - v_b)/R_{gain}$. Also, since no current can flow into the inverting terminal of the Op-Amp, $i_1 = i_2 = i_3$. With these currents know, we can now find the voltages v_1 and v_4 as

$$i_1 = (v_1 - v_a)/R_1 = (v_a - v_b)/R_{gain}$$
$$\Rightarrow v_1 = v_a + \frac{R_1}{R_{gain}}(v_a - v_b)$$
$$i_3 = (v_b - v_4)/R_2 = (v_a - v_b)/R_{gain} \tag{9.17}$$
$$\Rightarrow v_4 = v_b - \frac{R_2}{R_{gain}}(v_a - v_b)$$

Also, since no current can flow into the non-inverting terminal of the final Op-Amp, $i_4 = i_6$. Thus

$$i_4 = \frac{v_4 - v_6}{R_4} = i_6 = \frac{v_6}{R_6} \Rightarrow v_6\left(\frac{1}{R_6} + \frac{1}{R_4}\right) = \frac{v_4}{R_4}$$
$$\Rightarrow v_6 = v_4\left(\frac{R_6}{R_4 + R_6}\right) = \left(v_b - \frac{R_2}{R_{gain}}(v_a - v_b)\right)\left(\frac{R_6}{R_4 + R_6}\right) \tag{9.18}$$

Due to the high gain of the Op-Amp and feedback, $v_5 = v_6$. Also, since no current can flow into the inverting terminal, $i_5 = i_7$. Therefore, the final output voltage is given by

$$\frac{v_1 - v_5}{R_3} = \frac{v_5 - v_o}{R_5} \Rightarrow v_o = v_5\left(\frac{R_5}{R_3} + 1\right) - v_1\frac{R_5}{R_3}$$

$$v_o = \left(v_b - \frac{R_2}{R_{\text{gain}}}(v_a - v_b)\right)\left(\frac{1}{1 + \frac{R_4}{R_6}}\right)\left(\frac{R_5}{R_3} + 1\right) - \left(v_a + \frac{R_1}{R_{\text{gain}}}(v_a - v_b)\right)\left(\frac{R_5}{R_3}\right)$$

(9.19)

This output can be simplified significantly if $R_5 = R_3$ and $R_4 = R_6$ yielding

$$v_o = \left(v_b - \frac{R_2}{R_{\text{gain}}}(v_a - v_b)\right) - \left(v_a + \frac{R_1}{R_{\text{gain}}}(v_a - v_b)\right)$$

$$= v_b + \frac{R_2}{R_{\text{gain}}}(v_b - v_a) - v_a + \frac{R_1}{R_{\text{gain}}}(v_b - v_a)$$

(9.20)

$$= (v_b - v_a) + \left(\frac{R_2}{R_{\text{gain}}} + \frac{R_1}{R_{\text{gain}}}\right)(v_b - v_a) = \left(1 + \frac{R_1 + R_2}{R_{\text{gain}}}\right)(v_b - v_a)$$

The most common configuration would be to have $R_1 = R_2 = R_3 = R_4 = R_5 = R_6 = R$ as shown in Fig. 9.8. Under this condition, the output voltage would simply be given by

$$v_o = \left(1 + \frac{2R}{R_{\text{gain}}}\right)(v_b - v_a)$$

(9.21)

As stated previously, the goal of the instrumentation/differential amplifier is to amplify the difference between two signals while rejecting any signal component that the two signals have in common. In general, the output of a differential amplifier can be written as

$$v_o = G_d v_{\text{dm}} + G_{\text{cm}} v_{\text{cm}} + V_{\text{DC\_offset}}$$

(9.22)

where G_d is the differential mode gain, v_{dm} is the differential mode voltage, G_{cm} is the common mode gain, v_{cm} is the common mode voltage, and $V_{\text{DC\_offset}}$ is a DC offset voltage that can arise from using non-ideal Op-Amps. The differential mode and common mode voltages are given by

$$v_{\text{dm}} = (v_b - v_a)$$

$$v_{\text{cm}} = (v_b + v_a)/2$$

(9.23)

For the ideal instrumentation/differential amplifier shown in Fig. 9.8, the differential mode gain is given by $(1 + 2R/R_{\text{gain}})$ and the common mode gain is zero. However, if the resistors are not perfectly matched, a finite common mode gain can occur. Often the quality of the differential amplifier can be quantified by the Common Mode Rejection Ratio (CMRR) defined as

$$\text{CMRR} = 20 \cdot \log_{10}\left(\frac{|G_d|}{|G_{\text{cm}}|}\right) \text{ dB}$$

(9.24)

Ideally, the CMRR should be on the order of 100 dB for most applications.

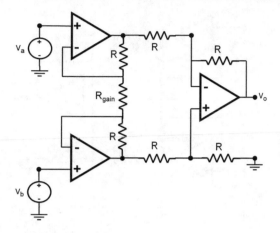

Fig. 9.8 Most common instrumentation/differential amplifier configuration

Lab Hint: Finding six resistors with exactly the same resistance values can be challenging given the allowable tolerances for most resistors. The key is to focus on matching the most critical resistors. For the instrumentation/differential amplifier shown in Fig. 9.7, this means having R_5 and R_3 closely matched as well as R_4 and R_6 closely matched. As long as each of these pairs is close to each other, the instrumentation amplifier will still effectively amplify the difference of the signals and the CMRR will be high.

Define dB: The CMRR in Eq. (9.24) is expressed in decibels or dB. Decibels are a common "unit" used when ratios are found. Examples include finding the voltage, current, or power gain of an amplifier or comparing the gains of different amplifiers. Converting a ratio to dB is done by taking $10 \log_{10}$ of the ratio of power quantities or $20 \log_{10}$ of the ratio of amplitude quantities. For example,

$$\#dB = \begin{cases} 10 \log_{10}\left(\frac{P_{out}}{P_{in}}\right) \\ 20 \log_{10}\left(\frac{v_{out}}{v_{in}}\right) \\ 20 \log_{10}\left(\frac{i_{out}}{i_{in}}\right) \end{cases}$$

Example 9.4 Design an instrumentation amplifier to have a differential mode gain of 21.

Solution The differential mode gain of the instrumentation amplifier is given by $\left(1 + 2R/R_{gain}\right)$. Therefore, if $R = 10 \text{ k}\Omega$ and $R_{gain} = 1 \text{ k}\Omega$, the amplifier will have a differential gain of 21.

Example 9.5 Now assume that one of the resistors from Example 9.4 is off in its resistance value by 10% (i.e., $R_5 = 11 \text{ k}\Omega$) as shown in Fig. 9.9. What is the differential mode gain, G_d, the common mode gain, G_{cm}, and the common mode rejection ratio, CMRR, for this amplifier?

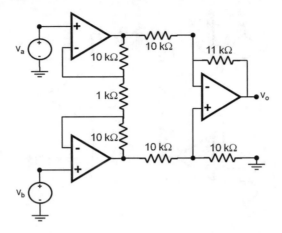

Fig. 9.9 Instrumentation amplifier for Example 9.5

Solution The output voltage for arbitrary resistor values was found in Eq. (9.19) and is given by

$$
v_o = \left(v_b - \frac{R_2}{R_{\text{gain}}} (v_a - v_b) \right) \left(\frac{1 + \frac{R_5}{R_3}}{1 + \frac{R_4}{R_6}} \right)
$$
$$
- \left(v_a + \frac{R_1}{R_{\text{gain}}} (v_a - v_b) \right) \left(\frac{R_5}{R_3} \right)
$$

Substituting in for our resistor values gives

$$
v_o = \left(v_b - \frac{10 \text{ k}\Omega}{1 \text{ k}\Omega} (v_a - v_b) \right) \left(\frac{1 + \frac{11 \text{ k}\Omega}{10 \text{ k}\Omega}}{1 + \frac{10 \text{ k}\Omega}{10 \text{ k}\Omega}} \right) - \left(v_a + \frac{10 \text{ k}\Omega}{1 \text{ k}\Omega} (v_a - v_b) \right) \left(\frac{11 \text{ k}\Omega}{10 \text{ k}\Omega} \right)
$$
$$
= 1.05(v_b + 10(v_b - v_a)) - 1.1(v_a - 10(v_b - v_a))
$$
$$
= 22.55 v_b - 22.6 v_a
$$

We now need to convert from v_a and v_b to v_{dm} and v_{cm}.

$$
\frac{2v_{\text{cm}} + v_{\text{dm}}}{2} = \frac{v_b + v_a + v_b - v_a}{2} = v_b
$$
$$
\frac{2v_{\text{cm}} - v_{\text{dm}}}{2} = \frac{v_b + v_a - v_b + v_a}{2} = v_a
$$

Therefore, v_o can be written as

$$
v_o = 22.55 \left(\frac{2v_{\text{cm}} + v_{\text{dm}}}{2} \right) - 22.6 \left(\frac{2v_{\text{cm}} - v_{\text{dm}}}{2} \right)
$$
$$
= (22.55 v_{\text{cm}} + 11.275 v_{\text{dm}}) + (-22.6 v_{\text{cm}} + 11.3 v_{\text{dm}})
$$
$$
= 22.575 v_{\text{dm}} - 0.05 v_{\text{cm}}
$$

Thus

$$
G_d = 22.575 \text{ V/V}
$$
$$
G_{\text{cm}} = -0.05 \text{ V/V}
$$

The CMRR is thus given by

$$\text{CMRR} = 20 \cdot \log_{10}\left(\frac{|G_d|}{|G_{cm}|}\right) \text{ dB} = 20 \cdot \log_{10}\left(\frac{22.575}{0.05}\right) = 53.0932 \text{ dB}$$

9.2.5 Generalized Op-Amp Analysis Assuming Ideal Op-Amps

In the previous sections, we considered some of the Op-Amp circuit topologies that are commonly encountered in measurement systems. However, many more topologies have been designed over the years, and new circuits can always be introduced. However, the ideal Op-Amp rules utilized previously can still be used to analyze all of these circuits. If you recall, for the ideal Op-Amp:

1. No current can flow in the inverting or non-inverting input ($i_p \cong i_n \cong 0$).
2. The voltages at the inverting and non-inverting input must be the same due to negative feedback ($v_p \cong v_n$).

The following examples illustrate how these rules can be used to find the voltages and currents for many different circuits consisting of multiple Op-Amps.

Example 9.6 Find i_o for the circuit shown in Fig. 9.10 assuming ideal Op-Amps.

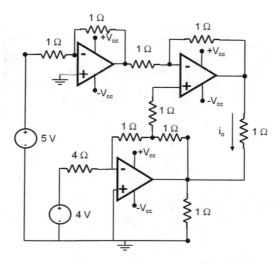

Fig. 9.10 Op-Amp circuit for Example 9.6

Solution To begin, we find the voltages at the inputs to the first two Op-Amps. Since $v_p \cong v_n$, these voltages must be zero. With the voltages known, we can find the currents flowing in the 1 Ω and 4 Ω resistors as shown. Also, since $i_p \cong i_n \cong 0$, this same current must also flow through the feedback resistor for each Op-Amp.

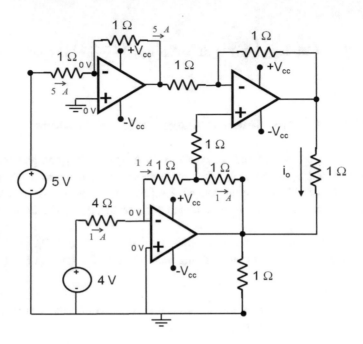

With the currents known and with $i_p \cong i_n \cong 0$, the voltages at the next set of nodes can be calculated.

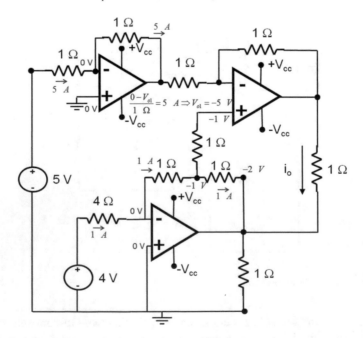

With this new set of voltages known, we can find the currents associated with the third Op-Amp. Lastly, we can find the output voltage of the last Op-Amp.

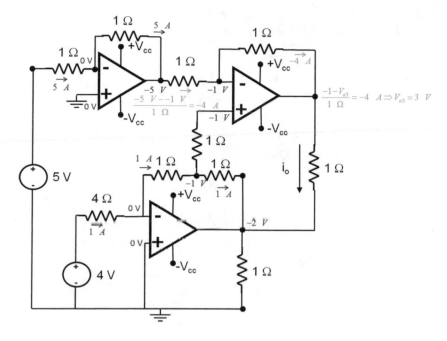

The current, i_o, is then given by $i_o = \frac{3\,\text{V} - 2\,\text{V}}{1\,\Omega} = 5$ A

Example 9.7 Find i_o for the circuit shown in Fig. 9.11 assuming ideal Op-Amps.

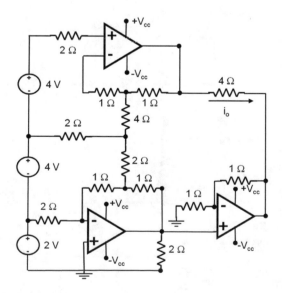

Fig. 9.11 Op-Amp circuit for Example 9.7

Solution Begin by finding the voltages at the left portion of the circuit. Also, since $i_p \cong i_n \cong 0$, the 2 Ω and 1 Ω resistors associated with the top most Op-Amp have no currents flowing in them. As a result, there will be no voltage drop across these resistors.

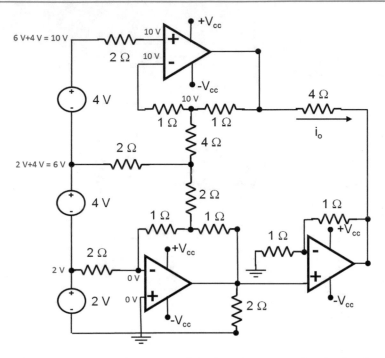

With the voltages known at these nodes, some of the branch currents for the Op-Amp on the lower left can be found along with the voltage between the two 1 Ω resistors.

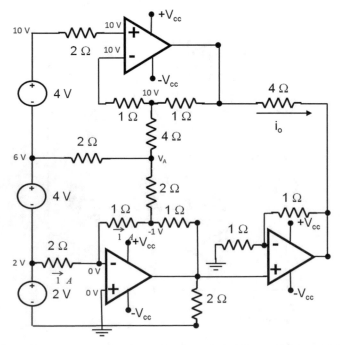

We can now use the node-voltage method to find the voltage V_A and the currents in the associated branches by using Kirchhoff's Current Law.

$$\frac{V_A - 6}{2\,\Omega} + \frac{V_A - -1}{2\,\Omega} + \frac{V_A - 10}{4\,\Omega} = 0 \Rightarrow 5V_A = 20 \Rightarrow V_A = 4\ \text{V}$$

$$\frac{V_A - 6}{2\,\Omega} = -1\ \text{A} \qquad \frac{V_A - -1}{2\,\Omega} = 2.5\ \text{A} \qquad \frac{V_A - 10}{4\,\Omega} = -1.5\ \text{A}$$

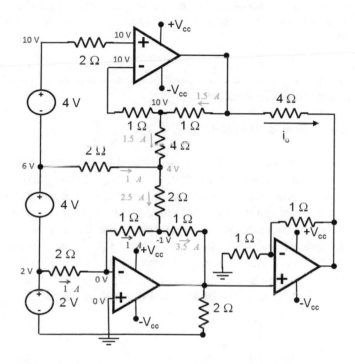

With the currents known, we can find additional node voltages at the output nodes of the first two Op-Amps. Also, the final Op-Amp is in the Non-inverting configuration, and therefore, its output is just 2x its input.

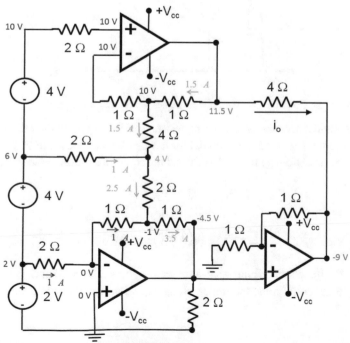

The current, i_o, is then just given by $i_o = \frac{11.5\,\text{V} - -9\,\text{V}}{4\,\Omega} = 5.125$ A.

> **Hint:** When solving problems with multiple Op-Amps it is generally a very bad idea to attempt to solve for all of the currents and voltages at the same time with one massive system of equations. Instead, one should walk through the problem solving for the unknowns as they are encountered. One should use what is already known to solve for voltages in nearby nodes and currents in adjacent branches. In other words, solve the problem as you would solve a puzzle.

9.3 Non-ideal Op-Amp Characteristics

9.3.1 Finite Impedances and Finite Gain

In the previous section, we analyzed circuits assuming the Op-Amps were ideal. Since most Op-Amps are designed to be very close to the ideal, the ideal Op-Amp approximations are normally sufficient when analyzing most circuits. However, there are non-idealities that can impact circuit performance and may limit the choice of Op-Amp for a specific application or impact the choice of resistors selected to achieve the desired gain. For example, unlike the ideal case, real Op-Amps have finite impedances and finite gains as shown in Fig. 9.12. For real Op-Amps, the input resistance, R_i, is on the order of $10^6\ \Omega$ (or higher) while the output resistance, R_o, is normally in the range from 10 to 50 Ω. The open-loop gain of the Op-Amp, A, is also typically on the order of 10^6 or higher. In the ideal case, R_i would be infinite, R_o would be zero, and A would be infinite.

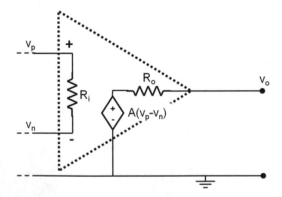

Fig. 9.12 Real model for an Op-Amp showing finite resistance values and finite gain

To illustrate the potential impact of the finite resistances and gain, let's refind the gain for the non-inverting configuration using the real model for the Op-Amp as shown in Fig. 9.13. To solve this circuit, we can use the node voltage method to write three equations for the node voltages v_p, v_n, and v_o.

$$\frac{v_n - v_p}{R_i} + \frac{v_n}{R_1} + \frac{v_n - v_o}{R_2} = 0$$

$$\frac{v_n - v_p}{R_i} = \frac{v_p - V_{Th}}{R_{Th}} \tag{9.25}$$

$$\frac{v_n - v_o}{R_2} = \frac{v_o - A(v_p - v_n)}{R_o} \Rightarrow v_o\left(1 + \frac{R_o}{R_2}\right) = A\left(v_p - v_n\right) + \frac{v_n R_o}{R_2}$$

Let's first focus on the impact of a finite open-loop gain, A, by assuming that R_i is very large and R_o is very small. Under these conditions, the node-voltage equations in (9.25) become

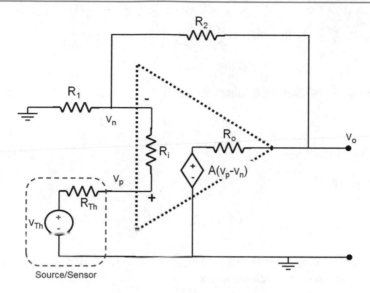

Fig. 9.13 Real model for an Op-Amp showing finite resistance values and finite gain in the noninverting configuration

$$\frac{v_n}{R_1} + \frac{v_n - v_o}{R_2} = 0 \Rightarrow v_o = v_n\left(1 + \frac{R_2}{R_1}\right)$$
$$0 = \frac{v_p - V_{Th}}{R_{Th}} \Rightarrow v_p = V_{Th} \tag{9.26}$$
$$v_o = A\left(v_p - v_n\right)$$

Therefore,

$$v_o = A\left(v_p - v_n\right) = A\left(V_{Th} - \frac{v_o}{\left(1 + \frac{R_2}{R_1}\right)}\right)$$

$$v_o + A\frac{v_o}{\left(1 + \frac{R_2}{R_1}\right)} = AV_{Th} \Rightarrow \text{Gain} = A_{CL} = \frac{v_o}{V_{Th}} = \frac{A}{1 + A\frac{1}{\left(1 + \frac{R_2}{R_1}\right)}} = \frac{A}{1 + A\beta} \tag{9.27}$$

where $\beta = 1/(1 + R_2/R_1)$ is the feedback factor and corresponds to the fraction of the output that is subtracted from the input, and A_{CL} is the closed-loop gain for the amplifier. For large values of open-loop gain, A, the amplifier gain, A_{CL}, is just given by $(1/\beta)$. However, as the open-loop gain is reduced, the amplifier gain is also reduced.

Example 9.8 An Op-Amp circuit in the non-inverting amplifier configuration has an infinite input resistance, R_i, and a zero output resistance, R_o. The feedback resistors used to set the gain of the amplifier are given by $R_2 = 49\ \text{k}\Omega$ and $R_1 = 1\ \text{k}\Omega$. The Thevenin resistance for the source at the input, R_{Th}, is $0\ \Omega$. Find the gain of the non-inverting amplifier as the open-loop gain, A, varies from 10^2 to 10^6 V/V.

Solution The gain of the non-inverting amplifier configuration is given by

$$\text{Gain} = A_{\text{CL}} = \frac{v_{\text{o}}}{V_{\text{Th}}} = \frac{A}{1 + A \frac{1}{\left(1 + \frac{R_2}{R_1}\right)}} = \frac{A}{1 + A\beta}$$

Therefore, we need to first find the feedback factor, β.

$$\beta = \frac{1}{\left(1 + \frac{R_2}{R_1}\right)} = \frac{1}{\left(1 + \frac{49 \text{ k}\Omega}{1 \text{ k}\Omega}\right)} = \frac{1}{50}$$

The gain is then given by

$$\text{Gain} = A_{\text{CL}} = \frac{A}{1 + \frac{A}{50}}$$

Therefore, if A = 10^2 V/V, the gain is 33.333 V/V, and if A = 10^6 V/V the gain is 49.9975 V/V. The gain plotted as a function of A is shown below.

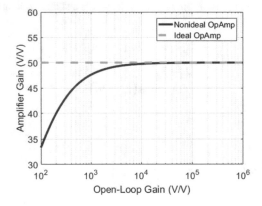

The impact of finite resistance on the gain of the amplifier can be found by directly solving for the three node voltage equations given in (9.25). Solving these equations and assuming that R_{o} is smaller than R_{i} will give a value for the gain of

$$\text{Gain} = A_{\text{CL}} = \frac{v_{\text{o}}}{V_{\text{Th}}} = \frac{A'}{1 + A' \frac{1}{\left(1 + \frac{R_2}{R_1}\right)}} = \frac{A'}{1 + A'\beta} \tag{9.28}$$

where A' is the effective open-loop gain given by

$$A' = A \left(\frac{R_i}{R_i + R_{\text{Th}} + \frac{R_2 R_1}{R_1 + R_2}}\right) \left(\frac{1}{1 + \frac{R_{\text{o}}}{R_1 + R_2}}\right) \tag{9.29}$$

Therefore, finite values for the input and output resistances will effectively lower the open loop gain of the Op-Amp circuit.

Example 9.9 An Op-Amp circuit in the non-inverting amplifier configuration has an input resistance, R_i, of 100 kΩ and an output resistance, R_o, of 100 Ω. The feedback resistors used to set the gain of the amplifier are given by $R_2 = 900$ MΩ and $R_1 = 100$ MΩ. The Thevenin resistance for the source at the input, R_{Th}, is 0 Ω. Find the gain of the non-inverting amplifier if the open-loop gain, A, is 10^5 V/V.

Solution The first step is to find the effective open-loop gain given by

$$A' = A \left(\frac{R_i}{R_i + R_{Th} + \frac{R_2 R_1}{R_1 + R_2}} \right) \left(\frac{1}{1 + \frac{R_o}{R_1 + R_2}} \right) = 10^5 \left(\frac{10^5}{10^5 + \frac{(900 \times 10^6)(100 \times 10^6)}{1000 \times 10^6}} \right) \left(\frac{1}{1 + \frac{100}{1000 \times 10^6}} \right)$$

$$\cong 10^5 \left(\frac{1}{1 + 900} \right) = 110.9878 \text{ V/V}$$

Thus, the amplifier gain is given by

$$\text{Gain} = A_{CL} = \frac{v_o}{V_{Th}} = \frac{A'}{1 + A' \frac{1}{\left(1 + \frac{R_2}{R_1} \right)}} = \frac{A'}{1 + A'\beta} = \frac{A'}{1 + \frac{A'}{10}} = 9.1735 \text{ V/V}$$

As a comparison, if the Op-Amp were ideal, the gain would be 10 V/V.

Example 9.10 Derive an expression relating the amplifier gain to the open-loop gain for an Op-Amp in the inverting configuration as shown in Fig. 9.14. Assume that the Op-Amp's input resistance, R_i, is infinite, the Op-Amp's output resistance, R_o, is zero.

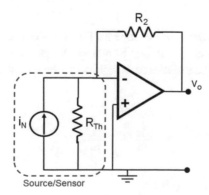

Fig. 9.14 Op-Amp in the inverting configuration for Example 9.10

Solution The first step is to draw the circuit with the open-loop gain expressed as a voltage-dependent voltage source as shown.

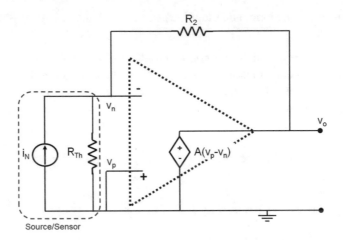

Source/Sensor

We can then write node-voltage equations for v_n to solve the circuit. The voltage v_p must be zero as it is connected to the ground node.

$$i_N = \frac{v_n}{R_{Th}} + \frac{v_n - A(0 - v_n)}{R_2}$$

$$= v_n \left(\frac{1}{R_{Th}} + \frac{(1+A)}{R_2} \right)$$

$$\Rightarrow v_o = -Av_n = -A \frac{i_N}{\left(\frac{1}{R_{Th}} + \frac{(1+A)}{R_2} \right)} = -A \frac{R_2 i_N}{\left(A + 1 + \frac{R_2}{R_{Th}} \right)}$$

$$\text{Gain} = A_{CL} = \frac{v_o}{i_N} = \frac{-AR_2}{\left(A + 1 + \frac{R_2}{R_{Th}} \right)} \ \text{V/A}$$

As can be seen from Example 9.10, a finite value for the open-loop gain, A, will also impact the gain of the amplifier in the inverting configuration. Similarly, the gain of the inverting amplifier will also be impacted by finite value for the resistances. Specifically, the gain for the circuit shown in Fig. 9.14 would be given by

$$\text{Gain} = A_{CL} = \frac{v_o}{i_N} = \frac{-A \left(\frac{R_2}{R_o + R_2} \right)}{\frac{1}{R_{Th}} + \frac{1}{R_i} + \frac{1}{R_2} + \left(\frac{A}{R_o + R_2} \right)} \tag{9.30}$$

where R_o and R_i are identified in Fig. 9.12.

9.3.2 Bias Currents and Offset Voltage

As was mentioned at the beginning of this Chapter, Op-Amps are active devices consisting of multiple transistors. To operate properly, bias voltages of $\pm V_{CC}$ need to be applied to the power supply terminals. These bias voltages place the transistors internal to the Op-Amp in the appropriate operating mode for the amplifier. However, when biasing these internal transistors, small currents are

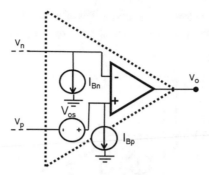

Fig. 9.15 Model of Op-Amp including bias currents and offset voltage

sometimes required at the inverting and/or non-inverting inputs. For most applications, these small bias currents can be ignored. However, there are cases when these currents can impact the operation of the Op-Amp. In addition to the currents, a DC offset voltage can also be generated.

Figure 9.15 shows the circuit model for the Op-Amp with the bias currents and offset voltages included. For most Op-Amps, the bias currents are on the order of nA while the offset voltage is on the order of ± 1 to ± 5 mV. The model also includes an ideal Op-Amp shown inside of the dotted triangle along with the DC voltage and current sources. The bias currents and offset voltages will produce DC voltages in the output of the amplifier that can mask the signal from a sensor of interest as is illustrated by the following examples.

Example 9.11 A non-inverting amplifier has been built using an Op-Amp as shown in Fig. 9.16. The Op-Amp has bias currents of $I_{Bn} = I_{Bp} = 500$ nA and an offset voltage of $V_{os} = 5$ mV. What is the voltage at the output, v_o, if the voltage at the input is $v_{in}(t) = 50 \sin(1000t)$ mV?

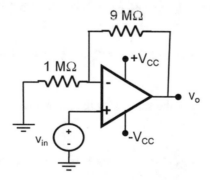

Fig. 9.16 Op-Amp in the circuit for Example 9.11

Solution This problem can be solved using superposition where the impact of the DC bias currents and offset voltage is solved separately from the AC signal. For the AC component, we have a simple non-inverting amplifier configuration. Therefore,

$$v_{o\_AC}(t) = v_{in}\left(1 + \frac{R_2}{R_1}\right) = 10 \cdot v_{in}$$
$$v_{o\_AC}(t) = 0.5 \sin(1000t) \text{ V}$$

For the DC output, we need to redraw the circuit with the sources included for the offset voltage and bias currents. We also eliminate v_{in} as we are now only solving for the DC component.

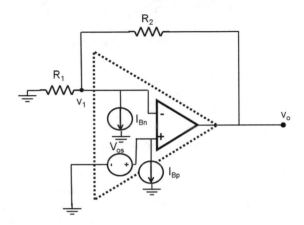

Assuming that the Op-Amp shown inside the dotted line is ideal, the voltage v_1 must be V_{os}. With v_1 known, we can sum the currents at the v_1 node to find v_o.

$$I_{\text{Bn}} + \frac{V_{\text{os}} - v_{\text{o\_DC}}}{R_2} + \frac{V_{\text{os}}}{R_1} = 0 \Rightarrow I_{\text{Bn}} R_2 + V_{\text{os}}\left(1 + \frac{R_2}{R_1}\right) = v_{\text{o\_DC}}$$

$$v_{\text{o\_DC}} = 4.55 \text{ V}$$

Therefore, the total output voltage is given by

$$v_\text{o}(t) = v_{\text{o\_AC}}(t) + v_{\text{o\_DC}} = 4.55 + 0.5 \sin(1000t) \text{ V}$$

Example 9.12 Repeat Example 9.11 with the 9 MΩ and 1 MΩ resistors replaced by 9 kΩ and 1 kΩ resistors respectively.

Solution The gain for the circuit is the same with the kΩ resistors. Therefore, the AC component will remain the same.

$$v_{\text{o\_AC}}(t) = v_{\text{in}}\left(1 + \frac{R_2}{R_1}\right) = 10 \cdot v_{\text{in}} = 0.5 \sin(1000t) \text{ V}$$

Hence, the only change will be in the DC component due to the bias currents and offset voltage.

$$I_{\text{Bn}} R_2 + V_{\text{os}}\left(1 + \frac{R_2}{R_1}\right) = v_{\text{o\_DC}} = 0.5045 \text{ V}$$

Therefore, the total output voltage is given by

$$v_\text{o}(t) = v_{\text{o\_AC}}(t) + v_{\text{o\_DC}} = 0.5045 + 0.5 \sin(1000t) \text{ V}$$

Hence, reducing the resistors used to set the gain significantly reduced the impact of the bias currents.

> **Lab Hint**: DC voltages in the output are not usually a concern in circuits as the DC component can be easily eliminated by a simple high pass filter. DC voltages are only a problem if they saturate the amplifiers (as will be discussed later in this chapter) OR if the signals being measured vary slowly with time.

Example 9.13 Use the output voltage, v_o, given in the circuits shown in Fig. 9.17 to find the bias currents and offset voltage for the Op-Amp. Assume that all three circuits are built with exactly the same Op-Amp.

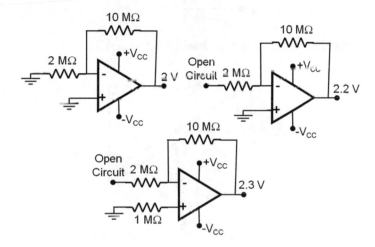

Fig. 9.17 Op-Amp circuits used to find bias currents and offset voltage in Example 9.13

Solution The first two circuits redrawn with sources representing the bias currents and offset voltages included are shown below. For both circuits, the voltage at the inverting terminal, v_1, is V_{os}. We can solve for I_{Bn} and V_{os} by summing the currents at v_1 for both circuits.

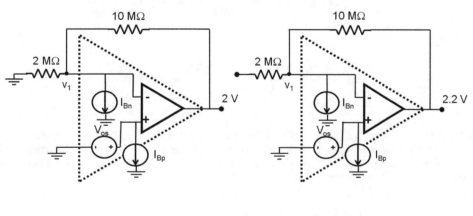

$$\frac{V_{os}}{2 \text{ M}\Omega} + I_{Bn} + \frac{V_{os} - 2 \text{ V}}{10 \text{ M}\Omega} = 0 \qquad I_{Bn} + \frac{V_{os} - 2.2 \text{ V}}{10 \text{ M}\Omega} = 0$$

Subtracting the two equations gives

$$\frac{V_{os}}{2 \text{ M}\Omega} + \frac{V_{os} - 2 \text{ V}}{10 \text{ M}\Omega} - \frac{V_{os} - 2.2 \text{ V}}{10 \text{ M}\Omega} = 0$$

$$\Rightarrow \frac{V_{os}}{2 \text{ M}\Omega} - \frac{2 \text{ V}}{10 \text{ M}\Omega} + \frac{2.2 \text{ V}}{10 \text{ M}\Omega} = 0$$

$$V_{os} = -40 \text{ mV}$$

Once, V_{os} is known, we can solve for I_{Bn}.

$$I_{Bn} = \frac{2.2 \text{ V} - V_{os}}{10 \text{ M}\Omega} = 224 \text{ nA}$$

With I_{Bn} and V_{os} known, we can then use the final circuit to find I_{Bp}.

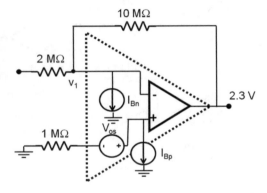

For this circuit, the voltage, v_1, is now given by $v_1 = -(1 \text{ M}\Omega)I_{Bp} + V_{os}$ as I_{Bp} forces a voltage drop across the 1 MΩ resistor. Summing the currents at v_1 now gives

$$I_{Bn} + \frac{-(1 \text{ M}\Omega)I_{Bp} + V_{os} - 2.3 \text{ V}}{10 \text{ M}\Omega} = 0$$

Therefore, I_{Bp} is given by

$$I_{Bp} = \frac{(10 \text{ M}\Omega)I_{Bn} + V_{os} - 2.3 \text{ V}}{(1 \text{ M}\Omega)} = -100 \text{ nA}$$

Lab Hint: The bias currents and offset voltages for an Op-Amp vary even for Op-Amps with the same part number from the same manufacturer. They can also vary with temperature. Therefore, measuring the bias currents and offset voltages for one Op-Amp will not give you the value for all Op-Amps or the value of a single Op-Amp under all usage conditions.

9.3.3 Saturation Limitations

While in many cases the bias currents and offset voltages have only a minimal impact on Op-Amp operation, amplifier saturation can easily distort a signal corrupting measured values. Since Op-Amps, and amplifiers in general, are active components, they need external power supplies to operate properly. For Op-Amps, the $+V_{CC}$ and $-V_{CC}$ applied at the positive and negative power supply terminals provide power to the Op-Amp as was described in Sect. 9.2.1. In addition to providing power, these terminals also set a limit on the maximum and minimum voltages that can be supplied by the Op-Amp. Specifically,

$$(-V_{CC} + \delta) \leq v_o \leq (+V_{CC} - \delta) \tag{9.31}$$

where v_o is the voltage at the output terminal of the Op-Amp and δ is a small offset voltage as v_o normally cannot reach exactly $\pm V_{CC}$. However, since δ is normally small, the saturation voltage is often approximated by $-V_{CC} \leq v_o \leq +V_{CC}$ as we did in Eq. (9.3).

Figure 9.18 illustrates the impact of saturation on a waveform as the saturation levels change. Initially, only the peaks of the waveform are impacted. However, as more of the waveform exceeds the saturation threshold, more of the waveform is clipped. For most amplifiers, the transition from the linear regime to the saturation regime is relatively gradual. Therefore, waveforms impacted by saturation in practice often appear "blunted" as shown in Fig. 9.19 rather than just clipped when they first begin to saturate. In either case, however, the waveform has been distorted by the amplifier and any measurement results gleaned from the waveform cannot be trusted.

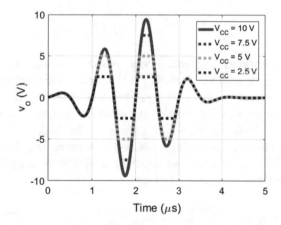

Fig. 9.18 Amplifier Saturation for different saturation levels

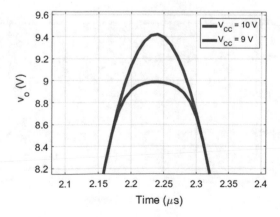

Fig. 9.19 "Blunting" of amplifier output when amplifier first begins to saturate

Example 9.14 If an Op-Amp in the non-inverting amplifier configuration has a gain of 10 and is biased with voltages of ±5 V, what is the largest and smallest input voltage allowed before the amplifier saturates?

Solution Since the Op-Amp will saturate at ±5 V, the voltage at the input of the amplifier must be

$$\frac{-5\text{ V}}{10} \le v_{\text{in}} \le \frac{+5\text{ V}}{10} \Rightarrow -0.5\text{ V} \le v_{\text{in}} \le +0.5\text{ V}$$

9.3.4 Op-Amp Frequency Response

Another non-ideality that must be considered when working with Op-Amp circuits is the frequency response of the Op-Amp(s). Cheaper Op-Amps, such as those typically used in undergraduate teaching labs, tend to have a very limited frequency response. As such, they cannot be used when working with higher frequency signals. The frequency limit results from the open-loop gain of the Op-Amp (discussed in Sect. 9.3.1) having a dependence on frequency. Many Op-Amps are designed to have a single dominate corner frequency such that

$$A(\omega) = \frac{A_{\text{o}}}{1 + j\frac{\omega}{\omega_{\text{o}}}} \tag{9.32}$$

where A_{o} is the low-frequency open-loop gain in V/V, ω is the operating frequency in rad/s, and ω_{o} is the corner frequency where the gain begins to drop dramatically when plotted in a log scale as shown in Fig. 9.20. In this plot, the gain is expressed in decibels (dB) as calculated from $20\log_{10}(|A(\omega)|)$. A factor of 20 is used instead of 10 since the gain is an amplitude quantity and not a power quantity.

As can be seen in Fig. 9.20, the open-loop gain is A_{o} for frequencies well below the corner frequency ω_{o}. However, after the corner frequency, the gain drops dramatically. In fact, if one goes from a frequency of 10–100 ω_{o}, the open-loop Op-Amp gain is reduced by 20 dB. A reduction of 20 dB corresponds to a reduction by a factor of 10 or

$$-20\text{ dB} = 20\log_{10}\left(\frac{\text{Gain}_2}{\text{Gain}_1}\right) \Rightarrow \log_{10}\left(\frac{\text{Gain}_2}{\text{Gain}_1}\right) = -1 \Rightarrow \frac{\text{Gain}_2}{\text{Gain}_1} = 10^{-1}$$

$$\Rightarrow \text{Gain}_2 = \frac{\text{Gain}_1}{10} \tag{9.33}$$

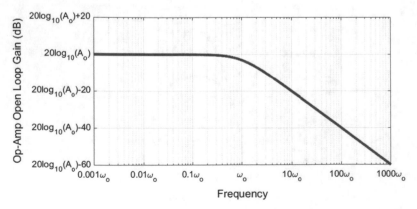

Fig. 9.20 Open-loop gain as a function of frequency for a typical Op-Amp

Therefore, the gain is reduced by 20 dB for every decade change in frequency past the corner frequency.

In addition to being illustrated graphically, the dependence of the gain on frequency can also be shown mathematically as

$$|A(\omega)| = \left| \frac{A_o}{1 + j\frac{\omega}{\omega_o}} \right| = \frac{A_o}{\sqrt{1 + \left(\frac{\omega}{\omega_o}\right)^2}}$$

$$20 \log_{10}(|A(\omega)|) = 20 \log_{10} \left(\frac{A_o}{\sqrt{1 + \left(\frac{\omega}{\omega_o}\right)^2}} \right) \tag{9.34}$$

$$= 20 \log_{10}(A_o) - 20 \log_{10} \left(\sqrt{1 + \left(\frac{\omega}{\omega_o}\right)^2} \right)$$

$$= 20 \log_{10}(A_o) - 10 \log_{10} \left(1 + \left(\frac{\omega}{\omega_o}\right)^2 \right)$$

If $\omega \ll \omega_o$, then Eq. (9.34) is approximately given by

$$20 \log_{10}(|A(\omega)|) = 20 \log_{10}(A_o) - 10 \log_{10} \left(1 + \overbrace{\left(\frac{\omega}{\omega_o}\right)^2}^{\text{Small}} \right) \tag{9.35}$$

$$\cong 20 \log_{10}(A_o) - 10 \log_{10}(1) = 20 \log_{10}(A_o)$$

Likewise, if $\omega \gg \omega_o$, then Eq. (9.34) is approximately given by

$$20 \log_{10}(|A(\omega)|) = 20 \log_{10}(A_o) - 10 \log_{10} \left(1 + \overbrace{\left(\frac{\omega}{\omega_o}\right)^2}^{\text{Dominates}} \right) \tag{9.36}$$

$$\cong 20 \log_{10}(A_o) - 20 \log_{10} \left(\frac{\omega}{\omega_o} \right)$$

where a factor of 10 change in frequency clearly results in a 20 dB change in the gain. When the frequency is exactly at the corner frequency, (i.e., $\omega = \omega_o$), then Eq. (9.34) becomes

$$20 \log_{10}(|A(\omega)|) = 20 \log_{10}(A_o) - 20 \log_{10} \left(\sqrt{1+1} \right) = 20 \log_{10}(A_o) - 3.0103 \text{ dB} \tag{9.37}$$

Therefore, the corner frequency, ω_o, is often called the -3 dB frequency since it is the frequency for which the gain has been reduced by approximately 3 dB (i.e., a factor of $\sqrt{2}$). ω_o is also called the -3 dB bandwidth for the amplifier as the gain of the amplifier does not drop below its maximum value by more than 3 dB over the range from 0 to ω_o.

Practical Hint: When determining if a quantity is much larger or much smaller than another quantity, usually a factor of 3–10 is sufficient. Therefore, the approximation given in (9.35) would be valid if ω is smaller than $\omega_o/3$, and the approximation given in (9.36) would be valid if ω is larger than $3\omega_o$.

Example 9.15 What is the highest frequency for which the open-loop gain of an Op-Amp is within 90% of its maximum value?

Solution The first step is to find the frequency corresponding to when the gain is $0.9A_o$.

$$|A(\omega)| = \left| \frac{A_o}{1 + j\frac{\omega}{\omega_o}} \right| = \frac{A_o}{\sqrt{1 + \left(\frac{\omega}{\omega_o}\right)^2}} = 0.9A_o \Rightarrow \sqrt{1 + \left(\frac{\omega}{\omega_o}\right)^2} = \frac{1}{0.9} = 1.1111$$

$$1 + \left(\frac{\omega}{\omega_o}\right)^2 = 1.23456 \Rightarrow \frac{\omega}{\omega_o} = 0.484322 \Rightarrow \omega < 0.484322\omega_o$$

Now that we have defined our terms for the frequency dependence of the Op-Amp, we can discuss the impact the frequency response of the Op-Amp has on the gain of the amplifier circuit. The exact impact will depend on the feedback configuration used in the Op-amp. Therefore, we will initially focus on Op-Amps in the noninverting configuration. As was shown in Eq. (9.27), the gain of the non-inverting amplifier when the Op-Amp has finite gain is given by

$$\text{Gain} = \frac{v_o}{V_{\text{Th}}} = A_{\text{CL}}(\omega) = \frac{A(\omega)}{1 + A(\omega)\frac{1}{\left(1 + \frac{R_2}{R_1}\right)}} = \frac{A(\omega)}{1 + A(\omega)\beta} = \frac{\frac{A_o}{\left(1 + j\frac{\omega}{\omega_o}\right)}}{1 + \frac{A_o}{\left(1 + j\frac{\omega}{\omega_o}\right)}\beta}$$

$$= \frac{A_o}{\left(1 + j\frac{\omega}{\omega_o}\right) + A_o\beta} = \frac{\frac{A_o}{1 + A_o\beta}}{1 + j\frac{\omega}{(1 + A_o\beta)\omega_o}}$$

(9.38)

Therefore, at low frequency, the non-inverting amplifier has a gain of $A_o/(1 + A_o\beta) \cong 1/\beta = (1 + R_2/R_1)$. However, at higher frequencies, the gain is reduced. The new corner frequency governing the transition between low and high frequency behavior is given by $\omega_o(1 + A_o\beta)$. It is critically important to notice that the product of the low-frequency amplifier gain and the -3 dB bandwidth is a constant for the Op-Amp as derived below.

$$\text{Gain} = \frac{A_o}{1 + A_o\beta} \cong (1 + R_2/R_1) \quad \text{Bandwidth} = (1 + A_o\beta)\omega_o = \left(1 + \frac{A_o}{(1 + R_2/R_1)}\right)\omega_o$$

(9.39)

$$\text{Gain} \cdot \text{Bandwidth} = A_o\omega_o = \omega_t$$

The gain-bandwidth product is also known as the unity gain frequency as this is the frequency at which the open-loop gain of the Op-Amp is reduced to one

$$|A(\omega_t)| = \frac{A_o}{\sqrt{1 + \left(\frac{A_o\omega_o}{\omega_o}\right)^2}} = \frac{A_o}{\sqrt{1 + A_o^2}} \cong 1$$

(9.40)

Since the product of the gain and the bandwidth is a constant for a particular Op-Amp, there is a design tradeoff between the gain and the usable frequency range for the Op-Amp circuit. If higher gains are needed, the usable frequency range will be reduced and vice versa.

Example 9.16 An Op-Amp with a unity gain frequency of 1 MHz is in the non-inverting configuration. The amplifier is designed to have a gain (i.e., $(1 + R_2/R_1)$) of 1, 10, 100, 1000, and 10000 V/V by varying the values of R_2 and R_1. What is the approximate low frequency gain in dB and usable frequency range corresponding to the -3 dB bandwidth of the amplifier for each case?

Solution The value of the gain in dB is given by $20 \log_{10}(1 + R_2/R_1)$. Therefore,

$$
\begin{aligned}
1 \text{ V/V} &\rightarrow & 0 \text{ dB} \\
10 \text{ V/V} &\rightarrow & 20 \text{ dB} \\
100 \text{ V/V} &\rightarrow & 40 \text{ dB} \\
1000 \text{ V/V} &\rightarrow & 60 \text{ dB} \\
10000 \text{ V/V} &\rightarrow & 80 \text{ dB}
\end{aligned}
$$

The bandwidth for each gain can be found from the gain-bandwidth product given by $f_t = 1$ MHz

$$
\begin{aligned}
\text{Bandwidth for 1 V/V} &= \frac{f_t}{\text{Gain}} = \frac{10^6 \text{ Hz}}{1 \text{ V/V}} = 1 \text{ MHz} \\
\text{Bandwidth for 10 V/V} &= \frac{f_t}{\text{Gain}} = \frac{10^6 \text{ Hz}}{10 \text{ V/V}} = 100 \text{ kHz} \\
\text{Bandwidth for 100 V/V} &= \frac{f_t}{\text{Gain}} = \frac{10^6 \text{ Hz}}{10 \text{ V/V}} = 10 \text{ kHz} \\
\text{Bandwidth for 1000 V/V} &= \frac{f_t}{\text{Gain}} = \frac{10^6 \text{ Hz}}{1000 \text{ V/V}} = 1000 \text{ Hz} \\
\text{Bandwidth for 10000 V/V} &= \frac{f_t}{\text{Gain}} = \frac{10^6 \text{ Hz}}{10000 \text{ V/V}} = 100 \text{ Hz}
\end{aligned}
$$

Example 9.17 Plot the exact gain of the non-inverting amplifiers described Example 9.16 as function of frequency in Hz if the open-loop gain of the Op-Amp is 100 V/mV.

Solution The exact gain as a function of frequency is given by

$$
|A_{\text{CL}}(\omega)| = \left| \frac{\frac{A_o}{1 + A_o\beta}}{1 + j\frac{\omega}{(1 + A_o\beta)\omega_o}} \right| = \frac{\frac{A_o}{1 + A_o\beta}}{\sqrt{1 + \left(\frac{\omega}{(1 + A_o\beta)\omega_o}\right)^2}} \Rightarrow |A_{\text{CL}}(f)| = \frac{\frac{A_o}{1 + A_o\beta}}{\sqrt{1 + \left(\frac{f}{(1 + A_o\beta)f_o}\right)^2}}
$$

Therefore, A_o, f_o, and β are all needed to generate the plot. β is just given by $1/(1 + R_2/R_1)$ while A_o and f_o can be found from.

$$
A_o = 100 \frac{\text{V}}{\text{mV}} \cdot \frac{1000 \text{ mV}}{\text{V}} = 10^5 \text{ V/V}
$$

$$
A_o f_o = f_t \Rightarrow f_o = \frac{f_t}{A_o} = \frac{10^6 \text{ Hz}}{10^5 \text{ V/V}} = 10 \text{ Hz}
$$

We can now plot the gain as a function of frequency for the different cases.

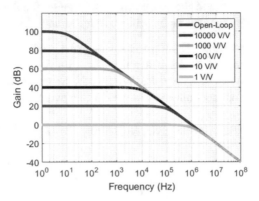

Clearly, as the amplifier gain decreases, the usable bandwidth of the amplifier increases.

Example 9.18 Determine the relationship between gain and bandwidth for an Op-Amp in the inverting configuration.

Solution The closed-loop gain for an Op-Amp in the inverting configuration is given in Eq. (9.30). Assuming R_i is very large and R_o is very small we get

$$A_{CL} = \frac{v_o}{i_N} = \frac{-A}{\frac{1}{R_{Th}} + \frac{1}{R_2} + \frac{A}{R_2}} = \frac{-AR_2}{A + 1 + \frac{R_2}{R_{Th}}} = \frac{\frac{-AR_2}{\left(1 + \frac{R_2}{R_{Th}}\right)}}{1 + \frac{A}{\left(1 + \frac{R_2}{R_{Th}}\right)}}$$

If we then include the frequency response of the Op-Amp, we get

$$A_{CL}(\omega) = \frac{\frac{-R_2}{(1 + R_2/R_{Th})}\left(\frac{A_o}{1 + j\frac{\omega}{\omega_o}}\right)}{1 + \frac{1}{(1 + R_2/R_{Th})}\left(\frac{A_o}{1 + j\frac{\omega}{\omega_o}}\right)} = \frac{\frac{-A_o R_2}{(1 + R_2/R_{Th})}}{\left(1 + j\frac{\omega}{\omega_o}\right) + \frac{A_o}{(1 + R_2/R_{Th})}} = \frac{\frac{-A_o R_2}{(1 + R_2/R_{Th})\left(1 + \frac{A_o}{(1 + R_2/R_{Th})}\right)}}{1 + j\frac{\omega}{\omega_o\left(1 + \frac{A_o}{(1 + R_2/R_{Th})}\right)}}$$

The gain, bandwidth, and gain-bandwidth product are thus given by

$$\text{Gain} = \frac{-A_o R_2}{(1 + R_2/R_{Th})\left(1 + \frac{A_o}{(1 + R_2/R_{Th})}\right)} \cong -R_2 \quad \text{Bandwidth} = \omega_o\left(1 + \frac{A_o}{(1 + R_2/R_{Th})}\right)$$

Notice that the bandwidth is very similar to the expression found for the bandwidth of the non-inverting configuration in Eq. (9.39).

9.3.5 Op-Amp Slew Rate Limitation

The frequency response of the Op-Amp discussed previously was a linear limitation to the operating range of the Op-Amp. The limitation was only based on the frequency of the signal and not the amplitude. However, the usable frequency range of an Op-Amp is also impacted by the slew rate. The

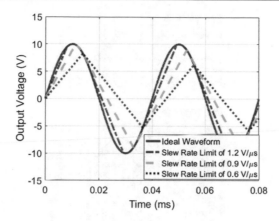

Fig. 9.21 Example of sinusoidal signal impacted by different slew rate values. The signal is at a frequency of 25 kHz and has an amplitude of 10 V

slew rate, *SR*, usually expressed in units of V/μs, is the maximum rate of change possible for the output of a real Op-Amp. Therefore, the magnitude of the slope of the output waveform cannot exceed the slew rate.

   A sinusoidal signal impacted by various slew rate limits is shown in Fig. 9.21. Initially, the sinusoidal signal increases faster than the slew rate and the output of the Op-Amp is not able to keep up. Eventually, the slope of the sinusoidal signal is reduced as it nears its peak. Since the desired output is still greater than the amplifier output, the output voltage continues to increase at the slew rate until it finally intersects with the desired output. It then attempts to follow the desired waveform until the slope of the decrease exceeds the slew rate. At this time, the amplifier output once again trails the input. For the smaller slew rates, the final output waveforms resembles a triangle wave rather than the desired sinusoidal waveform. Given the dependence on the slope of the waveform, both the frequency and the amplitude of the signal will impact the level of distortion. For sinusoidal signals, the slope is given by

$$v_o(t) = V_m \cos(\omega t + \theta_v) \Rightarrow \frac{dv_o}{dt} = -V_m \cdot \omega \cdot \sin(\omega t + \theta_v) \qquad (9.41)$$

However, the sin function varies from −1 to 1. Therefore, the largest slope that must be less than the slew rate is given by

$$\left(\frac{dv_o}{dt}\right)_{max} = V_m \cdot \omega = 2\pi f \cdot V_m = \frac{2\pi}{T} \cdot V_m < SR \qquad (9.42)$$

Example 9.19 Sort the waveforms shown in Fig. 9.22 from most impacted by slew rate limiting to least impacted by slew rate limiting assuming that all three waveforms have the same period as well as maxima and minima at the same locations with the same values.

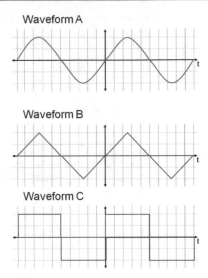

Fig. 9.22 Waveforms for Example 9.19

Solution Clearly, the square wave signal will be the most impacted by the slew rate as the slope of the ideal square wave is infinite. Therefore, all that remains is to determine the maximum slope of the triangle wave (Waveform B) relative to the sinusoidal wave (Waveform A). The triangle wave undergoes a change of $2V_m$ in $T/2$. Therefore, the slope is $\pm 4V_m/T$. This is less than the maximum/minimum slopes of the sinusoidal signal given by $\pm 2\pi V_m/T$. Therefore, the impact of slew rate on the sinusoidal waveform will be greater than the impact on a triangle waveform. Sorting the waveforms gives C, A, B as the order for most impacted to least impacted.

Example 9.20 A sensor for a specific application has an output of $v_s(t) = 20\cos\left(8 \times 10^5 t\right)$ mV. The sensor is connected to an amplifier in the non-inverting configuration. The op-amp used has a slew rate of 4 V/µs, a unity gain frequency of 40 MHz, and a saturation of ± 10 V. Find the largest gain value for the amplifier that will avoid the slew rate limit, saturation, and the bandwidth limit imposed by the unity gain frequency.

Solution The problem states that the amplifier will saturate at ± 10 V. Therefore, the output must be smaller than this value. Hence,

$$\text{Gain} < \frac{10 \text{ V}}{\max(v_s)} = \frac{10 \text{ V}}{20 \text{ mV}} = 500 \text{ V/V}$$

to avoid saturation. Similarly, in order to operate below the bandwidth limit, the frequency must be less than the bandwidth at the desired gain.

$$\text{Gain} < \frac{f_t}{\text{BW}} = \frac{40 \text{ MHz}}{(8 \times 10^5)/(2\pi) \text{ Hz}} = 314.159 \text{ V/V}$$

Lastly, to avoid distortion by the slew rate limit

$$\left(\frac{\mathrm{d}v_o}{\mathrm{d}t}\right)_{\max} < \mathrm{SR} \Rightarrow \left(\frac{\mathrm{d}}{\mathrm{d}t}\left(\mathrm{Gain} \cdot v_s(t)\right)\right)_{\max} < \mathrm{SR}$$

$$\frac{\mathrm{d}}{\mathrm{d}t}\left(\mathrm{Gain} \cdot v_s(t)\right) = -\mathrm{Gain} \cdot 20 \times 8 \times 10^5 \cdot \sin\left(8 \times 10^5 t\right) \mathrm{mV/s}$$

$$\left(\frac{\mathrm{d}}{\mathrm{d}t}\left(\mathrm{Gain} \cdot v_s(t)\right)\right)_{\max} = \mathrm{Gain} \cdot 16000 \, \mathrm{V/s} < 4 \, \mathrm{V/\mu s}$$

$$\mathrm{Gain} < \frac{4 \, \mathrm{V/\mu s}}{16000 \, \mathrm{V/s}} = 250 \, \mathrm{V/V}$$

All three of these limits on the gain must be satisfied. Therefore, the gain must be smaller than 250 V/V with the slew rate being the most limiting.

Example 9.21 For the sensor described in Example 9.20, find the unity gain frequency and slew rate required so that saturation was the only limit to the gain.

Solution The saturation limits the gain to 500 V/V. Therefore, we will assume that this is the gain for our Op-Amp. We also know that the band width for the Op-Amp must be at least 8×10^5 rad/s to accommodate our needed frequency. In practice, we would want a bandwidth slightly larger than 8×10^5 rad/s, but this will still give us a reasonable approximation. With this bandwidth and gain, the unity gain frequency for the Op-Amp would need to be

$$f_t = \mathrm{Gain} \cdot \mathrm{BW} = (500 \, \mathrm{V/V}) \cdot \left(\frac{8 \times 10^5 \, \mathrm{rad/s}}{2\pi}\right) = 63.662 \, \mathrm{MHz}$$

Likewise, the slew rate would need to be

$$\mathrm{SR} > V_m \cdot \omega = (500 \, \mathrm{V/V}) \cdot (20 \, \mathrm{mV}) \cdot \left(8 \times 10^5 \, \mathrm{rad/s}\right) = 8 \, \mathrm{V/\mu s}$$

Example 9.22 You need to design a non-inverting amplifier using one of the two following Op-Amps. Which Op-Amp should you use in your design if you want your output to have the largest amplitude possible without distortion or amplitude reduction due to the frequency limit of the Op-Amp and your input is $v_{in}(t) = 0.1 \sin\left(2\pi 10^6 \cdot t\right)$ V?

Op-Amp A	
Maximum saturation voltage	± 20 V
Slew Rate	10 V/μs
Unity Gain Frequency	50 MHz
Op-Amp B	
Maximum saturation voltage	± 15 V
Slew Rate	5 V/μs
Unity Gain Frequency	75 MHz

Solution To solve this problem, we need to find the limitation the slew rate and the unity gain frequency (i.e., Op-Amp frequency limit) places on the output of the amplifier. The unity gain frequency will limit the amount of gain that can be implemented with the Op-Amp and is given by

$$\text{Gain} = \frac{f_t}{\text{BW}} \Rightarrow \text{Gain}_A = \frac{50 \text{ MHz}}{1 \text{ MHz}} = 50 \text{ V/V} \quad \text{Gain}_B = \frac{75 \text{ MHz}}{1 \text{ MHz}} = 75 \text{ V/V}$$

Therefore, the maximum output would be 5 V or Op-Amp A and 7.5 V for Op-Amp B. Likewise, the slew rate would limit the output as

$$V_m < \frac{\text{SR}}{2\pi f} \Rightarrow V_{mA} = \frac{10 \text{ V/}\mu\text{s}}{2\pi \times 10^6 \text{ rad/s}} = 1.5915 \text{ V} \quad V_{mB} = \frac{5 \text{ V/}\mu\text{s}}{2\pi \times 10^6 \text{ rad/s}} = 0.7958 \text{ V}$$

Therefore, Op-Amp A with a maximum output amplitude of 1.5915 V would be the best choice between these two Op-Amps.

Lab Hint: When designing and constructing Op-Amp circuits, the slew rate is often the limiting factor. It is always possible to implement several amplification stages where the desired gain is spread out over several Op-Amp circuits to avoid the bandwidth limits introduced by the unity gain frequency. However, the limit introduced by the slew rate is tied directly to the output amplitude and frequency. In addition, the shape of the waveform can also have an impact even for waveforms with the same period. Therefore, knowledge of the expected waveforms and the potential impact of slew rate distortion should be clearly understood before selecting an Op-Amp for a particular application.

9.4 Problems

Problem 9.1: Classify the following amplifiers as either a voltage amplifier, a current amplifier, a transresistance amplifier, or a transconductance amplifier.

(a) $R_{in} = 20 \text{ }\Omega$ $R_{out} = 10 \text{ M}\Omega$ Gain $= 50 \text{ V/V}$

(b) $R_{in} = 20 \text{ }\Omega$ $R_{out} = 10 \text{ }\Omega$ Gain $= 50 \text{ V/V}$

(c) $R_{in} = 20 \text{ M}\Omega$ $R_{out} = 10 \text{ }\Omega$ Gain $= 50 \text{ V/V}$

(d) $R_{in} = 20 \text{ M}\Omega$ $R_{out} = 10 \text{ M}\Omega$ Gain $= 50 \text{ V/V}$

Problem 9.2: Design a voltage amplifier with a gain of 50 V/V using an Op-Amp.

Problem 9.3: A sensor has a Thevenin resistance of 200 kΩ and outputs a current of 80 μA when exposed to a light source. Design a transresistance amplifier to amplify this signal level so that $|v_o| = 0.5$ V using an Op-Amp.

Problem 9.4: A sensor for a specific application has an internal resistance of R_s of 100 kΩ and output of $i_s(t)$. The maximum voltage from the sensor is ± 1 V after which the sensor saturates.

(a) If the sensor is connected to the inverting amplifier as shown below using an ideal op-amp, what is the change in output voltage v_o if $i_s(t)$ changes from 50 to 80 µA? The value of R_1 is 500 Ω while the value of R_2 is 10 kΩ.

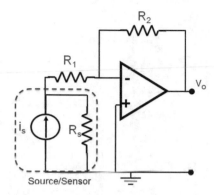

(b) If it is connected to the non-inverting amplifier as shown below using an ideal op-amp, what is the change in output voltage v_o if $i_s(t)$ changes from 50 to 80 µA? The value of R_1 is 500 Ω while the value of R_2 is 10 kΩ.

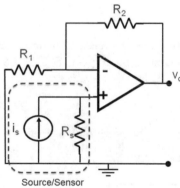

Problem 9.5: Design an instrumentation amplifier to have a differential mode gain of 100.

Problem 9.6: For this problem use the circuit shown below where $V_{in} = 5$ V, $R = 10$ kΩ, $R_0 = 500$ Ω, and $\Delta R = 1$ Ω. What value of R_{gain} would make $V_{out} = 500$ mV?

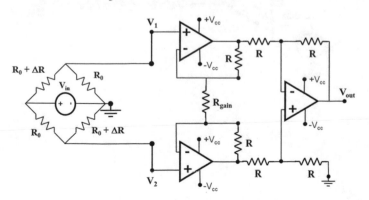

Problem 9.7: One of the resistors in an instrumentation amplifier is off in its resistance value as shown below. What is the differential mode gain, G_d, the common mode gain, G_{cm}, and the common mode rejection ratio, CMRR, for this amplifier?

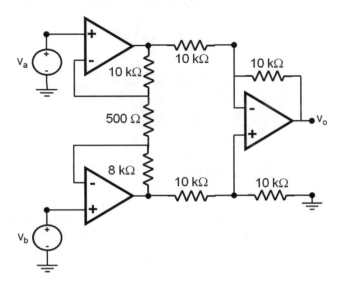

Problem 9.8: Several of the resistors in an instrumentation amplifier are off in their resistance values as shown below. What is the differential mode gain, G_d, the common mode gain, G_{cm}, and the common mode rejection ratio, CMRR, for this amplifier?

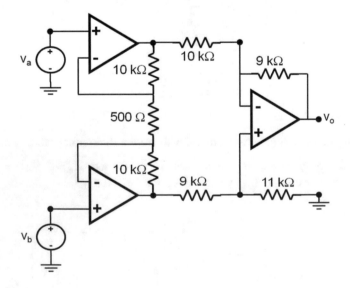

Problem 9.9: Find the gain, $G = v_o/v_s$, of the Op-Amp circuit shown below assuming the Op-Amp is ideal.

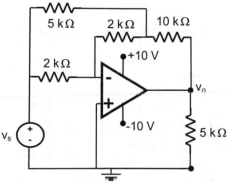

Problem 9.10: Find V_o assuming the Op-Amp is ideal.

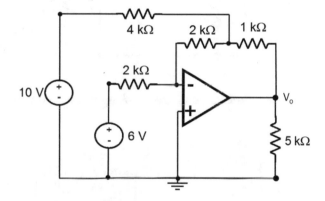

Problem 9.11: Find V_o assuming the Op-Amps are ideal.

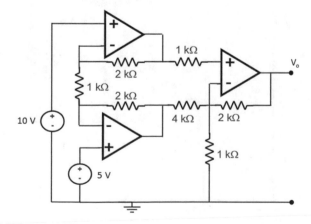

Problem 9.12: Find the output current, I_o, of the Op-Amp circuit shown below assuming ideal Op-Amps.

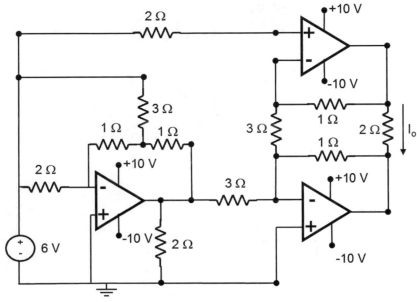

Problem 9.13: Find the output current, I_o, of the Op-Amp circuit shown below assuming ideal Op-Amps.

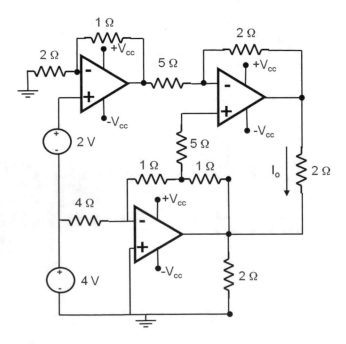

Problem 9.14: Find the output current, I_o, of the Op-Amp circuit shown below assuming ideal Op-Amps.

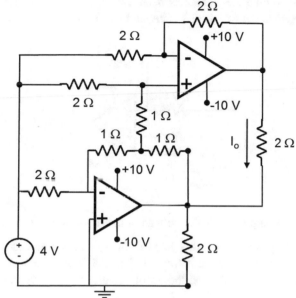

Problem 9.15: An Op-Amp circuit in the non-inverting amplifier configuration has an infinite input resistance, R_i, and a zero output resistance, R_o. The feedback resistors used to set the gain of the amplifier are given by $R_2 = 24$ kΩ and $R_1 = 1$ kΩ. The Thevenin resistance for the source at the input, R_{Th}, is 0 Ω. Find the gain of the non-inverting amplifier as the open-loop gain, A, varies as 100, 1000, and 10000 V/V.

Problem 9.16: An Op-Amp circuit in the non-inverting amplifier configuration has an input resistance, R_i, of 150 kΩ and an output resistance, R_o, of 200 Ω. The feedback resistors used to set the gain of the amplifier are given by $R_2 = 39$ MΩ and $R_1 = 1$ MΩ. The Thevenin resistance for the source at the input, R_{Th}, is 0 Ω. Find the effective open-loop gain of the non-inverting amplifier, A', as well as the closed loop gain if the open-loop gain, A, is 10^2, 10^4, and 10^6 V/V.

Problem 9.17: An Op-Amp circuit in the inverting amplifier configuration has an input resistance, R_i, of 400 kΩ and an output resistance, R_o, of 50 Ω. The feedback resistors used to set the gain of the amplifier are given by $R_2 = 10$ MΩ and $R_1 = 1$ MΩ. The Thevenin resistance for the source at the input, R_{Th}, is 100 kΩ. Find the closed loop gain $A_{CL} = v_o/i_n$ if the open-loop gain, A, is 10^2, 10^4, and 10^6 V/V.

Problem 9.18: A non-inverting amplifier has been built using an Op-Amp as shown below. The Op-Amp has bias currents of $I_{Bn} = I_{Bp} = 200$ nA and an offset voltage of $V_{os} = 35$ mV.

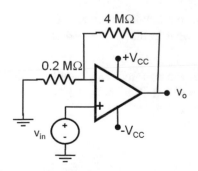

What is the voltage at the output, v_o, if the voltage at the input is $v_{in}(t) = 50\sin(1000t)$ mV?

Problem 9.19: A non-inverting amplifier has been built using an Op-Amp as shown below. The Op-Amp has bias currents of $I_{Bn} = I_{Bp} = 350$ nA and an offset voltage of $V_{os} = 10$ mV.

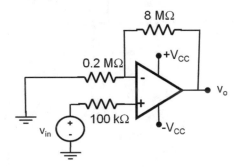

What is the voltage at the output, v_o, if the voltage at the input is $v_{in}(t) = 0.2\sin(10t)$ V?

Problem 9.20: An inverting amplifier with a gain of -50 V/V uses a non-ideal op-amp with 0.2 MΩ and 10 MΩ resistors. The output voltage is found to be $+9.5$ V when measured with the input open and $+9.2$ V with the input grounded. Find the offset voltage, V_{OS}, as well as the bias current I_{Bn}.

Problem 9.21: Use the output voltage, v_o, given in the circuits shown below to find the bias currents, I_{Bn} and I_{Bp}, and offset voltage for the Op-Amp. Assume that all three circuits are built with exactly the same Op-Amp.

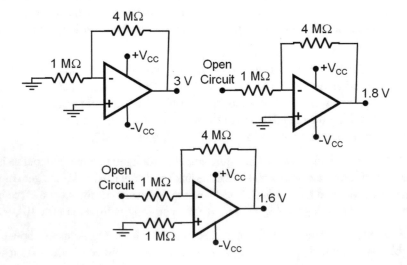

Problem 9.22: If an Op-Amp in the non-inverting amplifier configuration has a gain of 15 and is biased with voltages of ± 8 V, what is the largest and smallest input voltage allowed before the amplifier saturates?

Problem 9.23: If an Op-Amp in the non-inverting amplifier configuration has a gain of 20 and is biased with voltages of $+10$ V and -2 V, what is the largest and smallest input voltage allowed before the amplifier saturates?

Problem 9.24: What is the highest frequency for which the open-loop gain of an Op-Amp is within 75% of its maximum value?

Problem 9.25: The open loop gain of an Op-Amp at very low frequencies, A_o, is 90 dB and the -3 dB frequency is 100 Hz. When resistors are connected so that the Op-Amp is operating in the noninverting configuration, the gain is 40 dB. What is the unity gain frequency for the Op-Amp? What is the new -3 dB bandwidth with the 40 dB gain?

Problem 9.26: You need to build an amplifier with a gain of 40 dB (noninverting) under the design constraints that the -3 dB bandwidth of your amplifier needs to be at least 5 MHz.
(a) What unity gain frequency is needed for the Op-Amp in the circuit?
(b) If the open-loop -3 dB bandwidth of this Op-Amp is 10 kHz, what is the open-loop DC gain of the Op-Amp?

Problem 9.27: The open loop gain of an Op-Amp at very low frequencies, A_o, is 90 dB and the -3 dB frequency is 100 Hz. The Op-Amp is connected in the inverting configuration with $R_2 = 5$ kΩ and $R_{Th} = 50$ kΩ. What is the new -3 dB bandwidth for the inverting amplifier?

Problem 9.28: You have an op-amp with a unity gain frequency of 200 MHz, a slew rate of 300 V/µs, that is connected to voltage sources such that it saturates at ± 15 V. The Op-Amp is used in the non-inverting configuration with a gain of 100. The input to the amplifier is a sine wave.

(a) If the amplitude of the input is 0.1 V, what is the maximum frequency?
(b) If the frequency of the input is 1 MHz, what is the maximum amplitude for the input before the output distorts?

Problem 9.29: You have an op-amp with a unity gain frequency of 2 MHz, a slew rate of 1 V/µs, that is connected to voltage sources such that it saturates at ± 10 V. The Op-Amp is used in the non-inverting configuration with a gain of 10. The input to the amplifier is a sine wave.

(a) If the amplitude of the input is 0.5 V, what is the maximum frequency before the output distorts?
(b) If the frequency of the input is 20 kHz, what is the maximum amplitude before the output distorts?
(c) If the amplitude of the input is 50 mV, what is the usable frequency range for the amplifier?
(d) If the frequency of the input is 5 kHz, what is the useful input voltage range?

Problem 9.30: Assuming that the input waveform to a non-inverting amplifier with a gain of $+100$ V/V is always less than 0.5 V and you have an Op-Amp with a unity gain frequency of 100 MHz and a slew rate of 50 V/µs, find:

(a) What is the usable frequency range for your amplifier circuit?
(b) What would be the usable frequency range if your input was always less than 0.05 V?

Problem 9.31: A sensor for a specific application has an output of $v_s(t) = 3\cos(2 \times 10^5 t)$ mV. The sensor is connected to an amplifier in the non-inverting configuration. The Op-Amp used has a slew rate of 2 V/µs, a unity gain frequency of 50 MHz, and a saturation of ± 5 V. Find the largest gain value for the amplifier that will avoid the slew rate limit, saturation, and the bandwidth limit imposed by the unity gain frequency.

Problem 9.32: A non-inverting amplifier with a gain of 7 is to be used to amplify the signal from a sensor. The value of the voltage from the sensor ranges from 20 to 300 mV with frequency content from 50 to 200 kHz.

(a) What unity gain frequency is needed for your amplifier?
(b) If the typical waveform for v_s is sinusoidal (i.e., $v_s(t) = V_o\sin(\omega t)$), what slew rate is needed for the amplifier?

Problem 9.33: You need to design a non-inverting amplifier using one of the two following Op-Amps. Which Op-Amp should you use in your design if you want your output to have the largest amplitude possible without distortion or amplitude reduction due to the frequency limit of the Op-Amp and your input is $v_{in}(t) = 0.1\sin(20\pi 10^6 \cdot t)$ V?

Op-Amp A	
Maximum saturation voltage	± 10 V
Slew Rate	120 V/µs
Unity Gain Frequency	40 MHz
Op-Amp B	
Maximum saturation voltage	± 15 V
Slew Rate	60 V/µs
Unity Gain Frequency	80 MHz

Transformers

10

Efficient power delivery has long been an important problem in electrical engineering. Generating the power has its challenges, but fundamentally the power must reach the electrical devices that demand it. Efficient power delivery has become even more important with the growth of renewable energy sources. For example, the Midwest United States is a great source for wind power. However, most of the country's population centers are located in other regions. Therefore, the generated power may need to be transmitted over much greater distances than when relying on regional coal or nuclear power plants. Transformers are critical for efficient power transmission. In this chapter, we will discuss the fundamental concepts of transforms emphasizing their importance to power transmission.

10.1 Basic Transformer Operation

In order to understand the basic operation of transformers, it is useful to review the basic concepts of electromagnetics. A stationary charge on an object will result in a static electric field. If a constant electric field (i.e., voltage) is applied to the charge, the charge will move at a constant velocity creating a current known as a DC current. A DC current will produce a static magnetic field. However, if the electric voltage varies with time, then the motion of the charge will vary with time resulting in a time-varying current. As a result, the magnetic field produced by the current will also vary with time. According to Faraday's Law, however, a time-varying magnetic field can also be used to generate a second time-varying voltage if it passes through a looped conductor. Therefore, a transformer is a device that uses one set of coils to generate a time-varying magnetic field. The magnetic field is then directed through one or more additional coils where it is used to induce a new time-varying voltage.

As an example, consider the basic diagram of a transformer shown in Fig. 10.1. The primary winding will generate a time-varying magnetic field when a time-varying voltage, $v_1(t)$, is applied. The exact strength of the magnetic field will depend on both the current, $i_1(t)$, resulting from the voltage, $v_1(t)$, and the number of coils, N_1, used in the winding. Specifically, the magnetic flux, $\psi(t)$, in the core due to $i_1(t)$ is given by

$$\psi(t) = \frac{N_1 i_1(t)}{\Re} \tag{10.1}$$

© Springer Nature Switzerland AG 2020
T. A. Bigelow, *Electric Circuits, Systems, and Motors*,
https://doi.org/10.1007/978-3-030-31355-5_10

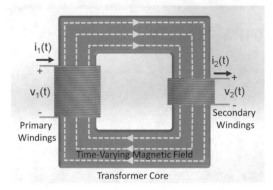

Fig. 10.1 Diagram of a transformer illustrating its basic operation

where $\Re$ is the magnetic reluctance of the core. The reluctance quantifies the difficulty in establishing the magnetic flux in the material and is approximately given by

$$\Re = \frac{l_c}{\mu A} \tag{10.2}$$

where μ is the magnetic permeability of the core, A is the cross-sectional area of the core, and l_c is the mean path length around the center of the core.

Once generated, the magnetic field/flux will be largely confined to the core of the transformer, usually iron, due to its magnetic properties. The magnetic field/flux will then pass through the coils of the secondary winding. Since the magnetic field is varying with time, a time-varying voltage will be induced in the secondary winding. The magnitude of the induced voltage depends on both the time-rate of change of the magnetic flux and the number of coils in the secondary winding. According to Lenz' Law, the polarity of the induced voltage is such that if the terminals were shorted, it would produce a current, $i_2(t)$, whose magnetic field would oppose the original magnetic field responsible for generating the voltage. Hence,

$$\text{Induced Voltage} = -N_2 \frac{d\psi(t)}{dt} \Rightarrow v_2(t) = +N_2 \frac{d\psi(t)}{dt} \tag{10.3}$$

For the moment, let's assume that no load is connected to the secondary winding. As a result, the current $i_2(t)$ would be zero, and the secondary winding would not be impacting the magnetic field in the core. Let's also assume that the resistance of the coils is negligible. Thus, the current flowing in the primary winding, $i_1(t)$, would be directly related to the voltage applied to the primary winding, $v_1(t)$, by

$$v_1(t) = L_m \frac{di_1(t)}{dt} \Rightarrow i_1(t) = \frac{1}{L_m} \int_{-\infty}^{t} v_1(\tau) d\tau \tag{10.4}$$

where L_m is the magnetization inductance for the transformer approximately given by

$$L_m = N_1^2 \frac{\mu A}{l_c} \tag{10.5}$$

Therefore,

$$v_2(t) = +N_2 \frac{d\psi(t)}{dt} = N_2 \frac{d}{dt}\left(N_1 \frac{\mu A}{l_c} i_1(t)\right) = N_2 N_1 \frac{\mu A}{l_c} \frac{di_1(t)}{dt} = N_2 N_1 \frac{\mu A}{l_c} \frac{v_1(t)}{L_m}$$

$$\Rightarrow v_2(t) \cong \frac{N_2}{N_1} v_1(t) \tag{10.6}$$

Hence, by adjusting the ratio of the number of turns in the secondary winding relative to the primary winding, we can scale the amplitude of the secondary voltage relative to the primary voltage.

One vitally important fact that must be remembered whenever one is working with transformers is that transformers cannot work at DC. The entire operation of the transformer is based on time-varying currents producing time-varying magnetic fields that in turn produces time-varying voltages. If the voltages/currents are not varying with time, the signals and energy cannot move from the primary winding to the secondary winding. In fact, both the primary winding and the secondary winding of the transformer will act as short circuits at DC (just like an inductor) with no electrical connection between the primary and secondary windings.

Example 10.1 A transformer has 1000 primary windings and 500 secondary windings. The core of the transformer has a mean path length of 4 cm, a cross-sectional area of 0.5 cm^2, and a magnetic permeability of 6×10^{-3} H/m. A voltage source with a Thevenin voltage of $v_{Th}(t) = 100\cos(400t)$ V and a Thevenin resistance of 50 Ω is connected across the primary winding of the transformer. The secondary winding is left open circuited (i.e., no load connected). What is the voltage across the secondary winding assuming the coils have negligible resistance?

Solution:
A sketch of the circuit is shown below.

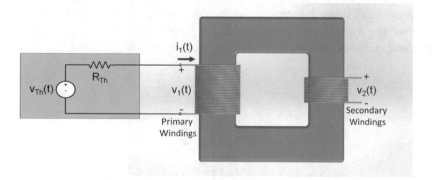

In order to find the voltage across the secondary winding, we need to first find the voltage across the primary winding. The voltage across the primary winding can be found from the inductance of the primary winding and the Thevenin resistance using a basic voltage divider. Hence,

$$v_2(t) = \frac{N_2}{N_1} v_1(t) \Rightarrow \tilde{V}_2 = \frac{N_2}{N_1} \tilde{V}_1 = \frac{N_2}{N_1} \cdot \left(\tilde{V}_{Th} \frac{j\omega L_m}{R_{Th} + j\omega L_m} \right)$$

where $\tilde{V}_1$, $\tilde{V}_2$, and $\tilde{V}_{Th}$ are all in the phasor domain. The inductance of the primary winding is given by

$$L_m = N_1^2 \frac{\mu A}{l_c} = 7.5\,\text{H}$$

Thus,

$$\tilde{V}_2 = \frac{500}{1000} \cdot \left((100\,\text{V}) \frac{j400 \cdot 7.5}{50 + j400 \cdot 7.5} \right) = (50\,\text{V}) \cdot \left(\frac{j3000}{50 + j3000} \right) = 49.99 \exp(j0.0167)\,\text{V}$$
$$\Rightarrow v_2(t) = 49.99 \cos(400t + 0.0167)\,\text{V}$$

Example 10.2 Repeat Example 10.1 if $v_{Th}(t) = 100\,\text{V}$ (i.e., DC source).

Solution:
Transformers do not work at DC. Therefore, there is no way for any voltage at the primary to reach the secondary winding. Therefore, the voltage across the secondary winding must be 0 V. In addition, since the primary winding will act as a short circuit, i_1 will be very large potentially damaging the transformer.

10.2   Circuit Symbol for the Transformer

The circuit symbol for a basic transformer is shown in Fig. 10.2. The relationship between the number of turns in the primary and secondary windings is given as a ratio and is normally written above the transformer. For example, if the primary winding had 2000 turns and the secondary winding had 200 turns, the ratio would be written as 10:1. The dots in the symbol shown adjacent to the windings are critically important as they provide the polarity of the voltage and the direction of the current. If the voltage at the dotted terminal of coil 1 is positive, then the voltage at the dotted terminal of coil 2 is also positive. Likewise, if a current is entering the dotted terminal of coil 1, then the

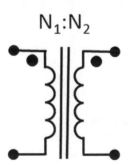

Fig. 10.2 Circuit symbol for a transformer

Lab Hint: Many transformers used in the lab have multiple primary and/or secondary windings wrapped around a common core. These windings are typically connected in series or in parallel to provide more flexibility in the current and voltage values. As an example, a circuit symbol for a transformer with two primary and three secondary windings is shown in Fig. 10.3.

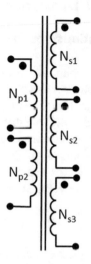

Fig. 10.3 Circuit symbol for a transformer with two primary and three secondary windings

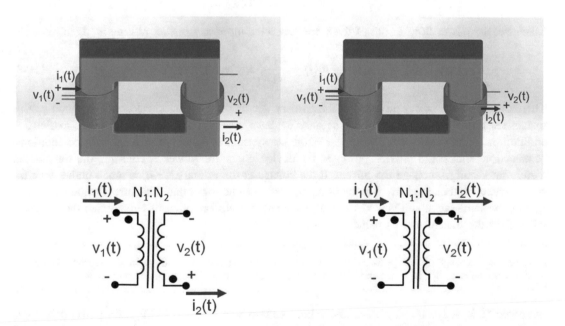

Fig. 10.4 Illustration of the impact the winding of the transformer coil has on the placement of the dot in the circuit symbol for the transformer

current is leaving the dotted terminal of coil 2. The position of the dot gives information on the direction the coil is wound as illustrated in Fig. 10.4. The secondary winding for the two transformers shown in the figure is wound in opposite directions. As a result, the dot in the circuit symbol is moved to reflect this change in coil direction.

10.3 Circuit Analysis with Ideal Transformers

10.3.1 Current and Voltage Relationships for Ideal Transformers

When analyzing transformer circuits, it is common to model the transformer as an ideal transformer at least in the initial analysis. The ideal transformer assumes that all of the magnetic flux passes through both the primary and secondary windings. Hence, there is no flux "leakage" outside of the transformer. The ideal transformer is also lossless meaning that there are no eddy current losses in the core and the primary/secondary windings have no resistance. Lastly, the magnetic reluctance of the core is relatively small allowing for easy establishment of the magnetic flux. This also means that the self-inductances of each winding, as well as the mutual inductance between the windings, are relatively high. Under these approximations, the voltages and currents in the transformer are given by

$$v_2(t) = \frac{N_2}{N_1} v_1(t) \quad i_2(t) = \frac{N_1}{N_2} i_1(t) \tag{10.7}$$

With these expressions, we can calculate the instantaneous power transfer through the transformer. The power flowing into the primary winding at any instant in time is given by

$$p_1(t) = i_1(t)v_1(t) \tag{10.8}$$

Likewise, the power flowing out of the secondary winding at any instant in time is given by

$$p_2(t) = i_2(t)v_2(t) = \frac{N_1}{N_2} i_1(t) \frac{N_2}{N_1} v_1(t) = \frac{N_1 N_2}{N_2 N_1} i_1(t)v_1(t) = i_1(t)v_1(t) = p_1(t) \tag{10.9}$$

Therefore, all of the power provided to the primary winding must reach the secondary winding. This makes sense for the ideal transformer because we have assumed that the transformer is lossless. In addition, no additional power can appear as the transformer is a passive element with no additional "connection" that could provide power to the device. Since the power is constant, the transformer merely interchanges voltage for current. If the voltage at the primary, $v_1(t)$, is much higher than the secondary, $v_2(t)$, then the current at the primary, $i_1(t)$, must be lower than the current at the secondary, $i_2(t)$ by the same factor and vice versa. This exchange or transformation of voltage into the current is what gives the transformer its name.

> **Note:** The exchange of voltage for current due to the different number of windings in a transformer is very similar to the exchange of torque and speed achieved using gears with different numbers of teeth.

Example 10.3 A 10,000 V_{rms} transmission line needs to be reduced to 120 V_{rms} for power delivery to a set of homes in a subdivision. What should be the turns ratio for the transformer?

Solution:
The voltage V_1 is 10 kV_{rms} while the voltage V_2 is 120 V_{rms}. Therefore,

$$\frac{N_1}{N_2} = \frac{v_1(t)}{v_2(t)} = \frac{10\,kV_{rms}}{120\,V_{rms}} = 83.3333$$

The turns ratio does not need to an integer number since multiple coils are used in each winding.

Example 10.4 Find all of the voltages and currents for the transformer circuit shown in Fig. 10.5 assuming that the transformer is ideal. Also, use $v_{in}(t) = 100\cos(200t)$ V, $R_{load} = 500\,\Omega$, and $N_2/N_1 = 8$.

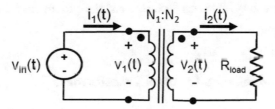

Fig. 10.5 Transformer circuit for Example 10.4

Solution:
Since the voltage source is connected directly across the primary winding $v_1(t) = v_{in}(t) = 100\cos(200t)$ V. The voltage across the secondary winding can then be found from the turns ratio as

$$v_2(t) = \frac{N_2}{N_1}v_1(t) = 8v_1(t) = 800\cos(200t)\,\text{V}$$

The voltage across the secondary winding is the same voltage that is across the load. Thus,

$$i_2(t) = i_{load}(t) = \frac{v_2(t)}{R_{load}} = 1.6\cos(200t)\,\text{A}$$

With the current flowing out of the secondary winding known, we can then use the turns ratio to find the current into the primary winding.

$$i_2(t) = \frac{N_1}{N_2}i_1(t) \Rightarrow i_1(t) = \frac{N_2}{N_1}i_2(t) = 8i_2(t) = 12.8\cos(200t)\,\text{A}$$

Notice that the voltage has been increased and the current has been decreased upon passing through the transformer.

Example 10.5 Repeat Example 10.4 with $v_{in}(t) = 100\cos(200t)$ V, $R_{load} = 50\,\Omega$, and $N_1/N_2 = 8$.

Solution Since the voltage source is still connected directly across the primary winding $v_1(t) = v_{in}(t) = 100\cos(200t)$ V. The voltage across the secondary winding can then be found from the turns ratio as

$$v_2(t) = \frac{N_2}{N_1}v_1(t) = \frac{v_1(t)}{8} = 12.5\cos(200t)\,\text{V}$$

The voltage across the secondary winding is the same voltage that is across the load. Thus,

$$i_2(t) = i_{\text{load}}(t) = \frac{v_2(t)}{R_{\text{load}}} = 0.25 \cos(200t) \text{ A}$$

With the current flowing out of the secondary winding known, we can then use the turns ratio to find the current into the primary winding.

$$i_2(t) = \frac{N_1}{N_2} i_1(t) \Rightarrow i_1(t) = \frac{N_2}{N_1} i_2(t) = \frac{i_2(t)}{8} = 31.25 \cos(200t) \text{ mA}$$

Notice that this time, the voltage has been decreased and the current has been increased upon passing through the transformer.

10.3.2   Equivalent Impedance for Ideal Transformers

When analyzing circuits with ideal transformers, it is often easier to replace the transformer by an equivalent impedance and then use traditional circuit analysis techniques (i.e., node-voltage) to solve for the currents and voltages in the circuit. This approach is very similar to the source transformation method of circuit analysis discussed in Chap. 3. Similar to our prior discussions of equivalent circuits, the two circuits shown in Fig. 10.6 are equivalent as long as $v_1(t)$ and $i_1(t)$ are the same for both circuits. This means that the transformer and impedance combination can be replaced by a single impedance value, Z'_L, when solving for $v_1(t)$ and $i_1(t)$. The value of Z'_L is given by

$$Z'_L = \frac{\widetilde{V}_1}{\widetilde{I}_1} = \frac{\left(\frac{N_1}{N_2}\widetilde{V}_2\right)}{\left(\frac{N_2}{N_1}\widetilde{I}_2\right)} = \left(\frac{N_1}{N_2}\right)^2 \frac{\widetilde{V}_2}{\widetilde{I}_2} = \left(\frac{N_1}{N_2}\right)^2 Z_L \tag{10.10}$$

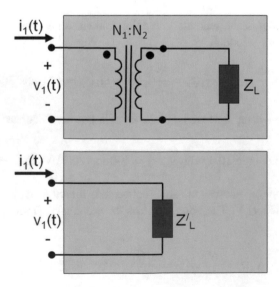

Fig. 10.6 Equivalent impedance for a transformer circuit

Of course, as was stated before, replacing circuit elements by their equivalents will hide the voltages and currents associated with the circuit elements that have been replaced by the new equivalent components. For transformers, this means that $v_2(t)$ and $i_2(t)$ have been hidden when the transformer is replaced by its equivalent impedance.

Example 10.6 Resolve for $i_1(t)$, the current flowing out of the source and into the primary winding, for the transformer circuit from Example 10.4 using the equivalent impedance of the transformer.

Solution The transformer and 500 Ω resistor can be replaced by a single resistor given by

$$Z_L' = \left(\frac{N_1}{N_2}\right)^2 Z_L = \left(\frac{1}{8}\right)^2 500\,\Omega = 7.8125\,\Omega$$

Thus, the new circuit is given by

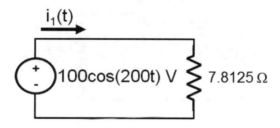

Thus,

$$i_1(t) = \frac{100\cos(200t)\text{ V}}{7.8125\,\Omega} = 12.8\cos(200t)\text{ A}$$

Example 10.7 Find the current between the transformers, $\tilde{I}_{\text{line}}$, as well as the voltage across the load, $\tilde{V}_{\text{load}}$, for the transformer circuit shown in Fig. 10.7 when $v_s(t) = 15\cos(2500t)$ V.

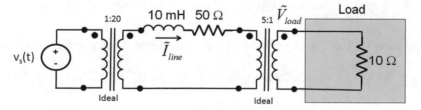

Fig. 10.7 Transformer circuit for Example 10.7

Solution Begin by replacing the inductor with its equivalent impedance given by

$$j\omega L = j2500 \cdot 10 \cdot 10^{-3} = j25\,\Omega$$

Also, since the source voltage, $v_s(t)$ is direct across the primary winding of the first transformer, we can determine the voltage across the secondary winding for this transformer as

$$v_2(t) = \frac{N_2}{N_1} v_1(t) \Rightarrow 20 \cdot v_s(t) = 300 \cos(2500t) \text{ V}$$

In addition, we can replace the second transformer and load with its equivalent impedance given by

$$Z_L' = \left(\frac{N_1}{N_2}\right)^2 Z_L \Rightarrow R_L' = \left(\frac{5}{1}\right)^2 (10\,\Omega) = 250\,\Omega$$

Therefore, the circuit can be redrawn as

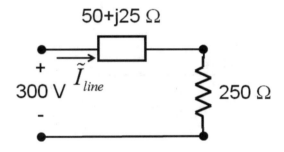

The line current is thus given by

$$\tilde{I}_{\text{line}} = \frac{300}{50 + j25 + 250} = 0.9965 \exp(-j0.0831) \text{ A}$$
$$\Rightarrow i_{\text{line}}(t) = 0.9965 \cos(2500t - 0.0831) \text{ A}$$

With the current in the line known, we can then find the current in the load using the transformer equation for current to yield

$$i_2(t) = \frac{N_1}{N_2} i_1(t) \Rightarrow \tilde{I}_{\text{load}} = \left(\frac{5}{1}\right)\tilde{I}_{\text{line}} = 4.9827 \exp(-j0.0831) \text{ A}$$

The voltage across the load can then be found from the current flowing in the load from

$$\tilde{V}_{\text{load}} = \tilde{I}_{\text{load}} \cdot R_{\text{load}} = (4.9827 \exp(-j0.0831) \text{ A}) \cdot (10\,\Omega) = 49.827 \exp(-j0.0831) \text{ V}$$

Example 10.8 Find the current between the transformers, $\tilde{I}_{\text{line}}$, as well as the voltage across the load, $\tilde{V}_{\text{load}}$, for the transformer circuit shown in Fig. 10.8 when $v_s(t) = 250 \cos(120\pi t)$ V.

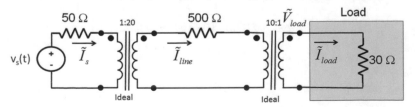

Fig. 10.8 Transformer circuit for Example 10.8

Solution Once again, we will replace the second transformer and its load by their equivalent impedance given by

$$Z'_L = \left(\frac{N_1}{N_2}\right)^2 Z_L \Rightarrow R'_L = \left(\frac{10}{1}\right)^2 (30\,\Omega) = 3000\,\Omega$$

However, we cannot reflect the source voltage across the 1st transformer as we did in Example 10.7 because the source voltage is NOT across the primary winding of the transformer. Instead, the source voltage is reduced by the voltage drop across the 50 Ω resistor before reaching the primary winding. Therefore, a better approach is to combine R'_L with the 500 Ω line resistance and then reflect the entire resistance across the 1st transformer as shown below.

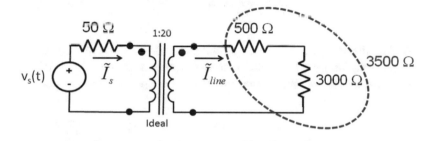

$$Z'_L = \left(\frac{N_1}{N_2}\right)^2 Z_L \Rightarrow R'_2 = \left(\frac{1}{20}\right)^2 (500\,\Omega + 3000\,\Omega) = 8.75\,\Omega$$

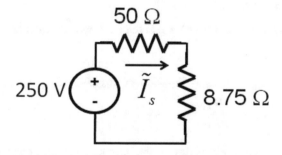

Therefore, the current from the source is given by

$$\tilde{I}_s = \frac{\tilde{V}_s}{50\,\Omega + 8.75\,\Omega} = \frac{250\text{ V}}{58.75\,\Omega} = 4.25532\text{ A}$$

The source current is also the same current that is flowing into the primary winding of the first transformer while the line current is the current flowing out of the secondary winding of the first transformer. The line current is thus given by

$$i_2(t) = \frac{N_1}{N_2} i_1(t) \Rightarrow \widetilde{I}_{\text{line}} = \left(\frac{1}{20}\right)\widetilde{I}_s = 0.212766 \text{ A}$$

Similarly, since the line current flows into the primary winding of the second transformer, the load current can be found from

$$i_2(t) = \frac{N_1}{N_2} i_1(t) \Rightarrow \widetilde{I}_{\text{load}} = \left(\frac{10}{1}\right)\widetilde{I}_{\text{line}} = 2.12766 \text{ A}$$

with the load voltage given by

$$\widetilde{V}_{\text{load}} = \widetilde{I}_{\text{load}} \cdot R_{\text{load}} = 63.8298 \text{ V}$$

Example 10.9 Find the current in the line, $\widetilde{I}_{\text{line}}$, the power flowing out of the source, and the turns ratio of the second transformer (N), for the transformer circuit shown in Fig. 10.9 given that the power delivered to the load is 800 W. Assume that most of the power from the source is delivered to the load.

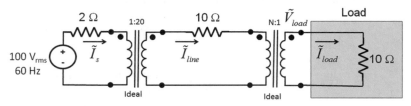

Fig. 10.9 Transformer circuit for Example 10.9

Solution Since we know the power flowing into the load, we can calculate the current flowing into the load.

$$P = \widetilde{V}_{\text{rms}}\widetilde{I}_{\text{rms}}^* \Rightarrow P_{\text{load}} = \widetilde{V}_{\text{load}}\widetilde{I}_{\text{load}}^* = \left(R_{\text{load}}\widetilde{I}_{\text{load}}\right)\widetilde{I}_{\text{load}}^* = R_{\text{load}}\left|\widetilde{I}_{\text{load}}\right|^2$$

$$\left|\widetilde{I}_{\text{load}}\right| = \sqrt{\frac{P_{\text{load}}}{R_{\text{load}}}} = 8.94427 \text{ A}_{\text{rms}}$$

However, since the circuit is purely resistive, the current must be purely real with no imaginary component (i.e., phase of 0 radians). Therefore, $\left|\widetilde{I}_{\text{load}}\right| = \widetilde{I}_{\text{load}} = 8.94427 \text{ A}_{\text{rms}}$. We can now reflect the load impedance and the load current across the 2nd transformer giving us

If we now reflect the currents and impedances across the first transformer we get

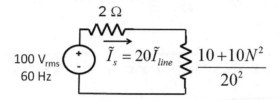

Therefore, the current from the source is given by both

$$\tilde{I}_s = 20\tilde{I}_{\text{line}} = \frac{178.88544}{N}$$

and

$$\tilde{I}_s = \frac{100\ V}{2 + \left(\frac{10+10N^2}{20^2}\right)}$$

Setting these currents equal gives

$$\tilde{I}_s = \frac{100\ V}{2 + \left(\frac{10+10N^2}{20^2}\right)} = \frac{178.88544}{N} \Rightarrow 100N = 357.77088 + \frac{1788.8544 + 1788.8544N^2}{400}$$

$$\Rightarrow 1788.8544N^2 - 40000N + 144897.2064 = 0$$

This quadratic has two roots corresponding to

$$N = 17.8136 \text{ or } N = 4.5471$$

In order to determine which root is correct, we need to use our assumption that most of the power from the source is delivered to the load. We already know that the power delivered to the load is 800 W, so we only need to find the power from the source for each turns ratio. The source current for each turns ratio is given by

$$\tilde{I}_s = \frac{178.88544}{N} = 10.0421\ A_{\text{rms}} \quad or \quad 39.3406\ A_{\text{rms}}$$

Therefore, the power from the source is given by

$$P = \tilde{V}_{\text{rms}} \tilde{I}_{\text{rms}}^* \Rightarrow P_s = \tilde{V}_s \tilde{I}_s^* = 1004.2\ W \quad or \quad 3934.1\ W$$

The 1004.2 W case corresponding to $N = 17.8136$ is the closest to the 800 W being delivered to the load. Therefore, this is the correct turns ratio for this problem. With the turns ratio known, we can now find the current in the line.

$$\widetilde{I}_{\text{line}} = \frac{\widetilde{I}_{\text{load}}}{N} = \frac{8.94427}{N} = 0.5021 \ A_{\text{rms}}$$

10.4 Improved Power Transfer Efficiency with Transformers

As was mentioned previously, the efficient transfer of power is just as critical as the efficient generation of power. Transformers greatly improve the efficiency of power delivery due to the ability to step up or increase the voltage before transmission and then step down or decrease the voltage where the electricity is needed. The improvement in efficiency achieved when using a transformer can be illustrated by the following examples.

Example 10.10 Determine the power efficiency when a load is connected directly to a transmission line as shown in Fig. 10.10. The load consists of 1000 light bulbs (60 W, 240 Ω) in parallel operating at a voltage of 120 V_{rms}. The transmission line is 5 km long (3.1 miles) and has a resistance of 0.06 Ω/km.

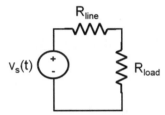

Fig. 10.10 Load connected directly to the transmission line for Example 10.10

Solution Since the load consists of 1000 light bulbs in parallel, the load resistance is given by

$$R_{\text{load}} \cong \frac{1}{\sum \frac{1}{R_i}} = \frac{R_i}{1000} = 0.24 \ \Omega$$

Since, the light bulbs require a voltage of 120 V_{rms}, the current flowing into the load is given by

$$\widetilde{I}_{\text{load}} \cong \frac{120 \ V_{\text{rms}}}{0.24 \ \Omega} = 500 \ A_{\text{rms}}$$

giving a power into the load of

$$P = \widetilde{V}_{\text{rms}}\widetilde{I}_{\text{rms}}^* \Rightarrow P_{\text{load}} = \widetilde{V}_{\text{load}}\widetilde{I}_{\text{load}}^* = (120 \ V_{\text{rms}})(500 \ A_{\text{rms}}) = 60 \ \text{kW}$$

The resistance of the transmission line is given by

$$R_{\text{line}} = (0.06 \ \Omega/\text{km})(5 \, \text{km}) \cong 0.3 \ \Omega$$

Since the load and the transmission line are in series, the load current is the same as the line current. Therefore, the power lost in the line is given by

$$P = \tilde{V}_{\text{rms}}\tilde{I}_{\text{rms}}^* \Rightarrow P_{\text{line}} = \tilde{V}_{\text{line}}\tilde{I}_{\text{line}}^* = \tilde{V}_{\text{line}}\tilde{I}_{\text{load}}^* = \left(R_{\text{line}}\tilde{I}_{\text{load}}\right)\tilde{I}_{\text{load}}^* = R_{\text{line}}\left|\tilde{I}_{\text{load}}\right|^2 = 75 \text{ kW}$$

Therefore, the total power supplied by the source is 60 kW + 75 kW = 135 kW. Therefore, the efficiency of power delivery is

$$\eta = 100\frac{P_{\text{load}}}{P_s} = 100\frac{P_{\text{load}}}{P_{\text{load}} + P_{\text{line}}} = 44.4444\%$$

Example 10.11 Repeat Example 10.10 with a transformer between the load and the transmission line as shown in Fig. 10.11. Assume that the transformer is ideal and the turns ratio, $N_1{:}N_2$ is 10:1.

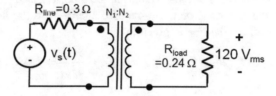

Fig. 10.11 Load connected to the transmission line through a transformer for Example 10.11

Solution Since the load and the voltage across the load are the same, the power being delivered to the load is still 60 kW and the current flowing in the load is still 500 A_{rms}. However, the line current is not the same as the load current due to the transformer. Instead, it is given by

$$i_2(t) = \frac{N_1}{N_2}i_1(t) \Rightarrow \tilde{I}_{\text{load}} = 10\tilde{I}_{\text{line}} \Rightarrow \tilde{I}_{\text{line}} = 50 \text{ } A_{\text{rms}}$$

Hence, the power lost in the line is

$$P_{\text{line}} = R_{\text{line}}\left|\tilde{I}_{\text{line}}\right|^2 = 750 \text{ W}$$

Therefore, the efficiency is given by

$$\eta = 100\frac{P_{\text{load}}}{P_s} = 100\frac{P_{\text{load}}}{P_{\text{load}} + P_{\text{line}}} = 98.7654\%$$

Hence, the use of the transformer dramatically improves the efficiency of power delivery.

Example 10.12 A power delivery system is needed to deliver 100 V DC to a 6 Ω load. Find the maximum length of the transmission line if 85% of the power from the source must be delivered to the load and the resistance of the transmission line is 9 Ω/km.

Solution The current at the load is given by

$$I_{\text{load}} = \frac{V_{\text{load}}}{R_{\text{load}}} = 16.667 \text{ A}$$

and the power delivered to the load is

$$P_{\text{load}} = I_{\text{load}}^2 R_{\text{load}} = 1666.7 \text{ W}$$

Given that the system must have an efficiency of 85%, the power from the source must be

$$P_s = \frac{P_{\text{load}}}{\eta} = \frac{1666.7 \text{ W}}{0.85} = 1960.8 \text{ W}$$

Therefore, the power lost in the line is

$$P_{\text{line}} = 0.15 P_s = 294.1 \text{ W}$$

Since the circuit is operating at DC, no transformers can be used as transformers do not work at DC. Therefore, the current in the line must be the same as the current in the load. This means that the maximum resistance allowed for the line is given by

$$P_{\text{line}} = I_{\text{line}}^2 R_{\text{line}} = I_{\text{load}}^2 R_{\text{line}} \Rightarrow R_{\text{line}} = \frac{P_{\text{line}}}{I_{\text{load}}^2} = 1.0588 \ \Omega$$

The maximum length of the transmission line is thus given by

$$Length = \frac{R_{\text{line}}}{9 \ \Omega/\text{km}} = 0.1176 \text{ km}$$

Therefore, the power plant must be practically next door.

Example 10.13 Repeat Example 10.12 utilizing a power delivery system with a transformer with a turns ratio of 13 to 1 as shown in Fig. 10.12. Assume this time the load needs to deliver an AC voltage of 100 V_{rms} to the load.

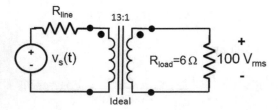

Fig. 10.12 Power delivery system for Example 10.13

Solution The current and power at the load are now given by

$$\widetilde{I}_{\text{load}} = \frac{\widetilde{V}_{\text{load}}}{R_{\text{load}}} = 16.667 \ A_{\text{rms}}$$

$$P = \widetilde{V}_{\text{rms}} \widetilde{I}_{\text{rms}}^* \Rightarrow P_{\text{load}} = \widetilde{V}_{\text{load}} \widetilde{I}_{\text{load}}^* = R_{\text{load}} \left| \widetilde{I}_E \right|^2 = 1666.7 \text{ W}$$

Since the power reaching the load is the same as before, the power from the source and the power delivered to the line must also be the same.

$$P_s = \frac{P_{\text{load}}}{\eta} = \frac{1666.7 \text{ W}}{0.85} = 1960.8 \text{ W}$$

$$P_{\text{line}} = 0.15 P_s = 294.1 \text{ W}$$

However, the line current and load current are not the same due to the use of the transformer. Specifically,

$$\tilde{I}_{\text{line}} = \frac{\tilde{I}_{\text{load}}}{13} = 1.282 \text{ } A_{\text{rms}}$$

Therefore, the maximum resistance of the line is given by

$$R_{\text{line}} = \frac{P_{\text{line}}}{\left|\tilde{I}_{\text{line}}\right|^2} = 178.9 \text{ } \Omega$$

The maximum length of the transmission line is thus given by

$$Length = \frac{R_{\text{line}}}{9 \text{ } \Omega/\text{km}} = 19.88 \text{ km}$$

This is 169 times farther (turns ratio squared) than when no transformer was used.

   Clearly, transformers have a significant impact on the efficiency of power delivery allowing power plants and consumers to be separated by a considerable distance. In fact, transformers are the primary reason our modern wall outlets operate at AC. In the 1880s, there was a "War of the Currents" between Thomas Edison, George Westinghouse, and Nikola Tesla. Thomas Edison was pushing for the adoption of DC power as the standard as it easily interfaced with the existing technology at the time. Westinghouse and Tesla, however, were encouraging the use of AC power due to its transmission efficiency with transformers even though there was very little existing technology at the time that operated at AC. Eventually, the need for efficient delivery won out and AC power became the standard.

10.5   Circuit Analysis with Non-ideal Transformers

Much of the prior analysis assumed that the transformers were ideal. However, real transformers have losses and core magnetic reluctances which must be considered. Figure 10.13 shows the circuit model for the non-ideal transformer. At its heart, the model has an ideal transformer to capture the step up or down of the voltage/current that is the primary purpose of the transformer. The model also has an inductor, L_m, to model the finite magnetic reluctance of the core, and a resistor, R_c, to model electrical losses in the core. The core losses result from the rearrangement of the magnetic domains in the core during each cycle (hysteresis losses) as well as losses due to the establishment of eddy currents in the core. The non-ideal transformer model also includes the coil resistances of the primary, R_1, and secondary, R_2, windings. Lastly, since some of the magnetic fields/flux will escape from the core, flux leakage for the primary and secondary windings is captured by L_1 and L_2, respectively. Typical values for the different components for a power transformer are given in Table 10.1.

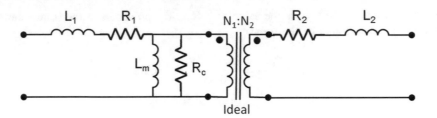

Fig. 10.13 Circuit model for non-ideal transformer

Table 10.1 Typical values for components in non-ideal transformer model for a transformer that might be used in the transmission of electrical power

Circuit quantity	Typical power transformer values	Ideal transformer values
Coil resistances (R_1 and R_2) (Depends on # of windings)	<5 Ω	0 Ω
Leakage inductances (L_1 and L_2) (Depends on # of windings)	<30 mH	0 H
Magnetizing reactance (L_m)	>25 H	∞ H
Core losses, (R_c)	>100 kΩ	∞ Ω

Example 10.14 Find the power from the source, power delivered to the load, and efficiency for the non-ideal transformer circuit shown in Fig. 10.14.

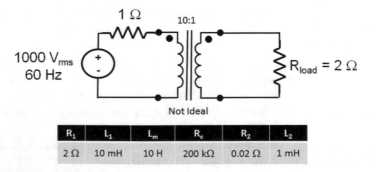

R_1	L_1	L_m	R_c	R_2	L_2
2 Ω	10 mH	10 H	200 kΩ	0.02 Ω	1 mH

Fig. 10.14 Non-ideal transformer circuit for Example 10.14

Solution The first step is to replace the non-ideal transformer with its equivalent circuit as shown below.

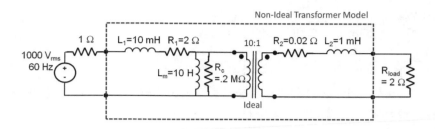

The next step is to convert the inductances into impedances (i.e., $j\omega L$) for the phasor domain analysis.

$$1 \text{ mH} \rightarrow j0.37699 \ \Omega$$
$$10 \text{ mH} \rightarrow j3.7699 \ \Omega$$
$$10 \text{ H} \rightarrow j3769.9 \ \Omega$$

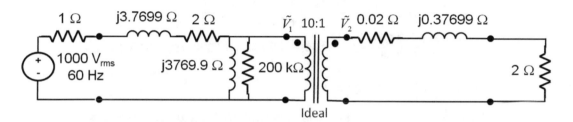

Combining impedances in series gives

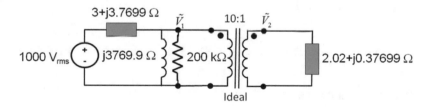

The transformer can then be replaced by its equivalent impedance by multiplying the impedance connected to the secondary by the turns ratio squared (i.e., $\times 100$) giving us

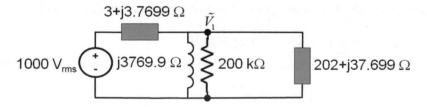

$\tilde{V}_1$ can then be solved using the node-voltage method.

$$\frac{\tilde{V}_1 - 1000\,V_{\text{rms}}}{3 + j3.7699} + \frac{\tilde{V}_1}{j3769.9} + \frac{\tilde{V}_1}{200k} + \frac{\tilde{V}}{202 + j37.699} = 0$$

$$\tilde{V}_1\left(\frac{1}{3 + j3.7699} + \frac{1}{j3769.9} + \frac{1}{200k} + \frac{1}{202 + j37.699}\right) = \frac{1000}{3 + j3.7699}$$

$$\tilde{V}_1\left((0.1292 - j0.1624) + (-j.00026526) + (5 \cdot 10^{-6}) + (0.0048 - j0.0009)\right) = (129.24 - j162.41)$$

$$\tilde{V}_1(0.1340 - j0.1636) = (129.24 - j162.41)$$

$$\tilde{V}_1 = \frac{(129.24 - j162.41)}{(0.1340 - j0.1636)}\frac{(0.1340 + j0.1636)}{(0.1340 + j0.1636)} = \frac{43.8877 - j0.6281}{0.0447} = 981.41 - j14.045$$

$$= 981.5115\exp(-j0.0143)\,V_{\text{rms}}$$

With the voltage $\tilde{V}_1$ known, the current from the source can be found from

$$\tilde{I}_s = \frac{1000\,V_{\text{rms}} - \tilde{V}_1}{3 + j3.7699} = \frac{1000\,V_{\text{rms}} - 981.41 - j14.045}{3 + j3.7699}$$

$$= \frac{18.589 + j14.0453}{3 + j3.7699} = 4.6836 - j1.2038 = 4.8358\exp(-j0.2516)\,A_{\text{rms}}$$

The power from the source is then given by

$$P_s = \left|\tilde{V}_s\right|\left|\tilde{I}_s\right|\cos(\theta_v - \theta_i) = 1000 \cdot 4.8358\cos(0 - -0.2516) = 4683.6\,\text{W}$$

In addition to finding the source power from $\tilde{V}_1$, we can also reflect the voltage across the ideal transformer in the model to find $\tilde{V}_2$.

$$v_2(t) = \frac{N_2}{N_1}v_1(t) \Rightarrow \tilde{V}_2 = \frac{\tilde{V}_1}{10} = 98.15115\exp(-j0.0143)\,V_{\text{rms}}$$

With $\tilde{V}_2$ known, $\tilde{I}_2 = \tilde{I}_{\text{load}}$ can be found from

$$\tilde{I}_2 = \tilde{I}_{\text{load}} = \frac{\tilde{V}_2}{2.02 + j0.37699} = 46.824 - j9.434 = 47.765\exp(-j0.1988)\,A_{\text{rms}}$$

The voltage drop across the load and power delivered to the load are thus given by

$$\tilde{V}_{\text{load}} = R_{\text{load}}\tilde{I}_{\text{load}} = R_{\text{load}}\tilde{I}_2 = 95.5299\exp(-j0.1988)\,V_{\text{rms}}$$

$$P_{\text{load}} = \left|\tilde{V}_{\text{load}}\right|\left|\tilde{I}_{\text{load}}\right|\cos(\theta_v - \theta_i) = 95.5299 \cdot 47.765\cos(-0.1988 - -0.1988) = 4563\,\text{W}$$

The efficiency is thus given by

$$\eta = 100\frac{P_{\text{load}}}{P_s} = 97.43\%$$

When solving non-ideal transformer circuits, the analysis is sometimes complicated by having the magnetizing reactance and core losses immediately adjacent to the ideal transformer in the model.

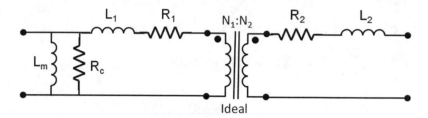

Fig. 10.15 Simplified circuit model for non-ideal transformer

The analysis can be simplified in some cases by moving L_m and R_c before the coil resistance and leakage inductance for the primary as shown in Fig. 10.15. In most cases, this simplification has a negligible effect on the final results.

Example 10.15 Find voltage across the load, $\tilde{V}_1$, for the non-ideal transformer circuit shown in Fig. 10.16 using the simplified model for the non-ideal transformer shown in Fig. 10.15.

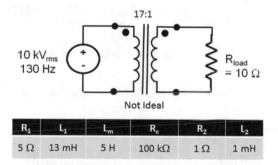

R_1	L_1	L_m	R_c	R_2	L_2
5 Ω	13 mH	5 H	100 kΩ	1 Ω	1 mH

Fig. 10.16 Non-ideal transformer circuit for Example 10.15

Solution We will once again first redraw the circuit with the non-ideal transformer replaced by its model.

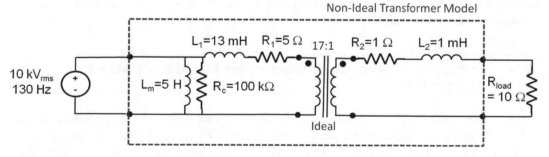

Replacing the inductors by their impedances and then reflecting the impedance from the secondary side of the transformer to the primary side gives

$$1 \text{ mH} \rightarrow j0.8168 \ \Omega$$
$$13 \text{ mH} \rightarrow j10.6186 \ \Omega$$
$$5 \text{ H} \rightarrow j4084.1 \ \Omega$$

$$Z_2 = R_{\text{load}} + R_2 + j\omega L_2 = 10\ \Omega + 1\ \Omega + j0.8168\ \Omega = 11 + j0.8168\ \Omega$$

$$Z_2' = \left(\frac{N_1}{N_2}\right)^2 Z_2 = 17^2 Z_2 = 3179 + j236.06\ \Omega$$

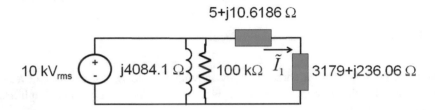

The current flowing into the primary side of the ideal transformer in the model, $\tilde{I}_1$, can then be found from

$$\tilde{I}_1 = \frac{10000\ V_{\text{rms}}}{(5 + j10.6186) + (3179 + j236.06)} = 3.122 - j0.2419 = 3.1313\exp(-j0.0773)\ A_{\text{rms}}$$

Once $\tilde{I}_1$ is known, the current flowing in the load can be found from

$$i_2(t) = \frac{N_1}{N_2} i_1(t) \Rightarrow \tilde{I}_2 = \tilde{I}_{\text{load}} = 17\tilde{I}_1 = 53.2324\exp(-j0.0773)\ A_{\text{rms}}$$

The voltage across the load is then given by

$$\tilde{V}_{\text{load}} = R_{\text{load}}\tilde{I}_{\text{load}} = R_{\text{load}}\tilde{I}_2 = 532.324\exp(-j0.0773)\ V_{\text{rms}}$$

As a comparison, if we were to use the complete model shown in Fig. 10.14, the load voltage would be given by $530.9216\exp(-j0.0762)\ V_{\text{rms}}$. Similarly, if we assumed the transformer was ideal, the voltage across the load would be $588.2353\ V_{\text{rms}}$. Therefore, the simplified model gives a more accurate estimate for the voltages and currents than the ideal case, but is much less computationally intensive in many situations.

10.6 Simultaneous Reflection of Sources and Impedances Across a Transformer

When working with transformers previously, we normally reflected the impedances from the secondary side of the transformer to the primary side before solving for the voltages or currents in the circuit. Once the voltage or current on the primary side was known, we then used the turns ratio to determine the voltage or current on the secondary side. This approach was utilized as it is relatively straightforward and typically results in less confusion when first learning transformer circuit analysis. However, both voltage sources and impedances can be reflected across the transformer when solving the circuit if done properly. When reflecting sources, the amplitude of the source is adjusted by the appropriate turns ratio as illustrated in the next example.

Example 10.8 Refind the current between the transformers, $\widetilde{I}_{line}$, for the transformer circuit shown in Fig. 10.8 when $v_s(t) = 250\cos(120\pi t)$ V by first reflecting the voltage source and all of the impedances to the center branch between the two transformers.

Solution Once again, we will replace the second transformer and its load by their equivalent impedance given by

$$Z'_L = \left(\frac{N_1}{N_2}\right)^2 Z_L \Rightarrow R'_L = \left(\frac{10}{1}\right)^2 (30\ \Omega) = 3000\ \Omega$$

We will then reflect both the voltage source, $v_s(t)$, and the 50 Ω resistance adjacent to the source across the primary winding. Both must be reflected or the voltage drop across the 50 Ω resistor will not be correctly captured. Therefore,

$$Z'_L = \left(\frac{N_1}{N_2}\right)^2 Z_L \Rightarrow (50\ \Omega) \rightarrow (50\ \Omega)\left(\frac{N_2}{N_1}\right)^2 = (50\ \Omega)(20)^2 = 20,000\ \Omega$$

$$v_2(t) = \frac{N_2}{N_1} v_1(t) \Rightarrow v_s(t) \rightarrow \frac{N_2}{N_1} v_s(t) = 20v_s(t)$$

Redrawing the circuit after reflecting the impedance and the voltage source gives

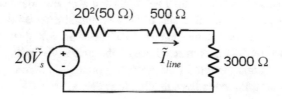

Therefore,

$$\widetilde{I}_{line} = \frac{20\widetilde{V}_s}{20000\ \Omega + 500\ \Omega + 3000\ \Omega} = \frac{5000\ \text{V}}{23500\ \Omega} = 0.212766\ \text{A}$$

Given that both voltage sources and impedances can be reflected across the transformer, the model for the non-ideal transformer can be redrawn as either of the cases shown in Fig. 10.17. These transformer models can simplify the analysis in some cases. However, they should be used with care as it is easy to forget to reflect one of the impedances or source values if you are not experienced in circuit analysis.

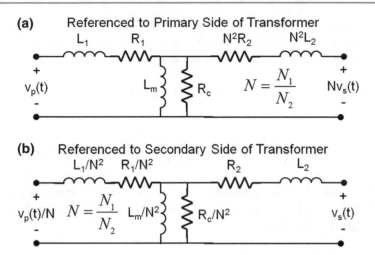

Fig. 10.17 Circuit model for the non-ideal transformer with everything reflected to the primary (**a**) or secondary (**b**) side of the transformer

10.7 Problems

Problem 10.1: A transformer has 500 primary windings and 100 secondary windings. The core of the transformer has a mean path length of 3 cm, a cross-sectional area of 0.15 cm², and a magnetic permeability of 6×10^{-3} H/m. A voltage source with a Thevenin voltage of $v_{Th}(t) = 15\cos(200t)$ V and a Thevenin resistance of 5 Ω is connected across the primary winding of the transformer. The secondary winding is left open circuited (i.e., no load connected). What is the voltage across the secondary winding assuming the coils have negligible resistance?

Problem 10.2: A transformer has 50 primary windings and 500 secondary windings. The core of the transformer has a mean path length of 2.5 cm, a cross-sectional area of 0.1 cm², and a magnetic permeability of 6×10^{-3} H/m. A voltage source with a Thevenin voltage of $v_{Th}(t) = 25\cos(700t)$ V and a Thevenin resistance of 2 Ω is connected across the primary winding of the transformer. The secondary winding is left open circuited (i.e., no load connected). What is the voltage across the secondary winding assuming the coils have negligible resistance?

Problem 10.3: A transformer has 200 primary windings and 500 secondary windings. The core of the transformer has a mean path length of 2 cm, a cross-sectional area of 0.25 cm², and a magnetic permeability of 6×10^{-3} H/m. A voltage source with a Thevenin voltage of $v_{Th}(t) = 80\cos(100t)$ V and a Thevenin resistance of 50 Ω is connected across the primary winding of the transformer. The secondary winding is left open circuited (i.e., no load connected). What is the voltage across the secondary winding assuming the coils have negligible resistance?

Problem 10.4: A 500 V_{rms} signal needs to be reduced to 120 V_{rms}. What should be the turns ratio assuming the transformer is ideal?

Problem 10.5: A 100 V_{rms} signal needs to be increased to 700 V_{rms}. What should be the turns ratio assuming the transformer is ideal?

Problem 10.6: An ideal transformer has a turns ratio of $(N_2/N_1) = 15$. If the time-varying current on the primary side has an amplitude of 15 A_{rms}, what is the current amplitude on the secondary side?

Problem 10.7: An ideal transformer has a turns ratio of $(N_1/N_2) = 20$. If the time-varying current on the primary side has an amplitude of 15 A_{rms}, what is the current amplitude on the secondary side?

Problem 10.8: An ideal transformer has a turns ratio of $(N_1/N_2) = 20$. If the time-varying current on the secondary side has an amplitude of 15 A_{rms}, what is the current amplitude on the primary side?

Problem 10.9: Find all of the voltages and currents for the transformer circuit shown below assuming that the transformer is ideal. Also, use $v_{in}(t) = 20\cos(100t)$ V, $R_{load} = 80\ \Omega$, and $N_2/N_1 = 10$.

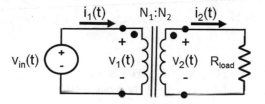

Problem 10.10: Find all of the voltages and currents for the transformer circuit shown below assuming that the transformer is ideal. Also, use $v_{in}(t) = 20\cos(100t)$ V, $R_{load} = 80\ \Omega$, and $N_1/N_2 = 10$.

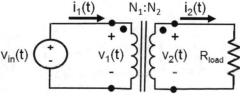

Problem 10.11: Find the equivalent impedance looking into the primary side of the transformer for the following ideal transformer circuits assuming $R_{load} = 80\ \Omega$, and $N_2/N_1 = 10$.

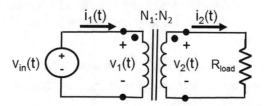

Problem 10.12: Find the equivalent impedance looking into the primary side of the transformer for the following ideal transformer circuits assuming $R_{load} = 80\ \Omega$, and $N_1/N_2 = 10$.

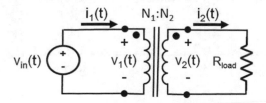

Problem 10.13: This problem deals with the following ideal transformer circuit where $v_s(t) = 150\cos(10000t)$ V.

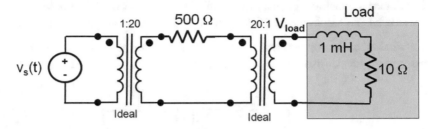

(a) Find the current in the 500 Ω resistor in phasor form.
(b) Find the voltage across the load, $v_{\text{load}}(t)$ in the time domain.

Problem 10.14: Find the current between the ideal transformers, $\widetilde{I}_{\text{line}}$, as well as the voltage across the load, $\widetilde{V}_{\text{load}}$ for the following transformer circuit when $v_s(t) = 150\cos(120\pi t)$ V, $R_{\text{line}} = 10$ Ω, and $R_{\text{load}} = 100$ Ω.

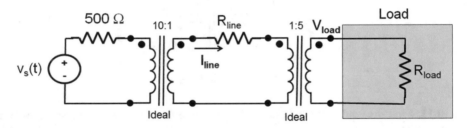

Problem 10.15: Could you use transformer to step down the 12.6 V from a car battery to a 4.2 V in order to charge a cell phone battery? If so, give the turns ratio needed for the transformer, if not explain why it would not work.

Problem 10.16: Find the current between the ideal transformers, $\widetilde{I}_{\text{line}}$, as well as the voltage across the load, $\widetilde{V}_{\text{load}}$ for the following transformer circuit when $v_s(t) = 135\cos(75\pi t)$ V and $R_{\text{load}} = 20$ Ω. Express your values in the phasor domain.

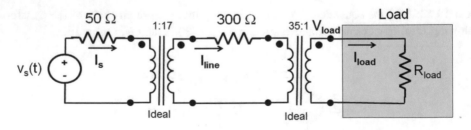

Problem 10.17: Find the current between the ideal transformers, $\tilde{I}_{line}$, as well as the voltage across the load, $\tilde{V}_{load}$ for the following transformer circuit when N = 15, $v_s(t) = 150\cos(120\pi t)$ V, $R_{line} = 100$ Ω, and $R_{load} = 10$ Ω.

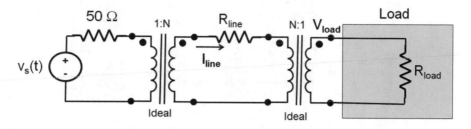

Problem 10.18: If $i_s(t) = 2\cos(200t)$ A, $R_{load} = 45$ Ω, and $v_s(t) = 500\cos(200t)$ V, find the turns ratio, N, for the second transformer for the circuit shown assuming ideal transformers.

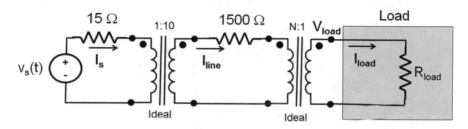

Problem 10.19: Find the current in the line, the power flowing out of the source, and turns ratio of the second transformer (N), given that the power delivered to the load is 500 W and the load resistance, shown as R_{load}, is 50 Ω for the following ideal transformer circuit. Assume that most of the power from the source is delivered to the load.

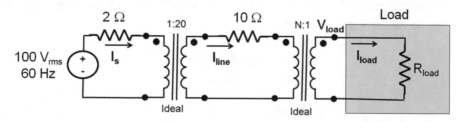

Problem 10.20: Find all possible turns ratios, N, for the first transformer and line currents, I_{line}, if the power delivered to the load is 235 W and $V_s(t) = 120\cos(100t)$ V_{rms} for the ideal transformer circuit shown if $R_1 = 10$ Ω, $R_2 = 500$ Ω, $R_{load} = 40$ Ω.

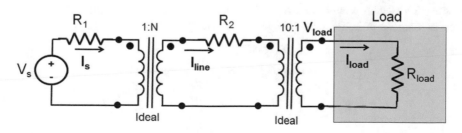

Problem 10.21: For the ideal transformer circuit shown below, find the voltage $\tilde{V}_1$ in phasor form, as well as the real power, reactive power, and apparent power delivered to the load.

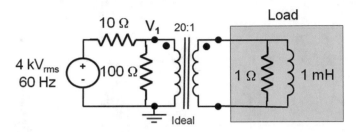

Problem 10.22: Assume a single ideal transformer is utilized between a transmission line and a load in order to improve efficiency as shown below. Given that we need 10 kW of power delivered to the load with an rms current of 50 A_{rms}, find the turns ratio which would achieve a 95% power efficiency while keeping the voltage at the source as small as possible.

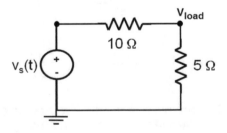

Problem 10.23: Your goal is to design a power delivery system that delivers power to a 5 Ω load. The load will be at the end of a 5 km long transmission line with a resistance of 2 Ω/km. The voltage needed at the load is 100 V_{rms}.

(a) Find the voltage required at the source and the power efficiency of the system, $\eta = 100 \frac{P_{load}}{P_{source}}$, when no transformers are used in the circuit.

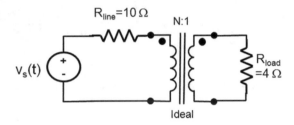

(b) Assuming a single ideal transformer is utilized between the transmission line and the load, find the turns ratio which would achieve a 90% power efficiency while keeping the voltage at the source as small as possible. The voltage at the load must still be 100 V_{rms}.

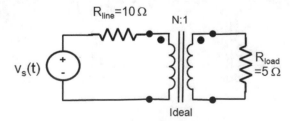

Problem 10.24: Your goal is to design a power delivery system shown below.

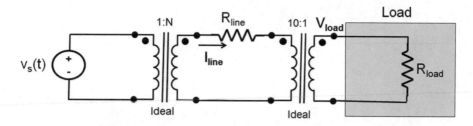

You know the following:

- The source voltage from your generator is $v_s(t) = 150 \cos(120\pi t)$ V
- The voltage needed at the load is $v_{load}(t) = 130 \cos(120\pi t)$ V
- The resistance of the load is $R_{load} = 5\ \Omega$
- 92% of the power from the source reaches the load.
- There are no inductors or capacitors in the system.

Find the resistance of the line, R_{line}, and the turns ratio, N, for the first transformer.

Problem 10.25: A non-ideal transformer is connected between a 10 k V_{rms} 130 Hz line and a 10 Ω load as shown below. Find the rms voltage across the load and the time average power coming from the source using the complete model for the non-ideal transformer.

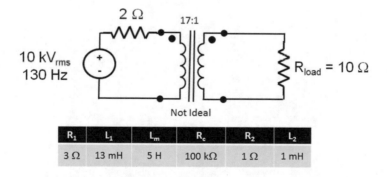

R_1	L_1	L_m	R_c	R_2	L_2
3 Ω	13 mH	5 H	100 kΩ	1 Ω	1 mH

Problem 10.26: A non-ideal transformer is connected between a 1 kV_{rms} 60 Hz line and a 2 Ω load as shown below. Find the rms voltage across the load using the simplified model for the non-ideal transformer.

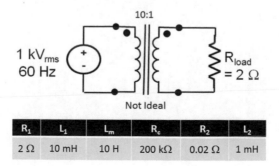

R_1	L_1	L_m	R_c	R_2	L_2
2 Ω	10 mH	10 H	200 kΩ	0.02 Ω	1 mH

Balanced Three-Phase Circuits

11

When driving in rural areas as well as in many cities, it is common to see power lines transmitting power to consumers. These power lines almost always have three relatively thick wires as shown in Fig. 11.1. Often, the power lines will also have a thinner wire accompanying the three primary wires as shown in Fig. 11.2. In addition, the power line might also have multiple sets of three wires, as shown in Fig. 11.3. However, in all cases, the three-wire layout is clearly seen.

Power lines consist of three dominate wires because power transmission is done in three different phases. A typical three-phase source in what is known as the wyc configuration is shown in Fig. 11.4. The source consists of three voltage sources whose voltages have the same amplitude but differ in phase by 120° (or $2\pi/3$ radians). For example,

$$
\begin{aligned}
v_{an}(t) &= V_o \cos(\omega t) \\
v_{bn}(t) &= V_o \cos(\omega t - 120°) \\
v_{cn}(t) &= V_o \cos(\omega t + 120°)
\end{aligned}
\tag{11.1}
$$

where for this case $v_{an}(t)$ sets the reference phase for the circuit. A plot of the different phases illustrating the 120° phase shift is shown in Fig. 11.5.

The nodes labeled a, b, and c in Fig. 11.4 would correspond to the three primary wires used in power transmission. The smaller fourth wire that is occasionally present would correspond to the neutral node labeled n. When the power system is working properly, very little (if any) current will

Fig. 11.1 Typical power lines consisting of three wires

© Springer Nature Switzerland AG 2020
T. A. Bigelow, *Electric Circuits, Systems, and Motors*,
https://doi.org/10.1007/978-3-030-31355-5_11

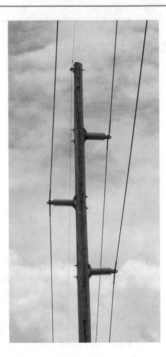

Fig. 11.2 Power lines with three primary wires and a smaller fourth wire

Fig. 11.3 Power lines consisting of multiple sets of three wires

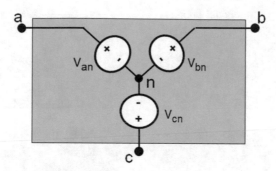

Fig. 11.4 Three-Phase voltage source in the Wye configuration

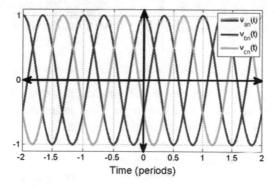

Fig. 11.5 Plot of three-phases given by Eq. (11.1)

flow along the wire associated with the n node. It is for this reason the optional fourth wire can be much smaller than the other three. In addition to its use in power transmission, three-phase power is also used in AC induction motors and synchronous generators described in Chap. 12. Therefore, this chapter will cover the basics of three-phased circuits.

11.1 Definition of Balanced Three-Phase Circuits

When working with three-phase circuits, it is normally assumed that all of the phases are balanced. This means that all of the voltage sources have the same amplitude and only differ in phase by exactly $120°$. In addition, the impedances from the source through the load are the same for each phase. This means that the impedance along each transmission line (a, b, and c) are all exactly the same. Lastly, the final load terminating each transmission line is also the same for each line. Part of the design process will include making the loads as balanced as possible.

Since three-phase power always involves three lines for the power flow, the load and source can be in either the delta configuration or the wye configuration. The concept of wye and delta configurations was originally introduced for resistive circuits in Chap. 1. However, each configuration is also shown in Fig. 11.6. The wye configuration is so named because the components are in the shape of a "Y" (or "T") with all three elements sharing a common node. Conversely, the delta configuration has all of the elements connected end to end in a triangle shape (greek letter Δ). Given that there are two possible configurations for the source (delta or wye) and two possible configurations for the load (delta or wye), there are a total of four possible configurations for balanced three-phase circuits (Wye-Wye, Delta-Delta, Wye-Delta, and Delta-Wye). Each of these configurations is shown in Fig. 11.7. Notice that the loads connected to or between the a, b, and c lines are all the same. Thus, all four circuit configurations are balanced.

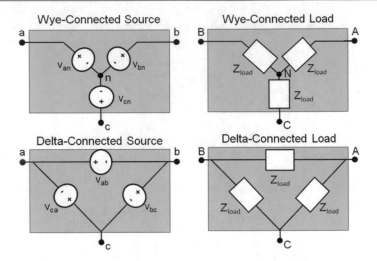

Fig. 11.6 Wye and delta connected sources and loads found in balanced three-phase circuits

When working with three-phase circuits, the normal convention is to use lower case subscripts when referencing the source side and upper case subscripts when referencing the load side. Hence, the line-to-line voltage from line a to b on the source side would be v_{ab} while the corresponding line-to-line voltage on the load side would be v_{AB} as shown in Fig. 11.7. The line currents flowing from the source to the load would have both upper and lower case subscripts (i.e., i_{aA}, i_{bB}, and i_{cC}) with the first subscript giving the node the current is flowing from and the second subscript giving the node the current is flowing to.

Example 11.1 How should the voltage drop across the line impedance for the a-line be specified?

Solution Since we anticipate the voltage at node a to be larger than the voltage at node A, a should be the first subscript and A should be the second subscript. Therefore, the voltage drop across the line impedance for the a-line should be written as v_{aA}.

One of the primary advantages of working with balanced three-phase circuits is that the currents and voltages associated with each line must mirror each other. Therefore, if we know the currents and voltages along one branch, then we will know the currents and voltages along all of the branches. Specifically, they will only differ in phase by 120° (or $2\pi/3$ radians) because the source voltages only differ in phase by 120° as given by

$$
\begin{aligned}
v_{an}(t) &= |V_{an}|\cos(\omega t + \theta_{V_{an}}) & v_{ab}(t) &= |V_{ab}|\cos(\omega t + \theta_{V_{ab}}) \\
v_{bn}(t) &= |V_{an}|\cos(\omega t + \theta_{V_{an}} - 120°) & v_{bc}(t) &= |V_{ab}|\cos(\omega t + \theta_{V_{ab}} - 120°) \\
v_{cn}(t) &= |V_{an}|\cos(\omega t + \theta_{V_{an}} + 120°) & v_{ca}(t) &= |V_{ab}|\cos(\omega t + \theta_{V_{ab}} + 120°)
\end{aligned}
\tag{11.2}
$$

For example, if

$$
\begin{aligned}
i_{aA}(t) &= |I_{aA}|\cos(\omega t + \theta_{I_{aA}}) \\
v_{AN}(t) &= |V_{AN}|\cos(\omega t + \theta_{V_{AN}}) \\
v_{AB}(t) &= |V_{AB}|\cos(\omega t + \theta_{V_{AB}}) \\
i_{AB}(t) &= |I_{AB}|\cos(\omega t + \theta_{I_{AB}}) \\
i_{ba}(t) &= |I_{ba}|\cos(\omega t + \theta_{I_{ba}})
\end{aligned}
\tag{11.3}
$$

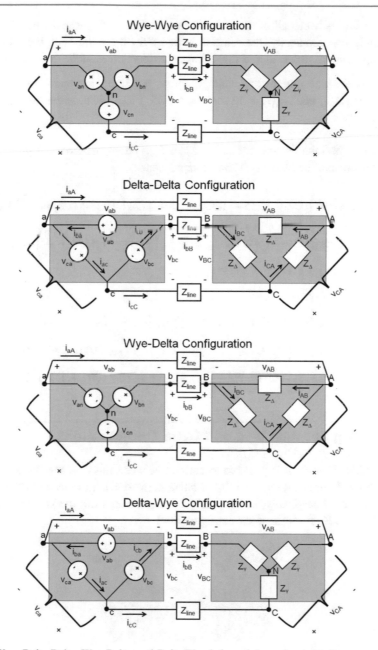

Fig. 11.7 Wye-Wye, Delta-Delta, Wye-Delta, and Delta-Wye balanced three-phase circuits

then

$$
\begin{aligned}
i_{\mathrm{bB}}(t) &= |I_{\mathrm{aA}}| \cos(\omega t + \theta_{I_{\mathrm{aA}}} - 120°) & i_{\mathrm{cC}}(t) &= |I_{\mathrm{aA}}| \cos(\omega t + \theta_{I_{\mathrm{aA}}} + 120°) \\
v_{\mathrm{BN}}(t) &= |V_{\mathrm{AN}}| \cos(\omega t + \theta_{V_{\mathrm{AN}}} - 120°) & v_{\mathrm{CN}}(t) &= |V_{\mathrm{AN}}| \cos(\omega t + \theta_{V_{\mathrm{AN}}} + 120°) \\
v_{\mathrm{BC}}(t) &= |V_{\mathrm{AB}}| \cos(\omega t + \theta_{V_{\mathrm{AB}}} - 120°) & v_{\mathrm{CA}}(t) &= |V_{\mathrm{AB}}| \cos(\omega t + \theta_{V_{\mathrm{AB}}} + 120°) \\
i_{\mathrm{BC}}(t) &= |I_{\mathrm{AB}}| \cos(\omega t + \theta_{I_{\mathrm{AB}}} - 120°) & i_{\mathrm{CA}}(t) &= |I_{\mathrm{AB}}| \cos(\omega t + \theta_{I_{\mathrm{AB}}} + 120°) \\
i_{\mathrm{cb}}(t) &= |I_{\mathrm{ba}}| \cos(\omega t + \theta_{I_{\mathrm{ba}}} - 120°) & i_{\mathrm{ac}}(t) &= |I_{\mathrm{ba}}| \cos(\omega t + \theta_{I_{\mathrm{ba}}} + 120°)
\end{aligned}
\tag{11.4}
$$

Since the relationship between all of the phases is known, the voltages and currents only need to be solved for a single line. Generally, the solution is found for the a-line with the a-line solution then shifted by the appropriate phase should the voltages/currents on another line be required.

Example 11.2 If the line current and voltage across the line impedance are given by $i_{aA}(t) = 4\cos(100\,t + 0.235)\,A_{rms}$ and $v_{aA}(t) = 12\cos(100\,t - 35°)\,V_{rms}$, what is the line current, $i_{bB}(t)$, and voltage, $v_{cC}(t)$?

Solution Let's first find the line current $i_{bB}(t)$. Since $i_{aA}(t)$ does not have a degree symbol with the phase term in the cosine, the phase is 0.235 radians. Hence,

$$i_{bB}(t) = 4\cos(100\,t + 0.235 - 120°)\,A_{rms} = 4\cos\left(100\,t + 0.235 - \frac{2\pi}{3}\right)A_{rms}$$

$$= 4\cos(100\,t - 1.8594)\,A_{rms} = 4\cos\left(100\,t - 1.8594\frac{180°}{\pi}\right)A_{rms}$$

$$= 4\cos(100\,t - 106.5355°)\,A_{rms}$$

Likewise, we can find $v_{cC}(t)$ by shifting the voltage $v_{aA}(t)$ by $+120°$.

$$v_{cC}(t) = 12\cos(100\,t - 35° + 120°)\,V_{rms} = 12\cos(100\,t + 85°)\,V_{rms}$$

$$= 12\cos\left(100\,t + 85°\frac{\pi}{180°}\right)V_{rms} = 12\cos(100\,t + 1.4835)\,V_{rms}$$

11.2 Converting Between Wye and Delta Configurations

As was discussed in Chap. 1, it is possible to convert between the wye and delta configurations. In Chap. 1, we restricted our attention to purely resistive networks, but the formulas can be generalized to impedances as well. Specifically, in Fig. 11.8, Z_1, Z_2, and Z_3 from the wye configuration can be related to Z_A, Z_B, and Z_C from the delta configuration by

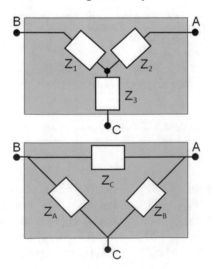

Fig. 11.8 Wye and Delta impedance networks

$$Z_A = \frac{Z_1 Z_2 + Z_1 Z_3 + Z_2 Z_3}{Z_1} \quad Z_1 = \frac{Z_B Z_C}{Z_A + Z_B + Z_C}$$
$$Z_B = \frac{Z_1 Z_2 + Z_1 Z_3 + Z_2 Z_3}{Z_2} \quad Z_2 = \frac{Z_A Z_C}{Z_A + Z_B + Z_C} \tag{11.5}$$
$$Z_C = \frac{Z_1 Z_2 + Z_1 Z_3 + Z_2 Z_3}{Z_3} \quad Z_3 = \frac{Z_A Z_B}{Z_A + Z_B + Z_C}$$

However, since we are only interested in balanced three-phase networks in this chapter, $Z_A = Z_B = Z_C = Z_\Delta$ and $Z_1 = Z_2 = Z_3 = Z_Y$. Therefore, Eq. (11.5) can be simplified as

$$Z_\Delta = 3Z_Y \tag{11.6}$$

As a result, the delta load can be converted to a wye load and vice versa without altering any of the other currents and voltages in the circuit by dividing or multiplying by a factor of 3.

In addition to exchanging delta and wye loads, it is possible to convert between the delta and wye source configurations. This is very similar to converting between Thevenin and Norton equivalent circuits as discussed in Chap. 3. Consider the wye connected sources shown in Fig. 11.7. The voltage between node a and node n is given by v_{an} while the voltage between node b and node n is v_{bn}. Therefore, by Kirchhoff's Voltage Law, the voltage v_{ab} must be given by $v_{ab}(t) = v_{an}(t) - v_{bn}(t)$. Converting this equation to the phasor domain and utilizing the fact that $\tilde{V}_{bn} = \tilde{V}_{an} \exp(-j120°)$ gives.

$$\tilde{V}_{ab} = \tilde{V}_{an} - \tilde{V}_{bn} = \tilde{V}_{an} - \tilde{V}_{an} \exp(-j120°) = \tilde{V}_{an}(1 - \exp(-j120°))$$
$$= \tilde{V}_{an}(1 - \cos(120°) + j\sin(120°)) = \tilde{V}_{an}\left(1 + 0.5 + j\frac{\sqrt{3}}{2}\right) \tag{11.7}$$
$$= \tilde{V}_{an} \cdot \sqrt{3} \exp\left(j\frac{\pi}{6}\right) = \tilde{V}_{an} \cdot \sqrt{3} \exp(j30°)$$

Repeating the derivation for the b and c lines gives

$$\tilde{V}_{bc} = \tilde{V}_{bn} \cdot \sqrt{3} \exp\left(j\frac{\pi}{6}\right) = \tilde{V}_{bn} \cdot \sqrt{3} \exp(j30°)$$
$$\tilde{V}_{ca} = \tilde{V}_{cn} \cdot \sqrt{3} \exp\left(j\frac{\pi}{6}\right) = \tilde{V}_{cn} \cdot \sqrt{3} \exp(j30°) \tag{11.8}$$

Therefore, wye and delta connected sources can be interchanged by scaling the amplitude by $\sqrt{3}$ and shifting the phase by 30° or π/6 radians. Since both the voltage sources and the loads can be converted between the delta and wye configuration, all the four configurations shown in Fig. 11.7 are equivalent after performing the appropriate source/load transformation.

Example 11.3 A wye connected source is connected to a wye connected load in a balanced three-phase circuit. The load consists of a 30 Ω resistor while the line impedance is a 5 Ω resistor as shown in Fig. 11.9. If $v_{an}(t) = 100\cos(100t)\ V_{rms}$, find the impedances and the source voltages in the phasor domain if the circuit is converted to the Delta-Delta configuration.

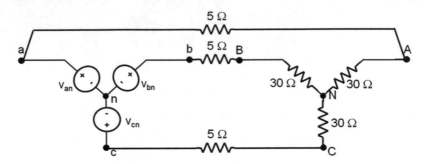

Fig. 11.9 Balanced three-phase circuit for Example 11.3

Solution The first step is to convert the load from the wye configuration to the delta configuration.

$$Z_\Delta = 3Z_Y = 3(30 \ \Omega) = 90 \ \Omega$$

Next, we need to convert the line-to-neutral voltage to a line-to-line voltage so that the wye connected source can be replaced by a delta connected source.

$$\tilde{V}_{ab} = \tilde{V}_{an} \cdot \sqrt{3} \exp\left(j\frac{\pi}{6}\right) = 100\sqrt{3} \exp\left(j\frac{\pi}{6}\right) V_{rms} = 173.2051 \exp(j0.5236) \ V_{rms}$$

Since, it is a balanced three-phase circuit, $\tilde{V}_{bn}$ and $\tilde{V}_{cn}$ are given by

$$\tilde{V}_{bc} = \left|\tilde{V}_{ab}\right| \exp\left(j\left(\theta_{V_{an}} - \frac{2\pi}{3}\right)\right) = 173.2051 \exp\left(-j\frac{\pi}{2}\right) V_{rms}$$
$$= 173.2051 \exp(-j1.5708) \ V_{rms}$$
$$\tilde{V}_{ca} = \left|\tilde{V}_{ab}\right| \exp\left(j\left(\theta_{V_{an}} + \frac{2\pi}{3}\right)\right) = 173.2051 \exp\left(j\frac{5\pi}{6}\right) V_{rms}$$
$$= 173.2051 \exp(j2.618) \ V_{rms}$$

The line impedance does not change. Therefore, the equivalent Delta-Delta circuit is given by

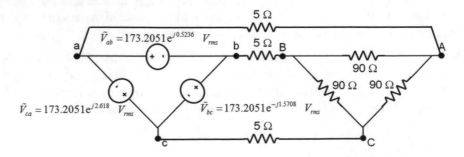

Example 11.4 A delta connected source is connected to a delta connected load in a balanced three-phase circuit. The load consists of a 30 Ω resistor in parallel with a 0.6667 mF capacitor while the line impedance is a 5 Ω resistor in series with a 200 mH inductor as shown in Fig. 11.10. If $v_{ab}(t) = 100 \cos(100 \, t + 2.92) \ V_{rms}$, find the impedances and the source voltages in the phasor domain if the circuit is converted to the Wye-Wye configuration.

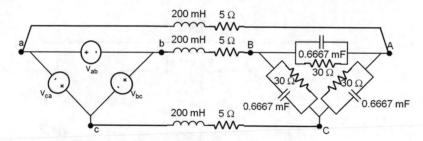

Fig. 11.10 Balanced three-phase circuit for Example 11.4

Solution The first step is to find the impedance along each branch of the load, Z_Δ.

$$Z_\Delta = \frac{1}{\frac{1}{30\,\Omega} + j\omega(0.6667\,\text{mF})} = \frac{1}{0.033333 + j0.06667} = 6 - j12\,\Omega$$

From this, we can find the impedance for the equivalent Y connected load.

$$Z_\Delta = 3Z_Y \Rightarrow Z_Y = \frac{Z_\Delta}{3} = 2 - j4\,\Omega$$

Likewise, the wye-connected voltage source is given by

$$\tilde{V}_{ab} = \tilde{V}_{an} \cdot \sqrt{3}\exp\left(j\frac{\pi}{6}\right) \Rightarrow \tilde{V}_{an} = \frac{\tilde{V}_{ab}}{\sqrt{3}}\exp\left(-j\frac{\pi}{6}\right) = \frac{100}{\sqrt{3}}\exp(j2.92)\exp\left(-j\frac{\pi}{6}\right)$$

$$\tilde{V}_{an} = 57.735\exp(j2.3964)\ V_{rms}$$

Since, it is a balanced three-phase circuit, $\tilde{V}_{bn}$ and $\tilde{V}_{cn}$ are given by

$$\tilde{V}_{bn} = \left|\tilde{V}_{an}\right|\exp\left(j\left(\theta_{V_{an}} - \frac{2\pi}{3}\right)\right) = 57.735\exp(j0.302)\ V_{rms}$$

$$\tilde{V}_{cn} = \left|\tilde{V}_{an}\right|\exp\left(j\left(\theta_{V_{an}} + \frac{2\pi}{3}\right)\right) = 57.735\exp(j4.4908)\ V_{rms}$$

$$= 57.735\exp(-j1.7924)\ V_{rms}$$

where we have subtracted 2π from the phase of $\tilde{V}_{cn}$ so that the phase term is between $-\pi$ and $+\pi$. The line impedances remains the same when converting between the different three-phase circuit configurations, but we still need to convert the inductor and resistor into their corresponding impedance values.

$$Z_{line} = 5\,\Omega + j\omega(200\,\text{mH}) = 5 + j20\,\Omega$$

Hence, the equivalent Wye-Wye configuration is given by

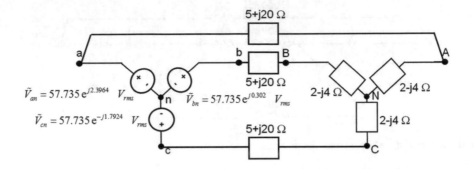

11.3 Solving for Voltages and Currents in Three-Phase Circuits

When solving most three-phase circuit problems, the first step is to convert the circuit into the Wye-Wye configuration as shown in Fig. 11.11. Node n will then serve as our reference ground node, and the voltage at node N relative to node n is denoted $\tilde{V}_N$. Summing all of the currents at node N gives

$$\tilde{I}_{aA} + \tilde{I}_{bB} + \tilde{I}_{cC} = 0 \Rightarrow \frac{\tilde{V}_{an} - \tilde{V}_N}{Z_{line} + Z_Y} + \frac{\tilde{V}_{bn} - \tilde{V}_N}{Z_{line} + Z_Y} + \frac{\tilde{V}_{cn} - \tilde{V}_N}{Z_{line} + Z_Y} = 0$$

$$\Rightarrow \tilde{V}_{an} - \tilde{V}_N + \tilde{V}_{bn} - \tilde{V}_N + \tilde{V}_{cn} - \tilde{V}_N = 0 \Rightarrow \tilde{V}_N = \frac{\tilde{V}_{an} + \tilde{V}_{bn} + \tilde{V}_{cn}}{3} \tag{11.9}$$

However, $\tilde{V}_{an}$, $\tilde{V}_{bn}$, and $\tilde{V}_{cn}$ only differ by a phase of $120°$. Therefore,

$$\tilde{V}_N = \frac{\tilde{V}_{an} + \tilde{V}_{an} \exp(-j120°) + \tilde{V}_{an} \exp(+j120°)}{3}$$

$$= \frac{\tilde{V}_{an}}{3} (1 + \cos(120°) - j\sin(120°) + \cos(120°) + j\sin(120°)) \tag{11.10}$$

$$= \frac{\tilde{V}_{an}}{3} (1 + 2\cos(120°)) = \frac{\tilde{V}_{an}}{3} \left(1 + 2\left(\frac{-1}{2}\right)\right) = 0$$

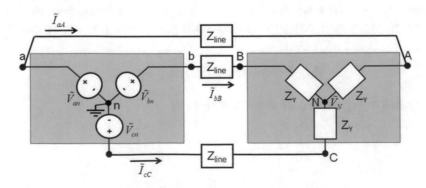

Fig. 11.11 Wye-Wye connected three-phase circuit used to find the voltage at $\tilde{V}_N$

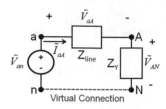

Fig. 11.12 Simplified circuit for "a" phase of Wye-Wye connected three-phase

Therefore, node N and node n are at the same voltage, and they must remain at the same voltage as long as the circuit is balanced. It is as if there is a virtual connection between the two nodes allowing them to be treated as the same node in the circuit. Occasionally, a wire will connect n and N for real circuits (the smaller wire shown in Fig. 11.2), but this wire will not be carrying any current if the phases are balanced.

Since node n and node N are virtually connected, we can focus on a single phase of the three-phase circuit as shown in Fig. 11.12 to find the line current $\tilde{I}_{aA}$, the line-to-neutral voltage at the load $\tilde{V}_{AN}$, and the voltage drop across the line impedance $\tilde{V}_{aA}$. Figure 11.12 is a fairly simple series circuit. Therefore,

$$\tilde{I}_{aA} = \frac{\tilde{V}_{an}}{Z_{line} + Z_Y}$$
$$\tilde{V}_{AN} = \tilde{I}_{aA} Z_Y = \frac{\tilde{V}_{an} Z_Y}{Z_{line} + Z_Y} \quad \tilde{V}_{aA} = \tilde{I}_{aA} Z_{line} = \frac{\tilde{V}_{an} Z_{Z_{line}}}{Z_{line} + Z_Y} \tag{11.11}$$

Once the line-to-neutral voltage at the load, $\tilde{V}_{AN}$, is known, the line-to-line voltage at the load, $\tilde{V}_{AB}$, can also be determined using the same steps shown in Eq. (11.7). Hence,

$$\tilde{V}_{AB} = \tilde{V}_{AN} \cdot \sqrt{3} \exp\left(j\frac{\pi}{6}\right) = \tilde{V}_{AN} \cdot \sqrt{3} \exp(j30°) \tag{11.12}$$

With $\tilde{V}_{AB}$ known, we can find $\tilde{I}_{AB}$

$$\tilde{I}_{AB} = \frac{\tilde{V}_{AB}}{Z_\Delta} \tag{11.13}$$

Equation (11.13) can also be rewritten to express $\tilde{I}_{AB}$ as a function of $\tilde{I}_{aA}$.

$$\tilde{I}_{AB} = \frac{\tilde{V}_{AB}}{Z_\Delta} = \frac{\tilde{V}_{AN} \cdot \sqrt{3} \exp\left(j\frac{\pi}{6}\right)}{3Z_Y} = \frac{\tilde{V}_{AN} \exp\left(j\frac{\pi}{6}\right)}{Z_Y \sqrt{3}} = \tilde{I}_{aA} \frac{\exp\left(j\frac{\pi}{6}\right)}{\sqrt{3}}$$
$$\Rightarrow \tilde{I}_{aA} = \tilde{I}_{AB} \sqrt{3} \exp\left(-j\frac{\pi}{6}\right) = \tilde{I}_{AB} \sqrt{3} \exp(-j30°) \tag{11.14}$$

The current in the b–a branch of the delta connected source, $\tilde{I}_{ba}$, can also be found in terms of $\tilde{I}_{aA}$ and is identical to $\tilde{I}_{AB}$. To see the equivalence, sum the currents at node a

$$\tilde{I}_{aA} = \tilde{I}_{ba} - \tilde{I}_{ac} = \tilde{I}_{ba} - \tilde{I}_{ba}\exp\left(j\frac{2\pi}{3}\right) = \tilde{I}_{ba}\left(1 - \exp\left(j\frac{2\pi}{3}\right)\right)$$

$$= \tilde{I}_{ba}\left(\frac{3}{2} - j\frac{\sqrt{3}}{2}\right) = \tilde{I}_{ba}\sqrt{3}\exp\left(-j\frac{\pi}{6}\right) = \tilde{I}_{ba}\sqrt{3}\exp(-j30°) \qquad (11.15)$$

$$\Rightarrow \tilde{I}_{ba} = \frac{\tilde{I}_{aA}\exp\left(j\frac{\pi}{6}\right)}{\sqrt{3}} = \tilde{I}_{AB}$$

Once the currents and voltages are known for the a-phase, the voltages and currents for all of the other phases can be found from Eq. (11.4) as they only differ by phases of 120°.

Example 11.5 Find all of the voltages and currents associated with all the phases for both the wye and delta configurations for the circuit shown in Fig. 11.9 if $v_{an}(t) = 100\cos(100t)$ V_{rms}. Express the values in the time domain with the angles expressed in radians.

Solution The circuit is already in the Wye-Wye configuration, so we can immediately simplify the analysis to a single phase.

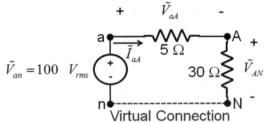

$$\tilde{I}_{aA} = \frac{\tilde{V}_{an}}{Z_{line} + Z_{Y}} = \frac{100\ V_{rms}}{5\ \Omega + 30\ \Omega} = 2.8571\ A_{rms}$$

$$\tilde{V}_{AN} = \tilde{I}_{aA}Z_{Y} = (2.8571\ A_{rms})(30\ \Omega) = 85.7143\ V_{rms}$$

$$\tilde{V}_{aA} = \tilde{I}_{aA}Z_{line} = (2.8571\ A_{rms})(5\ \Omega) = 14.2857\ V_{rms}$$

From $\tilde{V}_{AN}$ we can find $\tilde{V}_{AB}$

$$\tilde{V}_{AB} = \tilde{V}_{AN} \cdot \sqrt{3}\exp\left(j\frac{\pi}{6}\right) = 148.4615\exp\left(j\frac{\pi}{6}\right)\ V_{rms} = 148.4615\exp(j0.5236)\ V_{rms}$$

From $\tilde{V}_{AB}$ (or $\tilde{I}_{aA}$) we can find $\tilde{I}_{AB} = \tilde{I}_{ba}$

$$\tilde{I}_{AB} = \frac{\tilde{V}_{AB}}{Z_{\Delta}} = \frac{148.4615\exp\left(j\frac{\pi}{6}\right)\ V_{rms}}{90\ \Omega} = 1.6496\exp\left(j\frac{\pi}{6}\right)\ A_{rms} = 1.6496\exp(j0.5236)\ A_{rms}$$

$$\tilde{I}_{AB} = \tilde{I}_{ba} = \tilde{I}_{aA}\frac{\exp\left(j\frac{\pi}{6}\right)}{\sqrt{3}} = 1.6496\exp\left(j\frac{\pi}{6}\right)\ A_{rms} = 1.6496\exp(j0.5236)\ A_{rms}$$

The voltage $\tilde{V}_{ab}$ was found previously in Example 11.3 and is given by

$$\tilde{V}_{ab} = \tilde{V}_{an} \cdot \sqrt{3}\exp\left(j\frac{\pi}{6}\right) = 100\sqrt{3}\exp\left(j\frac{\pi}{6}\right)\ V_{rms} = 173.2051\exp(j0.5236)\ V_{rms}$$

Converting all of the values into the time domain gives

$$i_{aA}(t) = 2.8571\cos(100\,t)\;A_{rms} \qquad i_{AB}(t) = 1.6496\cos(100\,t + 0.5236)\;A_{rms}$$
$$v_{AN}(t) = 85.7143\cos(100\,t)\;V_{rms} \quad v_{AB}(t) = 148.4615\cos(100\,t + 0.5236)\;V_{rms}$$
$$v_{aA}(t) = 14.2857\cos(100\,t)\;V_{rms} \quad v_{ab}(t) = 173.2051\cos(100\,t + 0.5236)\;V_{rms}$$

We can now shift these values by 120° (or $2\pi/3$ radians) in phase to find the voltages and currents for the other phases.

$$v_{bn}(t) = 100\cos(100\,t - 2.0944)\;V_{rms} \quad v_{cn}(t) = 100\cos(100\,t + 2.0944)\;V_{rms}$$

$$i_{bB}(t) = 2.8571\cos(100\,t - 2.0944)\;A_{rms} \qquad i_{BC}(t) = 1.6496\cos(100\,t - 1.5708)\;A_{rms}$$
$$v_{BN}(t) = 85.7143\cos(100\,t - 2.0944)\;V_{rms} \quad v_{BC}(t) = 148.4615\cos(100\,t - 1.5708)\;V_{rms}$$
$$v_{bD}(t) = 14.2857\cos(100\,t - 2.0944)\;V_{rms} \quad v_{bc}(t) = 173.2051\cos(100\,t - 1.5708)\;V_{rms}$$

$$i_{cC}(t) = 2.8571\cos(100\,t + 2.0944)\;A_{rms} \qquad i_{CA}(t) = 1.6496\cos(100\,t + 2.618)\;A_{rms}$$
$$v_{CN}(t) = 85.7143\cos(100\,t + 2.0944)\;V_{rms} \quad v_{CA}(t) = 148.4615\cos(100\,t + 2.618)\;V_{rms}$$
$$v_{cC}(t) = 14.2857\cos(100\,t + 2.0944)\;V_{rms} \quad v_{ca}(t) = 173.2051\cos(100\,t + 2.618)\;V_{rms}$$

Example 11.6 A three-phase circuit in the Delta-Delta configuration has a source voltage $\tilde{V}_{ab}$ of 1000 V_{rms}, a line impedance of $Z_{line} = 0.5 + j0.5\;\Omega$ and a load impedance of $Z_\Delta = 24 + j6\;\Omega$. Find the line currents, the line-to-line voltage at the load, and the current in each phase of the delta connected load. Express the values in the phasor domain with the angles expressed in degrees.

Solution The first step is to convert the circuit into the Wye-Wye configuration.

$$\tilde{V}_{ab} = \tilde{V}_{an}\cdot\sqrt{3}\exp(j30°) \Rightarrow \tilde{V}_{an} = \frac{\tilde{V}_{ab}}{\sqrt{3}}\exp(-j30°) = 577.35\exp(-j30°)\;V_{rms}$$

$$Z_\Delta = 3Z_Y \Rightarrow Z_Y = \frac{Z_\Delta}{3} = 8 + j2\;\Omega$$

We can now focus on a single phase of the circuit

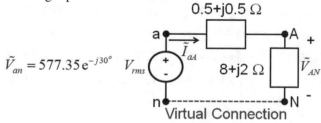

The line current and line-to-neutral voltage at the load are then given by

$$\tilde{I}_{aA} = \frac{\tilde{V}_{an}}{Z_{line} + Z_Y} = \frac{577.35\exp(-j30°)}{(0.5 + j0.5\;\Omega) + (8 + j2\;\Omega)} = 65.1635\exp(-j46.3895°)\;A_{rms}$$
$$\tilde{V}_{AN} = \tilde{I}_{aA}Z_Y = (65.1635\exp(-j46.3895°)\;A_{rms})(8 + j2\;\Omega) = 537.3522\exp(-j32.3553°)\;V_{rms}$$

We can now find the line-to-line voltage at the load as well as the current in the delta connected load.

$$\tilde{V}_{AB} = \tilde{V}_{AN} \cdot \sqrt{3} \exp(j30°) = 930.7212 \exp(-j2.3553°) \, V_{rms}$$

$$\tilde{I}_{AB} = \frac{\tilde{V}_{AB}}{Z_\Delta} = \frac{930.7212 \exp(-j2.3553°) \, V_{rms}}{24 + j6 \, \Omega} = 37.6222 \exp(-j16.3895°) \, A_{rms}$$

$$\tilde{I}_{AB} = \tilde{I}_{ba} = \tilde{I}_{aA} \frac{\exp(j30°)}{\sqrt{3}} = 37.6222 \exp(-j16.3895°) \, A_{rms}$$

We can now shift the current and voltage values by 120° to get the results for the other phases.

$$\tilde{I}_{aA} = 65.1635 \exp(-j46.3895°) \, A_{rms} \Rightarrow \begin{cases} \tilde{I}_{bB} = 65.1635 \exp(-j166.3895°) \, A_{rms} \\ \tilde{I}_{cC} = 65.1635 \exp(-j73.6105°) \, A_{rms} \end{cases}$$

$$\tilde{V}_{AB} = 930.7212 \exp(-j2.3553°) \, V_{rms} \Rightarrow \begin{cases} \tilde{V}_{BC} = 930.7212 \exp(-j122.3553°) \, V_{rms} \\ \tilde{V}_{CA} = 930.7212 \exp(j117.6447°) \, V_{rms} \end{cases}$$

$$\tilde{I}_{AB} = 37.6222 \exp(-j16.3895°) \, A_{rms} \Rightarrow \begin{cases} \tilde{I}_{BC} = 37.6222 \exp(-j136.3895°) \, A_{rms} \\ \tilde{I}_{CA} = 37.6222 \exp(j103.6105°) \, A_{rms} \end{cases}$$

11.4 Power Calculation in Three-Phase Circuits

Since three-phase circuits are common in power transfer applications, calculating the power delivered to the load and overall power efficiency of the circuit is often very important. Let's initially focus on the power associated with the Wye-Wye configuration as any three-phase circuit can be written in this form. The power from the wye connected source at any instant in time is given by

$$p(t) = v_{an}(t)i_{aA}(t) + v_{bn}(t)i_{bB}(t) + v_{cn}(t)i_{cC}(t) \tag{11.16}$$

Substituting in for the currents and voltages from Eqs. (11.2), (11.3), and (11.4) gives

$$\begin{aligned} p(t) = & |V_{an}| \cos(\omega t + \theta_{V_{an}})|I_{aA}| \cos(\omega t + \theta_{I_{aA}}) \\ & + |V_{an}| \cos(\omega t + \theta_{V_{an}} - 120°)|I_{aA}| \cos(\omega t + \theta_{I_{aA}} - 120°) \\ & + |V_{an}| \cos(\omega t + \theta_{V_{an}} + 120°)|I_{aA}| \cos(\omega t + \theta_{I_{aA}} + 120°) \\ = & |V_{an}||I_{aA}| \left(\begin{array}{l} \cos(\omega t + \theta_{V_{an}}) \cos(\omega t + \theta_{I_{aA}}) \\ + \cos(\omega t + \theta_{V_{an}} - 120°) \cos(\omega t + \theta_{I_{aA}} - 120°) \\ + \cos(\omega t + \theta_{V_{an}} + 120°) \cos(\omega t + \theta_{I_{aA}} + 120°) \end{array} \right) \end{aligned} \tag{11.17}$$

We can then apply the trigonometric identity

$$\cos(A)\cos(B) = \frac{\cos(A - B) + \cos(A + B)}{2} \tag{11.18}$$

to simplify each cosine product yielding

$$p(t) = \frac{|V_{an}||I_{aA}|}{2} \left(\begin{array}{l} \cos(\theta_{V_{an}} - \theta_{I_{aA}}) + \cos(2\omega t + \theta_{V_{an}} + \theta_{I_{aA}}) \\ + \cos(\theta_{V_{an}} - \theta_{I_{aA}}) + \cos(2\omega t + \theta_{V_{an}} + \theta_{I_{aA}} - 240°) \\ + \cos(\theta_{V_{an}} - \theta_{I_{aA}}) + \cos(2\omega t + \theta_{V_{an}} + \theta_{I_{aA}} + 240°) \end{array} \right) \tag{11.19}$$

If we then apply the second trigonometric identity

$$\cos(A \pm B) = \cos(A)\cos(B) \mp \sin(A)\sin(B) \tag{11.20}$$

to each cosine term in

$$\cos(2\omega t + \theta_{V_{an}} + \theta_{I_{aA}} - 240°) + \cos(2\omega t + \theta_{V_{an}} + \theta_{I_{aA}} + 240°) \tag{11.21}$$

we get

$$\begin{array}{l} \cos(2\omega t + \theta_{V_{an}} + \theta_{I_{aA}})\cos(240°) + \sin(2\omega t + \theta_{V_{an}} + \theta_{I_{aA}})\sin(240°) \\ + \cos(2\omega t + \theta_{V_{an}} + \theta_{I_{aA}})\cos(240°) - \sin(2\omega t + \theta_{V_{an}} + \theta_{I_{aA}})\sin(240°) \\ = -\cos(2\omega t + \theta_{V_{an}} + \theta_{I_{aA}}) \end{array} \tag{11.22}$$

Therefore, Eq. (11.19) becomes

$$p(t) = 3\frac{|V_{an}||I_{aA}|}{2}\cos(\theta_{V_{an}} - \theta_{I_{aA}}) \tag{11.23}$$

From Eq. (11.23), it is clear that power delivery in three-phase circuits is constant with respect to time. If the power flow along one of the phases is decreasing in time, the power flow on the other two phases will be increasing. This is very different from the power flow in the single-phase circuits discussed in Chap. 8. The instantaneous power in single phase is given by

$$p(t') = P + P\cos(2\omega t') - Q\sin(2\omega t') \tag{11.24}$$

where P was the time-average power and Q was the reactive power. Therefore, the power flow in single-phase circuits will come in bursts. The constant power flow in three-phase circuits will result in less vibrations and constant torque when applied to motors and generators.

The derivation leading to Eq. (11.23) can be repeated for the other three-phase circuit configurations as well as for the power lost in the line and delivered to the source. These derivations will show that the time-average power (i.e., real power) from the source, lost in the line, and delivered to the load are given by

$$\begin{aligned} P_{Source} &= 3\frac{|V_{an}||I_{aA}|}{2}\cos(\theta_{V_{an}} - \theta_{I_{aA}}) = 3|V_{an}|_{rms}|I_{aA}|_{rms}\cos(\theta_{V_{an}} - \theta_{I_{aA}}) \\ &= 3\frac{|V_{ab}||I_{ba}|}{2}\cos(\theta_{V_{ab}} - \theta_{I_{ba}}) = 3|V_{ab}|_{rms}|I_{ba}|_{rms}\cos(\theta_{V_{ab}} - \theta_{I_{ba}}) \\ P_{Line} &= 3\frac{|Z_{line}||I_{aA}|^2}{2}\cos(\theta_{Z_{line}}) = 3|Z_{line}||I_{aA}|^2_{rms}\cos(\theta_{Z_{line}}) \\ P_{Load} &= 3\frac{|V_{AN}||I_{aA}|}{2}\cos(\theta_{V_{AN}} - \theta_{I_{aA}}) = 3|V_{AN}|_{rms}|I_{aA}|_{rms}\cos(\theta_{V_{AN}} - \theta_{I_{aA}}) \\ &= 3\frac{|V_{AB}||I_{AB}|}{2}\cos(\theta_{V_{AB}} - \theta_{I_{AB}}) = 3|V_{AB}|_{rms}|I_{AB}|_{rms}\cos(\theta_{V_{AB}} - \theta_{I_{AB}}) \end{aligned} \tag{11.25}$$

where we have used $Z_{\text{line}} = |Z_{\text{line}}| \exp(j\theta_{Z_{\text{line}}})$ to simplify the equations for power lost in the line. Also, while not immediately clear from the total instantaneous power calculation, each phase of the three-phase circuit will still require reactive power. Therefore, the total reactive power associated with the source, line, and load will be given by

$$
\begin{aligned}
Q_{\text{Source}} &= 3\frac{|V_{\text{an}}||I_{\text{aA}}|}{2}\sin(\theta_{V_{\text{an}}} - \theta_{I_{\text{aA}}}) = 3|V_{\text{an}}|_{\text{rms}}|I_{\text{aA}}|_{\text{rms}}\sin(\theta_{V_{\text{an}}} - \theta_{I_{\text{aA}}}) \\
&= 3\frac{|V_{\text{ab}}||I_{\text{ba}}|}{2}\sin(\theta_{V_{\text{ab}}} - \theta_{I_{\text{ba}}}) = 3|V_{\text{ab}}|_{\text{rms}}|I_{\text{ba}}|_{\text{rms}}\sin(\theta_{V_{\text{ab}}} - \theta_{I_{\text{ba}}}) \\
Q_{\text{Line}} &= 3\frac{|Z_{\text{line}}||I_{\text{aA}}|^2}{2}\sin(\theta_{Z_{\text{line}}}) = 3|Z_{\text{line}}||I_{\text{aA}}|^2_{\text{rms}}\sin(\theta_{Z_{\text{line}}}) \\
Q_{\text{Load}} &= 3\frac{|V_{\text{AN}}||I_{\text{aA}}|}{2}\sin(\theta_{V_{\text{AN}}} - \theta_{I_{\text{aA}}}) = 3|V_{\text{AN}}|_{\text{rms}}|I_{\text{aA}}|_{\text{rms}}\sin(\theta_{V_{\text{AN}}} - \theta_{I_{\text{aA}}}) \\
&= 3\frac{|V_{\text{AB}}||I_{\text{AB}}|}{2}\sin(\theta_{V_{\text{AB}}} - \theta_{I_{\text{AB}}}) = 3|V_{\text{AB}}|_{\text{rms}}|I_{\text{AB}}|_{\text{rms}}\sin(\theta_{V_{\text{AB}}} - \theta_{I_{\text{AB}}})
\end{aligned}
\tag{11.26}
$$

Example 11.7 Find the total real and reactive power from the source, lost in the line, and delivered to the load for the circuit described in Example 11.5.

Solution When we solved for the currents and voltages in the circuit previously, we found that

$$
v_{\text{an}}(t) = 100\cos(100\,t)\ V_{\text{rms}} \quad v_{\text{AN}}(t) = 85.7143\cos(100\,t)\ V_{\text{rms}}
$$
$$
i_{\text{aA}}(t) = 2.8571\cos(100\,t)\ A_{\text{rms}}
$$

Therefore,

$$
\begin{aligned}
P_{\text{Source}} &= 3|V_{\text{an}}|_{\text{rms}}|I_{aA}|_{\text{rms}}\cos(\theta_{V_{\text{an}}} - \theta_{I_{aA}}) = 857.13\ \text{W} \\
P_{\text{Line}} &= 3|Z_{\text{line}}||I_{aA}|^2_{\text{rms}}\cos(\theta_{Z_{\text{line}}}) = 122.45\ \text{W} \\
P_{\text{Load}} &= 3|V_{\text{AN}}|_{\text{rms}}|I_{aA}|_{\text{rms}}\cos(\theta_{V_{\text{an}}} - \theta_{I_{aA}}) = 734.68\ \text{W}
\end{aligned}
$$

P_{Line} could also have been found from $P_{\text{Line}} = P_{\text{Source}} - P_{\text{Load}} = 122.45$ W. Also, since all of the circuit elements are purely resistive, all of the voltages and currents are in phase. Thus,

$$
Q_{\text{Source}} = Q_{\text{Line}} = Q_{\text{Load}} = 0\ VAR
$$

Example 11.8 Find the total real and reactive power from the source, lost in the line and delivered to the load for the circuit described in Example 11.6.

Solution When we solved for the currents and voltages in the circuit previously, we found that

$$
\begin{aligned}
\tilde{V}_{\text{ab}} &= 1000\ V_{\text{rms}} & \tilde{V}_{\text{AB}} &= 930.7212\exp(-j2.3553°)\ V_{\text{rms}} \\
\tilde{I}_{\text{aA}} &= 65.1635\exp(-j46.3895°)A_{\text{rms}} & \tilde{I}_{\text{AB}} &= 37.6222\exp(-j16.3895°)\ A_{\text{rms}}
\end{aligned}
$$

Also, $Z_{\text{line}} = 0.5 + j0.5\ \Omega = 0.7071\exp(j45°)\ \Omega$ and $\tilde{I}_{\text{ba}} = \tilde{I}_{\text{AB}}$. Therefore,

$$P_{\text{Source}} = 3|V_{ab}|_{\text{rms}}|I_{ba}|_{\text{rms}}\cos(\theta_{V_{ab}} - \theta_{I_{ba}})$$
$$= 3 \cdot 1000 \cdot 37.6222\cos(0 - -16.3895°) = 108.28 \text{ kW}$$
$$P_{\text{Line}} = 3|Z_{\text{line}}||I_{aA}|_{\text{rms}}^2\cos(\theta_{Z_{\text{line}}}) = 3 \cdot 0.7071 \cdot 65.1635\cos(45°) = 6.37 \text{ kW}$$
$$P_{\text{Load}} = 3|V_{AB}|_{\text{rms}}|I_{AB}|_{\text{rms}}\cos(\theta_{V_{AB}} - \theta_{I_{AB}})$$
$$= 3 \cdot 930.7212 \cdot 37.6222\cos(-2.3553° - -16.3895°) = 101.91 \text{ kW}$$

$$Q_{\text{Source}} = 3|V_{ab}|_{\text{rms}}|I_{ba}|_{\text{rms}}\cos(\theta_{V_{ab}} - \theta_{I_{ba}})$$
$$= 3 \cdot 1000 \cdot 37.6222\sin(0 - -16.3895°) = 31.85 \text{ } kVAR$$
$$Q_{\text{Line}} = 3|Z_{\text{line}}||I_{aA}|_{\text{rms}}^2\sin(\theta_{Z_{\text{line}}}) = 3 \cdot 0.7071 \cdot 65.1635\sin(45°) = 6.37 \text{ } kVAR$$
$$Q_{\text{Load}} = 3|V_{AB}|_{\text{rms}}|I_{AB}|_{\text{rms}}\sin(\theta_{V_{AB}} - \theta_{I_{AB}})$$
$$= 3 \cdot 930.7212 \cdot 37.6222\sin(-2.3553° \quad 16.3895°) - 23.47 \text{ } kVAR$$

11.5 Problems

Problem 11.1: If the line-to-line voltage at the source is given by $v_{ab}(t) = 10\cos(5t + 15°) \text{ } V_{\text{rms}}$, what are the line-to-line voltages $v_{bc}(t)$ and $v_{ca}(t)$ with the phases expressed in degrees?

Problem 11.2: If the line current is given by $i_{aA}(t) = 7\cos(200t - 0.4) \text{ } A_{\text{rms}}$, what are the line currents $i_{bB}(t)$ and $i_{cC}(t)$ with the phases expressed in radians?

Problem 11.3: If the line-to-neutral voltage at the source is given by $v_{bn}(t) = 100\cos(25t - 150°) \text{ } V$, what are the line-to-neutral voltages $v_{an}(t)$ and $v_{cn}(t)$ with the phases expressed in degrees?

Problem 11.4: A wye connected source is connected to a wye connected load in a balanced three-phase circuit. The impedance of the load is $Z_Y = 100 - j50 \text{ } \Omega$ while the impedance of the line is $Z_{\text{line}} = 20 + j20 \text{ } \Omega$. If the line-to-neutral voltage at the source is given by $v_{an}(t) = 80\cos(50t) \text{ } V$, what is the line-to-neutral voltage at the load, $v_{AN}(t)$, and the line current $i_{aA}(t)$?

Problem 11.5: If the line-to-line voltage at the source is given by $v_{ab}(t) = 9\cos(3t) \text{ } V_{\text{rms}}$, what is the line-to-neutral voltage $v_{an}(t)$ with the phase expressed in radians?

Problem 11.6: A wye connected source is connected to a delta connected load in a balanced three-phase circuit. The impedance of the load is $Z_\Delta = 12 + j15 \text{ } \Omega$ while the impedance of the line is $Z_{\text{line}} = 3 + j3 \text{ } \Omega$. If the line-to-neutral voltage at the source is given by $\tilde{V}_{an} = 100 \text{ } V_{\text{rms}}$, what is the line current $\tilde{I}_{aA}$ and the line-to-line voltage $v_{AB}(t)$ given that the circuit is operating at 60 Hz?

Problem 11.7: A delta connected source is connected to a delta connected load in a balanced three-phase circuit. The impedance of the load is $Z_\Delta = 12 \text{ } \Omega$ while the impedance of the line is $Z_{\text{line}} = 2 \text{ } \Omega$. If the line-to-line voltage at the source is given by $v_{ab}(t) = 120\cos(100t) \text{ } V_{\text{rms}}$, find

(a) What is the line-to-neutral voltage at the source, $\tilde{V}_{an}$?
(b) What is the line current, $\tilde{I}_{aA}$?
(c) What is the line-to-line voltage at the load, $\tilde{V}_{AB}$?

Problem 11.8: A delta connected source is connected to a delta connected load in a balanced three-phase circuit. The impedance of the load is $Z_\Delta = 15 + j9$ Ω while the impedance of the line is $Z_{\text{line}} = 1 + j1$ Ω. If the line-to-line voltage at the source is given by $\tilde{V}_{ab} = 325 \exp(j\pi/3)$ V_{rms}, what is the line current $\tilde{I}_{aA}$, the line-to-line voltage $\tilde{V}_{AB}$, and the total time average power delivered to the load?

Problem 11.9: A delta connected source is connected to a delta connected load in a balanced three-phase circuit. The impedance of the load is $Z_\Delta = 130 - j150$ Ω while the impedance of the line is $Z_{\text{line}} = 10 + j20$ Ω. If the line-to-line voltage at the source is given by $\tilde{V}_{ab} = 140$ V_{rms}, find $\tilde{V}_{AB}$ and $\tilde{I}_{AB}$?

Problem 11.10: A delta connected source is connected to a delta connected load in a balanced three-phase circuit. The impedance of the line is $Z_{\text{line}} = 16$ Ω while $v_{ab}(t) = 171 \cos(100\pi t)$ V_{rms} and $i_{aA}(t) = 1.7 \cos(100\pi t - 0.6)$ A_{rms}.

(a) What is $\tilde{V}_{AB}$?
(b) What is $\tilde{I}_{AB}$?
(c) What is Z_Δ?

Problem 11.11: A delta connected source is connected to a delta connected load in a balanced three-phase circuit. The impedance of the load is $Z_\Delta = 132 - j667$ Ω while the impedance of the line is $Z_{\text{line}} = 166 + j88$ Ω. If the line-to-line voltage at the load is $\tilde{V}_{AB} = 94$ V_{rms}, what is the line current $\tilde{I}_{aA}$, the line-to-line voltage at the source $\tilde{V}_{AB}$, and the time average power lost in the line?

AC Motors and Generators

12

As was stated in Chap. 5, translating electrical energy into motion is one of the oldest applications of electrical engineering. Both AC and DC motors have been utilized for this purpose. Previously, the universal motor (series connected DC motor) was introduced as a cheap motor that can operate on AC power. Other AC motors include the induction motor and the synchronous motor. However, the brushes of the universal motor and most synchronous motors reduce their reliability. Conversely, AC induction motors have no brushes and are, therefore, the primary motor in many industrial and commercial applications. When it comes to generating electric power, however, synchronous machines are the first choice. Therefore, in this chapter, we will cover the basics of AC induction motors and synchronous generators.

12.1 Three-Phase Induction Motors Basic Operation

The basic structure of an induction motor is shown in Fig. 12.1. Power is provided to the motor via the stator windings. The rotor also has a series of conductors which are often aluminum bars connected to rings that short the ends together (squirrel-cage induction motor). Shorting the ends will ultimately maximize the induced current flow in the rotor bars. The bars are often formed by pouring molten aluminum into slots on the laminated iron rotor.

The basic operation of all induction motors involves the establishment of a rotating magnetic field in the interior of the motor that drives the rotation of the machine. For the three-phase induction motor, the rotating magnetic field is established by driving the motor with three-phase AC power as was discussed in Chap. 11. The three-phases are connected to three different sets of windings on the stator as illustrated in Fig. 12.2 for a simple 2-pole machine. The number of poles for the AC induction motor refers to the number of poles of the rotating magnetic field and must always be an even number. In addition to being excited electrically by a different phase, the windings are also positioned at different angles around the stator. For a 2-pole machine, each phase has only one set of windings, so the windings are positioned every 120°. A 4-pole induction motor would have two sets of windings for each phase, so the windings at different phases would be positioned every 60° as shown in Fig. 12.3.

© Springer Nature Switzerland AG 2020
T. A. Bigelow, *Electric Circuits, Systems, and Motors*,
https://doi.org/10.1007/978-3-030-31355-5_12

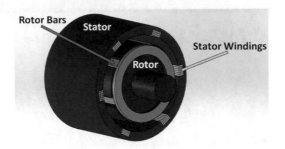

Fig. 12.1 Basic structure of an AC induction motor

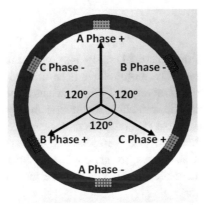

Fig. 12.2 Excitation and position of stator windings for a 2-pole three-phase AC induction motor

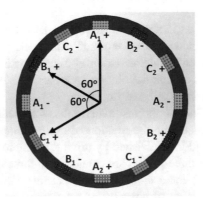

Fig. 12.3 Excitation and position of stator windings for a 4-pole three-phase AC induction motor

Returning to the simple 2-pole induction motor, the magnetic fields corresponding to each phase are given by

$$
\begin{aligned}
B_A(t) &= K \cdot i_A(t) \cos(\theta_s) \\
B_B(t) &= K \cdot i_B(t) \cos(\theta_s - 120°) \\
B_C(t) &= K \cdot i_C(t) \cos(\theta_s + 120°)
\end{aligned}
\tag{12.1}
$$

where K is a constant related to the number of windings in each phase and θ_s is the angle corresponding to our location on the stator as shown in Fig. 12.4. The magnetic field corresponding to the a-phase will have its maximum at $\theta_s = 0$. Therefore, the magnetic fields for the b and c phases will have their maximums at $\theta_s = +120°$ and $\theta_s = -120°$, respectively, as the windings are physically shifted during motor construction by 120° for the 2-pole induction motor. The spatial location, θ_s, should not be confused with the phase of the electrical signal being applied to a particular phase of the motor.

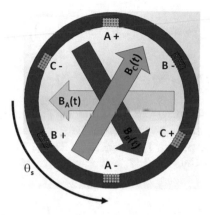

Fig. 12.4 Magnetic fields corresponding to each phase of a 2-pole three-phase AC induction motor

The currents $i_A(t)$, $i_B(t)$, and $i_C(t)$ are the currents flowing through each phase of the windings and correspond to either $i_{AN}(t)$, $i_{BN}(t)$, and $i_{CN}(t)$ or $i_{AB}(t)$, $i_{BC}(t)$, and $i_{CA}(t)$ depending on if the motor windings are connected in the wye or delta configuration. In either case, the currents in each phase of the stator differ by 120°. Therefore, the magnetic fields in Eq. (12.1) can be written as

$$B_A(t) = K \cdot I_o \cos(\omega_e t) \cos(\theta_s) = B_o \cos(\omega_e t) \cos(\theta_s)$$
$$B_B(t) = K \cdot I_o \cos(\omega_e t - 120°) \cos(\theta_s - 120°) = B_o \cos(\omega_e t - 120°) \cos(\theta_s - 120°) \quad (12.2)$$
$$B_C(t) = K \cdot I_o \cos(\omega_e t + 120°) \cos(\theta_s + 120°) = B_o \cos(\omega_e t + 120°) \cos(\theta_s + 120°)$$

where ω_e is the electric frequency being applied to the stator windings. In previous chapters, this frequency was just denoted as ω. However, there are multiple frequency terms associated with three-phase induction motors. Therefore, it can be useful to distinguish the electrical frequency of excitation from the other frequencies by the use of an additional subscript. As a result, the total magnetic field generated by the stator windings at any instant in time is given by

$$B_{\text{total}}(t) = B_A(t) + B_B(t) + B_C(t)$$
$$= B_o \cdot \left[\begin{array}{c} \cos(\omega_e t) \cos(\theta_s) + \cos(\omega_e t - 120°) \cos(\theta_s - 120°) \\ + \cos(\omega_e t + 120°) \cos(\theta_s + 120°) \end{array} \right] \quad (12.3)$$

However,

$$\cos(\omega_e t) \cos(\theta_s) = \tfrac{1}{2} (\cos(\omega_e t - \theta_s) + \cos(\omega_e t + \theta_s))$$
$$\cos(\omega_e t \pm 120°) \cos(\theta_s \pm 120°) = \tfrac{1}{2} (\cos(\omega_e t - \theta_s) + \cos(\omega_e t + \theta_s \pm 240°)) \quad (12.4)$$
$$= \tfrac{1}{2} (\cos(\omega_e t - \theta_s) + \cos(\omega_e t + \theta_s) \cos(240°) \mp \sin(\omega_e t + \theta_s) \sin(240°))$$

Therefore, Eq. (12.3) becomes

$$
\begin{aligned}
B_{total}(t) &= \frac{B_o}{2} \cdot \left[\begin{array}{l}
\cos(\omega_e t - \theta_s) + \cos(\omega_e t + \theta_s) \\
+ \cos(\omega_e t - \theta_s) + \cos(\omega_e t + \theta_s)\cos(240^\circ) - \sin(\omega_e t + \theta_s)\sin(240^\circ) \\
+ \cos(\omega_e t - \theta_s) + \cos(\omega_e t + \theta_s)\cos(240^\circ) + \sin(\omega_e t + \theta_s)\sin(240^\circ)
\end{array} \right] \\
&= \frac{B_o}{2} \cdot \left[3\cos(\omega_e t - \theta_s) + \cos(\omega_e t + \theta_s)\overset{=0}{\overbrace{\left(1 + 2\cos(240^\circ)\right)}} \right] \\
&= \frac{3B_o}{2}\cos(\omega_e t - \theta_s) = B_{max}\cos(\omega_e t - \theta_s)
\end{aligned}
\tag{12.5}
$$

Figure 12.5 shows a sketch of the magnetic field from Eq. (12.5) for different time values. As time increases, the magnetic field spins in the counterclockwise direction. The magnetic field makes one complete rotation in one electric period, T_e, where $\omega_e = 2\pi/T_e$. Therefore, the frequency of rotation for the magnetic field is given by $f_s = 1/T_e = f_e \Rightarrow \omega_s = \omega_e$ for a 2-pole induction motor. The

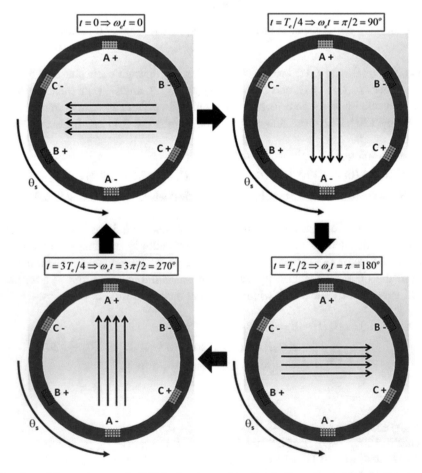

Fig. 12.5 Rotation of the total magnetic field in the air gap for a 2-pole three-phase AC induction motor

frequency f_s in Hz or ω_s in radians/sec is also known as the synchronous speed for the motor. As we will see shortly, the motor will always try to spin at the same speed as the magnetic field (i.e., in sync) and, therefore, this speed is called the synchronous speed.

As was stated previously, the number of poles for the induction motor can be increased by adding more windings around the stator. As an example, Fig. 12.6 shows a sketch of the magnetic fields at some instant in time for the 4-pole case. The total magnetic field has the magnetic field lines going into the center at two locations and out of the stator at two locations giving rise to two virtual North and South poles. The virtual north and south poles will always be opposite each other and will rotate about the stator as time increases due to the phase difference between the windings. However, this time, the magnetic field will only rotate 180° for every period of the electrical frequency due to the closer proximity of the windings. As a result, the frequency of rotation for the magnetic field (i.e., synchronous speed) would be given by $f_s = f_e/2 \Rightarrow \omega_s = \omega_e/2$. Generalizing these results to an induction motor with any number of poles gives a synchronous speed of

$$\omega_s = \frac{2\omega_e}{P} \quad n_s = \frac{120 f_e}{P} \tag{12.6}$$

where ω_s is the speed in radians/sec and n_s is the speed in revolutions per minute (rpm). Recall from Chap. 5 that speeds in rad/sec are related to speeds in rpm by $n_m = 60\omega_m/2\pi \Rightarrow n_s = 60\omega_s/2\pi$.

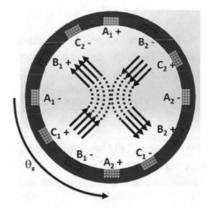

Fig. 12.6 Total magnetic field at some instant in time for a 4-pole three-phase AC induction motor

Now that we have established that the induction motor has a spinning magnetic field due to the excitation of the stator windings, we can consider the interaction of this magnetic field with the conducting bars on the rotor. For the moment, let's assume that the rotor is stationary. The spinning magnetic field lines would then be passing through the conducting bars. From Chap. 5, we know that when a bar cuts through magnetic field lines a voltage is induced on the bar. However, if we alter our point of reference, magnetic field lines passing through a bar will also induce a voltage across the bar as illustrated in Fig. 12.7. As a result, the spinning magnetic field will induce voltages on the rotor bars. The voltages on the rotor bars will then induce current flow in the rotor bars. The current in the rotor bars will then establish a magnetic field which will then attempt to align with the rotating magnetic field causing the rotor to spin. The motor will then move at the synchronous speed if the rotor is able to spin at the same speed as the rotating magnetic field. Since there is no torque that can drive the motor faster than the synchronous speed, the synchronous speed is also the maximum speed (or no load speed) for the motor.

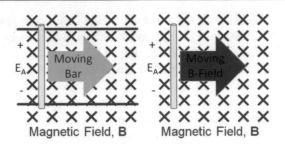

Fig. 12.7 Induced voltage due to relative motion between a conducting bar and a magnetic field

Example 12.1 What is the fastest speed a three-phase induction motor can rotate if it is connected to a power source that operates at 60 Hz?

Solution:
The fastest speed of the motor would be the synchronous speed which is governed by the number of poles and the electrical frequency. Since the electrical frequency is 60 Hz, and the smallest possible number of poles is 2, the fastest speed is given by

$$n_s = \frac{120 f_e}{P} = \frac{120 \cdot 60}{2} = 3600 \text{ rpm}$$

As was stated previously, when no load is connected to the motor, it will spin at the synchronous speed. However, as the torque demand from the motor increases, the motor will slow down. For induction motors, the reduction in speed with increasing load is quantified by the slip, s, given by

$$s = \frac{\omega_s - \omega_m}{\omega_s} = \frac{n_s - n_m}{n_s} \tag{12.7}$$

Therefore, if the motor is not moving, the slip would be 1, and if the motor is spinning at the synchronous speed the slip would be 0. For other values of slip, the motor speed can be calculated from

$$\omega_m = (1 - s)\omega_s \quad n_m = (1 - s)n_s \tag{12.8}$$

AC induction motors are normally operated at slip values close to 0 with speeds close to the synchronous speed.

Example 12.2 A three-phase induction motor is being driven at a frequency of 75 Hz, and the motor speed is 1200 rpm. How many poles does the motor probably have and what is the slip?

Solution:
The motor must have an even number of poles. Therefore, the potential values for the synchronous speed are given by

$$n_s = \frac{120 f_e}{P} = \frac{120 \cdot 75}{P} = \frac{9000}{P}$$

$$= \underbrace{4500 \text{ rpm}}_{P=2} \quad \underbrace{2250 \text{ rpm}}_{P=4} \quad \underbrace{1500 \text{ rpm}}_{P=6} \quad \underbrace{1125 \text{ rpm}}_{P=8} \quad \underbrace{900 \text{ rpm}}_{P=10} \quad \underbrace{750 \text{ rpm}}_{P=12}$$

We know that the motor speed MUST be less than the synchronous speed, but it is usually close to the synchronous speed. The synchronous speed closest to the stated motor speed of 1200 rpm would be the 1125 rpm associated with P = 8. However, this speed is smaller than the given motor speed. Hence, 1500 rpm associated with P = 6 is the correct choice as this is the speed that is closest to the given motor speed while still being larger. Therefore, the motor probably has 6 poles and a slip of

$$s = \frac{n_s - n_m}{n_s} = \frac{1500 - 1200}{1500} = 0.2$$

12.2   Speed-Torque Curve for Three-Phase Induction Motor

The typical speed-torque curve for a three-phase induction motor is shown in Fig. 12.8. As was stated previously, induction motors normally operate near the synchronous speed, n_s. As the torque increases away from the no-load case, the slip increases and the motor slows. The reduction in speed with torque is relatively minor in the normal operating range of the motor which spans from the no-load to the full-load torque. The full-load torque would be the maximum rated torque for the motor. Extended operation at loads higher than the full-load torque could damage the motor. Unlike the previous motors considered in Chap. 5, induction motors have maximum in their speed-torque curves known as the break-over or pullout torque as shown in Fig. 12.8. This peak is normally outside of the normal operating range for the motor. Lastly, the starting torque is the torque provided by the motor when power is first applied to the motor. If the initial torque demanded by the load is greater than the starting torque, then the motor will never start spinning.

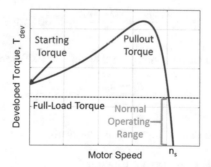

Fig. 12.8 Speed-Torque curve for a three-phase induction motor

Practical Hint: When power is first applied to the motor, the motor will accelerate until the torque demanded by the load equals the torque supplied by the motor. If the torques from the motor and the load balance too soon, the motor will operate on the wrong side of the pullout torque peak. This could potentially burn-out the motor. Therefore, induction motors should not be operated outside of their rated values.

Figure 12.9 shows the developed torque plotted as a function of slip instead of speed. From this plot, it is clear that the developed torque and slip are proportional when the motor is operating in its normal operating range. This relationship between slip and torque can be used to find the speed of the motor as the torque changes in the normal operating range for the motor.

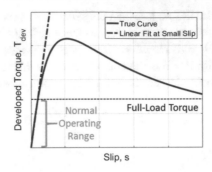

Fig. 12.9 Slip-Torque curve for a three-phase induction motor

---

Hint: When operating in the normal operating range, it is the torque and the slip that are proportional NOT the torque and speed. It is very common for students to solve problems incorrectly by making the speed and the torque proportional. However, if the torque increases, then the speed should decrease NOT increase.

Example 12.3 A three-phase induction motor with 6 poles is being driven at 60 Hz and is running in its normal operating range. When connected to a certain load, the motor rotates at a speed of 1070 rpm and outputs 2000 W. What is the slip and developed torque? What is the new motor speed if the developed torque is reduced to 5 Nm?

Solution:
First, we need to find the synchronous speed for the motor.

$$n_s = \frac{120 f_e}{P} = \frac{120 \cdot 60}{6} = 1200 \text{ rpm}$$

Then, we can calculate the slip corresponding to a speed of 1070 rpm.

$$s = \frac{n_s - n_m}{n_s} = 0.1083$$

Likewise, the developed torque can be found from the power as

$$P_{\text{dev}} = T_{\text{dev}} \cdot \omega_m \text{ (From Chapter 5)}$$
$$\Rightarrow T_{\text{dev}} = \frac{P_{\text{dev}}}{\omega_m} = \frac{2000 \text{ W}}{\frac{2\pi}{60} n_m} = \frac{2000 \text{ W}}{112.0501 \text{ rad/s}} = 17.8492 \text{ Nm}$$

When operating in its normal operating range, the slip and torque should be proportional. Therefore,

$$\frac{T_{\text{dev1}}}{T_{\text{dev2}}} = \frac{s_1}{s_2} \Rightarrow s_2 = s_1 \frac{T_{\text{dev2}}}{T_{\text{dev1}}} = (0.1083) \frac{5 \text{ Nm}}{17.8492 \text{ Nm}} = 0.0303$$

Once the new slip is known, the new speed would be given by

$$n_m = (1 - s)n_s \Rightarrow n_{m2} = 1163.6 \text{ rpm}$$

Example 12.4 A three-phase induction motor with 4 poles is being driven at 50 Hz and is running in its normal operating range. When connected to a certain load, the motor rotates at a speed of 1300 rpm and outputs 1800 W. What would be the new speed if a new load is connected that demands 2000 W of power? Also, determine the developed torque demanded by both loads.

Solution:
The developed torque demanded by the first load would be given by

$$P_{dev} = T_{dev} \cdot \omega_m \text{ (From Chapter 5)}$$
$$\rightarrow T_{dev} = \frac{P_{dev}}{\omega_m} = \frac{1800 \text{ W}}{\frac{2\pi}{60} n_m} = \frac{1800 \text{ W}}{136.1357 \text{ rad/s}} = 13.2221 \text{ Nm}$$

Since the power is different for the new load, the torque would also be different. Hence,

$$P_{dev1} = T_{dev1}\omega_{m1}$$
$$P_{dev2} = T_{dev2}\omega_{m2}$$
$$\frac{P_{dev1}}{P_{dev2}} = \frac{T_{dev1}\omega_{m1}}{T_{dev2}\omega_{m2}} = \frac{T_{dev1}}{T_{dev2}}\frac{\omega_s(1 - s_1)}{\omega_s(1 - s_2)} = \frac{T_{dev1}}{T_{dev2}}\frac{(1 - s_1)}{(1 - s_2)}$$

However, the torques should still be proportional to the slip. Therefore,

$$\frac{T_{dev1}}{T_{dev2}} = \frac{s_1}{s_2}$$
$$\Rightarrow \frac{P_{dev1}}{P_{dev2}} = \frac{s_1(1 - s_1)}{s_2(1 - s_2)} =$$
$$\Rightarrow s_2(1 - s_2) = \frac{P_{dev2}}{P_{dev1}}s_1(1 - s_1) = 0.12836$$
$$\Rightarrow 0 = s_2^2 - s_2 + 0.12836 \Rightarrow s_2 = 0.8487 \quad or \quad 0.1513$$

The slip must be small to be running in the normal operating range. Hence, $s_2 = 0.1513$. To get the new motor speed, we need the synchronous speed given by

$$n_s = \frac{120 f_e}{P} = \frac{120 \cdot 50}{4} = 1500 \text{ rpm}$$

Thus, the new motor speed would be given by

$$n_{m2} = n_s(1 - s_2) = n_s(1 - 0.1513) = 1273.1 \text{ rpm}$$

12.3   Equivalent Circuit for Three-Phase Induction Motor

The operation of the induction motor is very similar to the operation of a transformer as was discussed in Chap. 10. The voltage applied to the stator windings results in the generation of a changing magnetic field which in turn induces a voltage on the rotor bars. For the transformer, the voltage applied to the primary winding generates a time-varying magnetic field in the core that passes through the secondary winding. The changes of the magnetic field with respect to time then produce a voltage across the secondary winding. Given the similarity between the induction motor and the transformer, the circuit models for the two devices are also very similar. Figure 12.10 shows an approximate circuit model for a single-phase of the induction motor. In this model, $\tilde{V}_s$ and $\tilde{I}_s$ are the voltage and current for a single-phase of the stator windings and could be either $\tilde{V}_{AN}$ and $\tilde{I}_{AN}$ OR $\tilde{V}_{AB}$ and $\tilde{I}_{AB}$ (or equivalent) depending on if the stator is wired in the wye or delta configuration.

The resistive losses of the stator windings and rotor bars are given by R_s and R_r, respectively. Likewise, L_s and L_r are the self-inductances of the stator windings and rotor bars. L_m is included to model the reluctance of the rotor core and air gap between the rotor and the stator. The presence of the air gap makes the value of L_m much smaller than would normally be encountered in a transformer. The resistor, R_c, is also included to model electrical losses in the rotor cores. However, the impact of these losses are normally minor, so R_c is frequently neglected in the model. Lastly, an ideal transformer is included to model the transfer of energy from the stator to the rotor. However, unlike for a basic transformer, the transfer cannot be expressed as a simple turn's ratio due to the motion of the rotor relative to the stator.

To understand the transfer of energy across the air gap, let's focus on the rotor side of the circuit shown in Fig. 12.11. The voltage generated on the rotor bars, $\tilde{V}_r$, will depend on magnetic fields established by the stator and the cutting of the magnetic field lines by the rotor. Therefore, it will depend on the slip with the highest voltage induced when the rotor is stationary. As a result, $\tilde{V}_r$ can be written as

$$\tilde{V}_r = s\tilde{V}_{ro} \tag{12.9}$$

The motion of the rotor bars relative to the changing magnetic field will also mean that the frequency of $\tilde{V}_r$ will not be the same as the electrical frequency being applied to the stator winding. Instead, the time-variation of the magnetic field will appear different than if the rotor bars were stationary. Hence, the frequency of the electric currents and voltages for the rotor bars will be given by

$$\omega_r = s \cdot \omega_e \tag{12.10}$$

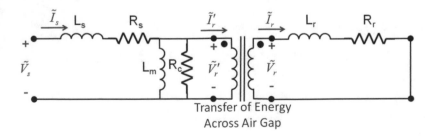

Fig. 12.10 Basic model of single-phase of an induction motor

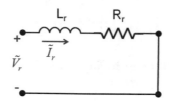

Fig. 12.11 Circuit model for the rotor in an induction motor

The rotor current is then found from

$$\tilde{I}_r = \frac{\tilde{V}_r}{R_r + j\omega_r L_r} = \frac{s\tilde{V}_{ro}}{R_r + js \cdot \omega_e L_r} = \frac{\tilde{V}_{ro}}{\frac{R_r}{s} + j\omega_e L_r} \qquad (12.11)$$

However, the voltage $\tilde{V}_{ro}$ would only depend on the magnetic field established by the stator and thus must be related to the voltage $\tilde{V}'_r$ by

$$\tilde{V}_{ro} = \frac{\tilde{V}'_r}{N_{\mathrm{eff}}} \qquad (12.12)$$

where N_{eff} is the effective turns ratio relating the stator windings and rotor bars. Likewise,

$$\tilde{I}_r = N_{\mathrm{eff}}\tilde{I}'_r \qquad (12.13)$$

and

$$Z'_{\mathrm{eq}} = \frac{\tilde{V}'_r}{\tilde{I}'_r} = \frac{N_{\mathrm{eff}}\tilde{V}_{ro}}{\left(\frac{\tilde{I}_r}{N_{\mathrm{eff}}}\right)} = N^2_{\mathrm{eff}}\frac{\tilde{V}_{ro}}{\tilde{I}_r} = N^2_{\mathrm{eff}}\left(\frac{R_r}{s} + j\omega_e L_r\right) = \frac{R'_r}{s} + jX'_r \qquad (12.14)$$

Therefore, the transformer and rotor impedances from Fig. 12.10 can be replaced by the impedance from Eq. (12.14) as shown in Fig. 12.12.

The model shown in Fig. 12.12 can be useful for finding the impedance of a single-phase of an induction motor during operation. However, translating the currents into torque for a particular motor speed is challenging as the resistance R'_r/s captures both conductive losses in the rotor bars and power going into the mechanical motion of the rotor. Specifically, the total time-average power crossing the air gap is given by

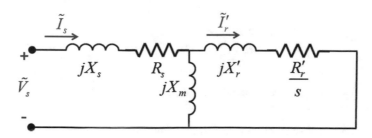

Fig. 12.12 Simplified model of single-phase of an induction motor

$$P_{ag} = \frac{3}{2}\left|\tilde{I}'_r\right|^2 \frac{R'_r}{s} = 3\left|\tilde{I}'_r\right|^2_{\mathrm{rms}} \frac{R'_r}{s} \tag{12.15}$$

where the factor of three corresponds to having three phases. Likewise, the power lost due to conductive losses in the rotor bars is given by

$$P_r = \frac{3}{2}\left|\tilde{I}_r\right|^2 R_r = \frac{3}{2}\left|\tilde{I}'_r\right|^2 N^2_{\mathrm{eff}} R_r = \frac{3}{2}\left|\tilde{I}'_r\right|^2 R'_r = 3\left|\tilde{I}'_r\right|^2_{\mathrm{rms}} R'_r \tag{12.16}$$

The difference between these two powers must be the power going into the mechanical motion. Hence,

$$P_{\mathrm{dev}} = P_{ag} - P_r = \frac{3}{2}\left|\tilde{I}'_r\right|^2 \left(\frac{R'_r}{s} - R'_r\right) = \frac{3}{2}\left|\tilde{I}'_r\right|^2 \left(\frac{1-s}{s}\right)R'_r = 3\left|\tilde{I}'_r\right|^2_{\mathrm{rms}} \left(\frac{1-s}{s}\right)R'_r \tag{12.17}$$

As a result, the resistance R'_r/s can be split into two different resistances to independently calculate the power going into conductive losses and the power going into mechanical motion as shown in Fig. 12.13.

With the circuit model now complete, we can generate the power flow diagram for the three phase induction motor shown in Fig. 12.14. The powers, P_{in} and P_s are given by

$$P_{\mathrm{in}} = \tfrac{3}{2}\left|\tilde{V}_s\right|\left|\tilde{I}_s\right|\cos(\theta_{vs} - \theta_{is}) = 3\left|\tilde{V}_s\right|_{\mathrm{rms}}\left|\tilde{I}_s\right|_{\mathrm{rms}}\cos(\theta_{vs} - \theta_{is})$$
$$P_s = \tfrac{3}{2}\left|\tilde{I}_s\right|^2 R_s = 3\left|\tilde{I}_s\right|^2_{\mathrm{rms}} R_s \tag{12.18}$$

while the other power quantities can be found from Eqs. (12.15) through (12.17). The output torque and developed torque can be calculated from the power expressions as

$$T_{\mathrm{out}} = \frac{P_{\mathrm{out}}}{\omega_m}$$
$$T_{\mathrm{dev}} = \frac{P_{\mathrm{dev}}}{\omega_m} = \frac{(1-s)P_{ag}}{(1-s)\omega_s} = \frac{P_{ag}}{\omega_s} \tag{12.19}$$

Notice that the starting developed torque for the motor corresponding to a slip of 1 is given by

$$T_{\mathrm{start}} = \frac{P_{ag}}{\omega_s} = \frac{3}{2}\frac{\left|\tilde{I}'_r\right|^2 R'_r}{\omega_s} = 3\frac{\left|\tilde{I}'_r\right|^2_{\mathrm{rms}} R'_r}{\omega_s} \tag{12.20}$$

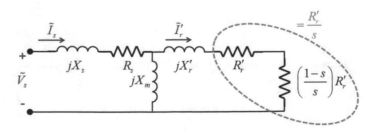

Fig. 12.13 Simplified model of single-phase of an induction motor with separate resistors to account for conductive losses and mechanical motion

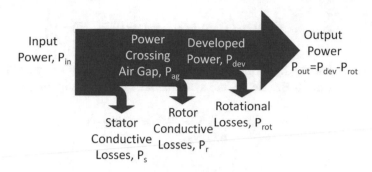

Fig. 12.14 Power flow diagram for a three-phase induction motor

Example 12.5 A three-phase delta-connected induction motor with 4 poles is connected to a wye-connected source as shown in Fig. 12.15. The impedance of the transmission line between the source and the load is given by $Z_{line} = 1 + j2 \, \Omega$. The circuit model for the motor has $R_s = 1 \, \Omega$, $R'_r = 0.5 \, \Omega$, $X_s = 1 \, \Omega$, $X'_r = 0.5 \, \Omega$, and $X_m = 50 \, \Omega$. In addition, $v_{an}(t) = 120 \cos(100\pi t) \, V_{rms}$ and the motor has 100 W of rotational losses. The motor is spinning at a speed of 1400 rpm. What is the impedance of each phase of the motor, Z_{Motor}, and the output power and torque for the motor?

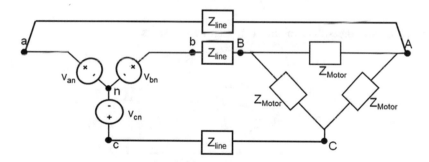

Fig. 12.15 Wye–Delta-connected three-phase induction motor for Example 12.5

Solution:
The circuit model for a single-phase of the motor is given by

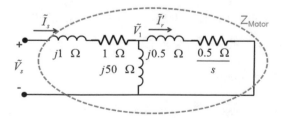

Since the motor is delta-connected, the source voltages and currents would be given by $\tilde{V}_s = \tilde{V}_{AB}$ and $\tilde{I}_s = \tilde{I}_{AB}$ assuming we are focusing on the a-phase. To find $\tilde{V}_{AB}$, we need to first find Z_{motor} and then solve the three-phase circuit. Z_{Motor} is given by

$$Z_{\text{Motor}} = 1 + j1 + \left((j50) \text{ in parallel w/} \left(\frac{0.5}{s} + j0.5 \right) \right)$$

Therefore, we need to find the slip. We know that the motor speed is 1400 rpm and that the electrical frequency is 100π rad/s or 50 Hz. Hence, the synchronous speed and slip are given by

$$n_s = \frac{120 f_e}{P} = \frac{120 \cdot \frac{100\pi}{2\pi}}{4} = \frac{120 \cdot 50}{4} = 1500 \text{ rpm}$$
$$s = \frac{n_s - n_m}{n_s} = 0.066667$$

Z_{Motor} is thus given by

$$Z_{\text{Motor}} = 1 + j1 + \left(\frac{1}{\frac{1}{j50} + \frac{1}{\left(\frac{0.5}{s} + j0.5 \right)}} \right) = 1 + j1 + \left(\frac{1}{\frac{1}{j50} + \frac{1}{(7.5 + j0.5)}} \right)$$
$$= 1 + j1 + 7.1936 + j1.5634 = 8.1936 + j2.5634 \ \Omega$$

With Z_{Motor} known, we can convert the delta-connected load to a wye-connected load.

$$Z_Y = \frac{Z_{\text{Motor}}}{3} = 2.7312 + j0.8545 \ \Omega$$

A single branch of the three-phase circuit can then be drawn as

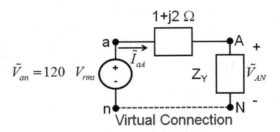

As a result, $\tilde{I}_{aA}$ and $\tilde{V}_{AN}$ are given by

$$\tilde{I}_{aA} = \frac{\tilde{V}_{an}}{Z_{\text{line}} + Z_Y} = \frac{120 \ V_{\text{rms}}}{1 + j2 + 2.7312 + j0.8545} = \frac{120 \ V_{\text{rms}}}{3.7312 + j2.8545} = 25.5436 e^{-j0.6531} \ A_{\text{rms}}$$

$$\tilde{V}_{AN} = \tilde{I}_{aA} Z_Y = 73.0989 e^{-j0.3498} \ V_{\text{rms}}$$

Yielding a $\tilde{I}_s = \tilde{I}_{AB}$ and $\tilde{V}_s = \tilde{V}_{AB}$ of

$$\tilde{V}_{AB} = \tilde{V}_{AN} \cdot \sqrt{3} \exp\left(j\frac{\pi}{6} \right) = 126.6111 e^{+j0.1738} \ V_{\text{rms}}$$

$$\tilde{I}_{AB} = \tilde{I}_{aA} \frac{\exp\left(j\frac{\pi}{6} \right)}{\sqrt{3}} = 14.7476 e^{-j0.1295} \ A_{\text{rms}}$$

Either of these values can be used to generate a node voltage equation for $\tilde{V}_1$. We will use the current $\tilde{I}_s = \tilde{I}_{AB}$.

$$\tilde{I}_{AB} = 14.7476e^{-j0.1295} \; A_{\text{rms}} = \frac{\tilde{V}_1}{j50 \; \Omega} + \underbrace{\frac{\tilde{V}_1}{\dfrac{0.5}{s} + j0.5 \; \Omega}}_{7.5 + j0.5 \; \Omega}$$

$$\Rightarrow 14.7476e^{-j0.1295} \; A_{\text{rms}} = \tilde{V}_1(0.1327 - j0.0288)$$

$$\Rightarrow \tilde{V}_1 = 108.5644e^{+j0.0846} \; V_{\text{rms}}$$

Once $\tilde{V}_1$ is known, $\tilde{I}'_r$ can be found from

$$\tilde{I}'_r = \frac{\tilde{V}_1}{\underbrace{\dfrac{0.5}{s} + j0.5 \; \Omega}_{7.5 + j0.5 \; \Omega}} = 14.4432e^{+j0.018} \; A_{\text{rms}}$$

The developed power and output power are then given by

$$P_{\text{dev}} = 3|\tilde{I}'_r|^2_{\text{rms}} \left(\frac{1-s}{s}\right) R'_r = 4380.7 \; W$$

$$P_{\text{out}} = P_{\text{dev}} - P_{\text{rot}} = 4280.7 \; W$$

Finally, the output torque is given by

$$T_{\text{out}} = \frac{P_{\text{out}}}{\omega_m} = \frac{P_{\text{out}}}{n_m \frac{2\pi}{60}} = 29.1985 \; \text{Nm}$$

Example 12.6 Find the starting torque for the motor described in Example 12.5.

Solution:
The starting torque corresponds to a slip value of 1. As a result,

$$\frac{0.5}{s} + j0.5 \; \Omega \rightarrow 0.5 + j0.5 \; \Omega$$

This impedance is very different from the impedance value of $7.5 + j0.5 \; \Omega$ found in Example 12.5. As a result, all of the previously derived currents, voltages, and impedance are now different, and the problem must be solved completely over from the beginning. Therefore, we will once again find a new value for Z_{Motor}.

$$Z_{\text{Motor}} = 1 + j1 + \left(\frac{1}{\frac{1}{j50} + \frac{1}{(0.5 + j0.5)}}\right) = 1.4901 + j1.4999 \; \Omega$$

$$\Rightarrow Z_Y = 0.4967 + j0.5 \; \Omega$$

As a result, $\tilde{I}_{aA}$ and $\tilde{V}_{AN}$ are given by

$$\tilde{I}_{aA} = \frac{\tilde{V}_{an}}{Z_{line} + Z_Y} = 41.1840e^{-j1.0313} \; A_{rms}$$

$$\tilde{V}_{AN} = \tilde{I}_{aA}Z_Y = 29.0246e^{-j0.2427} \; V_{rms}$$

Yielding a $\tilde{I}_s = \tilde{I}_{AB}$ and $\tilde{V}_s = \tilde{V}_{AB}$ of

$$\tilde{V}_{AB} = \tilde{V}_{AN} \cdot \sqrt{3} \exp\left(j\frac{\pi}{6}\right) = 50.2721e^{+j0.2809} \; V_{rms}$$

$$\tilde{I}_{AB} = \tilde{I}_{aA} \frac{\exp\left(j\frac{\pi}{6}\right)}{\sqrt{3}} = 29.0246e^{j0.8045} \; A_{rms}$$

Moving on to find $\tilde{V}_1$ and $\tilde{I}_r'$ yields

$$\tilde{I}_{AB} = \frac{\tilde{V}_1}{j50 \; \Omega} + \frac{\tilde{V}_1}{0.5 + j0.5 \; \Omega} \Rightarrow \tilde{V}_1 = 16.646e^{+j0.2876} \; V_{rms}$$

$$\tilde{I}_r' = \frac{\tilde{V}_1}{0.5 + j0.5 \; \Omega} = 23.541e^{-j0.4978} \; A_{rms}$$

Notice that these voltages and currents are completely different from the values found in Example 12.5. The starting torque is then given by

$$T_{start} = \frac{P_{ag}}{\omega_s} = 3\frac{|\tilde{I}_r'|^2_{rms}R_r'}{\omega_s} = 5.292 \; Nm$$

12.4   Circuit Analysis of Three-Phase Induction Motors with Negligible Line Impedances

As was illustrated by Examples 12.5 and 12.6, solving circuit problems involving induction motors can be very long and tedious. However, the solution process can be simplified considerably if the line impedances, Z_{line}, is negligible. Under this condition, there would be no voltage drop across the line and the voltage $\tilde{V}_s$ would be known directly from the source voltages. Since the voltage across the stator windings is known and independent of the current in the stator windings, we can replace the elements in the box in Fig. 12.16 by their Thevenin equivalent. The open-circuit voltage (Thevenin Voltage) would be given by

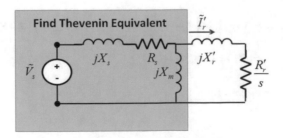

Fig. 12.16 Circuit model for single-phase of an induction motor when the line impedance in negligible

$$\tilde{V}_{Th} = \tilde{V}_s \frac{jX_m}{R_s + j(X_s + X_m)} \tag{12.21}$$

Likewise, the short circuit current would be given by

$$\tilde{I}_{SC} = \frac{\tilde{V}_s}{R_s + jX_s} \tag{12.22}$$

Hence, the Thevenin impedance would be given by

$$Z_{Th} = \frac{\tilde{V}_{Th}}{\tilde{I}_{SC}} = \frac{jX_m(R_s + jX_s)}{R_s + j(X_s + X_m)} \tag{12.23}$$

However, for induction motors, $R_s \ll X_m$. Therefore, the Thevenin voltage and Thevenin resistance can be rewritten as

$$\tilde{V}_{Th} \cong \tilde{V}_s \left(\frac{X_m}{X_s + X_m}\right)$$

$$Z_{Th} = R_{Th} + jX_{Th} \stackrel{\sim}{=} (R_s + jX_s)\frac{X_m}{(X_s + X_m)} \Rightarrow \begin{cases} R_{Th} = R_s \left(\frac{X_m}{X_s + X_m}\right) \\ X_{Th} = X_s \left(\frac{X_m}{X_s + X_m}\right) \end{cases} \tag{12.24}$$

As a result, the circuit shown in Fig. 12.16 can be simplified yielding the circuit shown in Fig. 12.17. In this simplified version, the current $\tilde{I}'_r$ can be found directly from Ohm's Law and is given by

$$\tilde{I}'_r \cong \frac{\tilde{V}_{Th}}{Z_{Th} + jX'_r + \frac{R'_r}{s}}$$

$$|\tilde{I}'_r|^2 = \frac{|\tilde{V}_{Th}|^2}{\left|R_{Th} + jX_{Th} + jX'_r + \frac{R'_r}{s}\right|^2} = \frac{|\tilde{V}_{Th}|^2}{\left(R_{Th} + \frac{R'_r}{s}\right)^2 + \left(X_{Th} + X'_r\right)^2} \tag{12.25}$$

As a result, the power crossing the air gap, P_{ag}, is given by

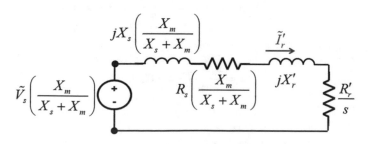

Fig. 12.17 Simplified circuit model for single-phase of an induction motor when the line impedance is negligible

$$P_{ag} = 3|\tilde{I}'_r|^2_{\text{rms}} \frac{R'_r}{s} \cong 3 \frac{|\tilde{V}_{Th}|^2_{\text{rms}}}{\left(\left(R_{Th} + \frac{R'_r}{s}\right)^2 + \left(X_{Th} + X'_r\right)^2\right)} \frac{R'_r}{s} \qquad (12.26)$$

Hence, the developed torque provided by the motor would be given by

$$T_{\text{dev}} = \frac{P_{ag}}{\omega_s} \cong \frac{3|\tilde{V}_{Th}|^2_{\text{rms}}}{\omega_s \left(\left(R_{Th} + \frac{R'_r}{s}\right)^2 + \left(X_{Th} + X'_r\right)^2\right)} \frac{R'_r}{s}$$

$$\cong \frac{3|\tilde{V}_s|^2_{\text{rms}} \left(\frac{X_m}{X_s + X_m}\right)^2 \frac{R'_r}{s}}{\omega_s \left(\left(R_s \left(\frac{X_m}{X_s + X_m}\right) + \frac{R'_r}{s}\right)^2 + \left(X_s \left(\frac{X_m}{X_s + X_m}\right) + X'_r\right)^2\right)} \qquad (12.27)$$

> **Note**: The approximations based on $R_s \ll X_m$ are reasonably accurate when calculating the torque and output power for the induction motor. However, the approximation leads to relatively large errors in the estimation of the conductive losses in the stator. Therefore, the approximation should only be used to calculate the motor output and not the losses in the circuit.

Example 12.7 A 3-phase induction motor with 4 poles is connected to a voltage source with $v_s(t) = 147\cos(90\pi t)$ V_{rms}. The circuit has negligible line impedance. The motor has $R_s = 2\,\Omega$, $R'_r = 3\,\Omega$, $X_s = 1\,\Omega$, $X'_r = 3\,\Omega$, and $X_m = 500\,\Omega$. In addition, the motor is spinning at a speed of 1300 rpm. What is the output torque of the motor assuming negligible rotational losses?

Solution:
Clearly, $R_s \ll X_m$. Therefore, the expression derived in Eq. (12.27) can be utilized. The first step is to find the slip. The synchronous speed for this motor would be given by

$$n_s = \frac{120 f_e}{P} = \frac{120 \cdot \frac{90\pi}{2\pi}}{4} = \frac{120 \cdot 45}{4} = 1350 \text{ rpm}$$

yielding a slip of

$$s = \frac{n_s - n_m}{n_s} = 0.0370$$

Also,

$$\left(\frac{X_m}{X_s + X_m}\right) = \frac{500}{1 + 500} = 0.998$$

The developed torque, which would be the same as the output torque since the rotational losses are negligible would then be given by

$$T_{dev} \cong \frac{3|\tilde{V}_s|^2_{\text{rms}}\left(\frac{X_m}{X_s+X_m}\right)^2\frac{R'_r}{s}}{\omega_s\left(\left(R_s\left(\frac{X_m}{X_s+X_m}\right)+\frac{R'_r}{s}\right)^2+\left(X_s\left(\frac{X_m}{X_s+X_m}\right)+X'_r\right)^2\right)}$$

$$=\frac{3\cdot(147\ V_{\text{rms}})^2(0.998)^2\frac{3}{0.0370}}{\frac{2\cdot90\pi}{4}\left((2(0.998)+\frac{3}{0.0370})^2+((0.998)+3)^2\right)}=5.3582\ \text{Nm}$$

Example 12.8 Find the starting torque for the motor described in Example 12.7.

Solution:

When starting, the slip would be 1. Therefore, the starting torque would be given by

$$T_{dev} \cong \frac{3|\tilde{V}_s|^2_{\text{rms}}\left(\frac{X_m}{X_s+X_m}\right)^2\frac{R'_r}{s}}{\omega_s\left(\left(R_s\left(\frac{X_m}{X_s+X_m}\right)+\frac{R'_r}{s}\right)^2+\left(X_s\left(\frac{X_m}{X_s+X_m}\right)+X'_r\right)^2\right)}$$

$$=\frac{3\cdot(147\ V_{\text{rms}})^2(0.998)^2 3}{\frac{2\cdot90\pi}{4}\left((2(0.998)+3)^2+((0.998)+3)^2\right)}=33.4648\ \text{Nm}$$

Example 12.9 A three-phase induction motor with 4 poles is connected to a voltage source with an amplitude of 100 V_{rms} and a frequency of 70 Hz. The circuit has negligible line impedance. The motor has $R_s = 0.6\ \Omega$, $R'_r = 0.3\ \Omega$, $X_s = 0.8\ \Omega$, $X'_r = 0.4\ \Omega$, and $X_m = 20\ \Omega$. In addition, the motor is providing a T_{dev} of 25 Nm while operating close to its normal operating range. What is the speed of the motor in rpm?

Solution:

Begin by finding the synchronous speed

$$\omega_s = \frac{2\omega_e}{P} = \frac{4\pi f_e}{P} = 219.9\ \text{rad/sec}$$

The relationship between torque and slip is then approximately given by

$$T_{dev} \cong \frac{3|\tilde{V}_s|^2_{\text{rms}}\left(\frac{X_m}{X_s+X_m}\right)^2\frac{R'_r}{s}}{\omega_s\left(\left(R_s\left(\frac{X_m}{X_s+X_m}\right)+\frac{R'_r}{s}\right)^2+\left(X_s\left(\frac{X_m}{X_s+X_m}\right)+X'_r\right)^2\right)}$$

Therefore,

$$\left(R_s\left(\frac{X_m}{X_s+X_m}\right)+\left(\frac{R'_r}{s}\right)\right)^2+\left(X_s\left(\frac{X_m}{X_s+X_m}\right)+X'_r\right)^2\cong\left[3\frac{|\tilde{V}_s|^2_{\text{rms}}}{\omega_sT_{\text{dev}}}\left(\frac{X_m}{X_s+X_m}\right)^2\right]\left(\frac{R'_r}{s}\right)$$

$$\Rightarrow\left(\frac{R'_r}{s}\right)^2+\left[\frac{2R_s\left(\frac{X_m}{X_s+X_m}\right)}{-3\frac{|\tilde{V}_s|^2_{\text{rms}}}{\omega_sT_{\text{dev}}}\left(\frac{X_m}{X_s+X_m}\right)^2}\right]\left(\frac{R'_r}{s}\right)+\left[\begin{array}{c}\left(R_s\left(\frac{X_m}{X_s+X_m}\right)\right)^2\\+\left(X_s\left(\frac{X_m}{X_s+X_m}\right)+X'_r\right)^2\end{array}\right]=0$$

$$\Rightarrow\left(\frac{R'_r}{s}\right)^2+[1.1538-5.0451]\left(\frac{R'_r}{s}\right)+[0.3328+1.3671]=0$$

$$\Rightarrow\left(\frac{R'_r}{s}\right)^2-3.8912\left(\frac{R'_r}{s}\right)+1.6999=0$$

This is a quadratic with two possible roots given by

$$\left(\frac{R'_r}{s}\right)=3.3897\quad or\quad 0.5015\Rightarrow s=0.0885\quad or\quad 0.5982$$

Since we are told that the motor is operating in its normal operating range, the smaller of the two possible slip values should be assumed. Hence, the motor speed can then be found from the synchronous speed and the slip by

$$n_s=\frac{120f_e}{P}=2100\text{ rpm}$$

$$n_m=(1-s)n_s=1914.1\text{ rpm}$$

The solution can also be represented graphically as shown below.

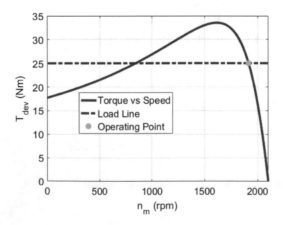

Returning now to Eq. (12.27), consider the limiting case when the slip is very small. Under this condition, $(R_{Th}+R'_r/s)^2+(X_{Th}+X'_r)^2\approx(R'_r/s)^2$ as (R'_r/s) will be much larger than the other impedance terms. Hence,

$$(T_{\text{dev}})_{\text{Small }s}\approx\frac{3|\tilde{V}_{Th}|^2_{\text{rms}}}{\omega_s\left(\frac{R'_r}{s}\right)^2}\frac{R'_r}{s}=\frac{3|\tilde{V}_{Th}|^2_{\text{rms}}}{\omega_sR'_r}s\tag{12.28}$$

Therefore, the torque is proportional to the slip at small slip as was stated previously. Also, the range over which the proportionality approximation is valid will depend on the values of the impedances in the circuit.

The maximum torque or pullout torque shown in Fig. 12.8 is also strongly influenced by the impedances in the circuit. The slip corresponding to the maximum torque can be found by taking the derivative of developed torque with respect to (R'_r/s) and setting it equal to zero. Hence,

$$\frac{\partial T_{\text{dev}}}{\partial (R'_r/s)} = \frac{3|\tilde{V}_{Th}|^2_{\text{rms}}}{\omega_s\left(\left(R_{Th}+\frac{R'_r}{s}\right)^2+\left(X_{Th}+X'_r\right)^2\right)} - \frac{6|\tilde{V}_{Th}|^2_{\text{rms}}\left(R_{Th}+\frac{R'_r}{s}\right)\frac{R'_r}{s}}{\omega_s\left(\left(R_{Th}+\frac{R'_r}{s}\right)^2+\left(X_{Th}+X'_r\right)^2\right)^2} = 0$$

$$\Rightarrow \left(\left(R_{Th}+(R'_r/s)\right)^2+\left(X_{Th}+X'_r\right)^2\right) = (R'_r/s)\left(2\left(R_{Th}+(R'_r/s)\right)\right) \qquad (12.29)$$

$$\Rightarrow \frac{R'_r}{s} = \sqrt{R^2_{Th}+\left(X_{Th}+X'_r\right)^2} \Rightarrow s_{\text{max}} = \frac{R'_r}{\sqrt{R^2_{Th}+\left(X_{Th}+X'_r\right)^2}}$$

Substituting this value of slip into the equation for the developed torque yields

$$T_{\text{max}} \cong \frac{3|\tilde{V}_{Th}|^2_{\text{rms}}}{2\omega_s\left(\sqrt{R^2_{Th}+\left(X_{Th}+X'_r\right)^2}+R_{Th}\right)} \qquad (12.30)$$

Example 12.10 Find the maximum torque (pullout torque) and speed corresponding to this torque for the three-phase induction motor described in Example 12.9.

Solution:

The first step is to find V_{Th}, R_{Th}, and X_{Th}

$$\tilde{V}_{Th} = \tilde{V}_s\left(\frac{X_m}{X_s+X_m}\right) = 100(0.96154) = 96.1538\ V_{\text{rms}}$$

$$R_{Th} = R_s\left(\frac{X_m}{X_s+X_m}\right) = 0.6(0.96154) = 0.5769\ \Omega$$

$$X_{Th} = X_s\left(\frac{X_m}{X_s+X_m}\right) = 0.8(0.96154) = 0.7692\ \Omega$$

The maximum torque and corresponding slip are then given by

$$T_{\text{max}} \cong \frac{3|\tilde{V}_{Th}|^2_{\text{rms}}}{2\omega_s\left(\sqrt{R^2_{Th}+\left(X_{Th}+X'_r\right)^2}+R_{Th}\right)} = 33.5311\ \text{Nm}$$

$$s = \frac{R'_r}{\sqrt{R^2_{Th}+\left(X_{Th}+X'_r\right)^2}} = \frac{0.3}{\sqrt{R^2_{Th}+\left(X_{Th}+0.4\right)^2}} = 0.2301$$

The speed can then be found from the slip as

$$n_m = (1-s)n_s = (1-0.2301)2100 = 1616.8\ \text{rpm}$$

12.5 Single-Phase Induction Motors

Three-phase induction motors are common when three-phase power is available. However, single-phase power, such as that found in standard wall outlets, is far more accessible. Therefore, it is also important to have motors that can operate from single-phase sources. The universal motor described in Chap. 5 is one example of a motor that can operate from a standard wall outlet. However, the brushes and commutators in the universal motor negatively impact its reliability. Therefore, other single-phase motors are needed.

Figure 12.18 shows the magnetic fields being generated by a two-pole single-phase induction motor as a function of time.

Since the motor has only a single-phase, the magnetic field grows and shrinks over time as given by

$$B_{\text{total}}(t) = B_o \cdot \cos(\omega_e t) \cos(\theta_s) \tag{12.31}$$

However, if we use the trigonometric identity given in Eqs. (12.4) and (12.31) can also be written as

$$B_{\text{total}}(t) = \frac{B_o}{2} \left[\cos(\omega_e t - \theta_s) + \cos(\omega_e t + \theta_s) \right] \tag{12.32}$$

The $\cos(\omega_e t - \theta_s)$ term would correspond to a magnetic field spinning in the counterclockwise direction while the $\cos(\omega_e t + \theta_s)$ term would correspond to a magnetic field spinning in the clockwise direction. Hence, the total magnetic field could be interpreted as two magnetic fields spinning in opposite directions as is also illustrated in Fig. 12.18. The rotor can be made to spin in sync with one of these two spinning magnetic fields if it is started properly. At low speeds, however, the motor would be incapable of providing any torque as the rotor would not be tracking either magnetic field. Hence, the speed torque curve for a single-phase AC induction motor is given by Fig. 12.19.

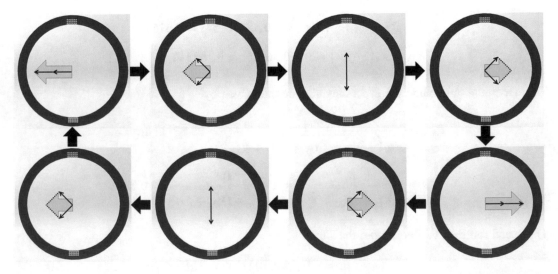

Fig. 12.18 Total magnetic field in the air gap for a 2-pole single-phase AC induction motor illustrating the decomposition of the pulsating magnetic field into the sum of two magnetic fields rotating in opposite directions

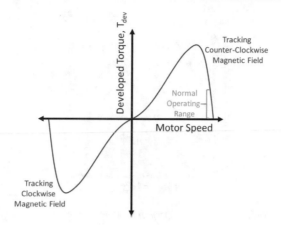

Fig. 12.19 Speed-Torque curve for a single-phase AC induction motor

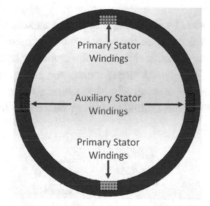

Fig. 12.20 Position of auxiliary windings in single-phase induction motor

Since the starting torque of the single-phase induction motor is zero, the primary challenge is to get the motor started. This is accomplished by introducing auxiliary windings positioned perpendicular to the primary windings as shown in Fig. 12.20. The phase of the AC signal in the auxiliary windings is then delayed relative to the primary windings to provide the necessary starting torque.

12.5.1   Split-Phase Windings

The simplest approach to create a phase shift between the primary and auxiliary windings is to use wire with a smaller diameter in the auxiliary winding. This will result in a higher ratio of resistance to inductance altering the phase of the impedance of the winding. If the phase of the impedance is different between the two windings, the phase of the magnetic field will also be different giving rise to a starting torque on the motor. Once the motor has started to spin at a sufficient speed, a centrifugal switch is triggered that disconnects the auxiliary winding. The motor then continues to spin from the power provided by the primary windings. The speed-torque curve for the basic split-phase motor is shown in Fig. 12.21. Split-phase motors tend to have moderate starting torques, but they tend to have pulsating vibrations during operation due to the pulsating magnetic field.

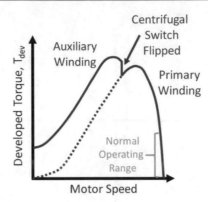

Fig. 12.21 Speed-Torque curve for split-phase single-phase induction motor

12.5.2   Capacitor-Start Windings

The starting torque can be increased by including a capacitor in series with the auxiliary windings as this will place the phase difference between the auxiliary and primary windings much closer to 90°. Once again, after the motor reaches a sufficient speed, the auxiliary winding with the capacitor will be disconnected from the circuit. The speed-torque curve for a capacitor start motor is shown in Fig. 12.22. Capacitor start motors have relatively high starting torques, but they still have pulsating vibrations during their operation in the normal operating range due to the pulsating magnetic field.

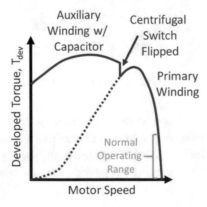

Fig. 12.22 Speed-Torque curve for a capacitor-start single-phase induction motor

12.5.3   Capacitor-Run Windings

In some applications, the vibrations introduced by the pulsating magnetic field are undesirable. Therefore, it is possible to leave the capacitor and associated auxiliary windings permanently attached to the circuit without the use of a centrifugal switch. However, since the goal of the auxiliary windings/capacitor is to now provide the phase shift during normal operation, the starting torques are lower than those provided by the capacitor start motors. The operation of the capacitor run motor is very similar to a three-phase induction motor with a constantly rotating magnetic field.

12.5.4   Capacitor-Start, Capacitor-Run Windings

It is also possible to have some capacitance connected to the auxiliary windings at all times and a second capacitance that is only connected when the motor is operating at a low speed. These motors are called capacitor start, capacitor run motors. The starting capacitance is designed to maximize the starting torque. Once the motor gains sufficient speed, a centrifugal switch changes the capacitance to optimize the phase difference during normal operation. Capacitor start, capacitor run motors have both high starting torques and relatively smooth motion during operation. However, they are more complicated than either the capacitor run or capacitor start motors.

12.6   Synchronous Generators

Synchronous generators, or alternators, are one of the most common methods of turning mechanical motion into electrical power. In many ways, they operate similar to an induction motor only in reverse. For an induction motor, the three-phases on the stator generate a rotating magnetic field. The rotating magnetic field cuts across the rotor bars creating voltages and currents in the rotor bars. For an alternator, we begin by establishing a magnetic field on the rotor. This can be accomplished by either using permanent magnets or by flowing DC currents on windings in the rotor. The DC power can either come from an external source connected via slip rings and brushes or the power can come from a small generator mounted directly on the shaft of the rotor. Once the magnetic field has been established, the prime mover which provides the mechanical motion will cause the magnets/magnetic field to rotate. The rotating magnetic field from the rotor will then cut across the windings on the stator inducing a three-phase time-varying voltage as illustrated in Fig. 12.23.

When working with three-phase induction motors, we noted that the synchronous speed of the rotating magnetic field was directly related to the number of poles and the electrical frequency of the signal applied to the stator windings. In a similar fashion, the electrical frequency, f_e, being produced by the synchronous generator is given by

$$f_e = \frac{n_m P}{120} \tag{12.33}$$

where n_m is the rotational speed of the rotor in rpm and P is the number of poles in the generator. Hence, the frequency produced by the generator will depend on the speed of rotation.

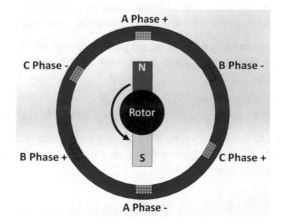

Fig. 12.23   Basic operation of a 2-Pole synchronous generator

Example 12.11 A small 1.5 m diameter turbine spins at 720 rpm in 30 km/h (18.64 mph) wind. How many poles are needed to produce a 60 Hz signal?

Solution:
The electrical frequency and number of poles are related by

$$f_e = \frac{n_m P}{120} \Rightarrow P = \frac{120 \cdot f_e}{n_m} = 10$$

Example 12.12 For the same generator described in Example 12.11, what would be the electrical frequency produced if the turbine were to spin at 1250 rpm in a 50 km/h wind?

Solution:
The electrical frequency would be given by

$$f_e = \frac{n_m P}{120} = 104.2 \text{ Hz}$$

For traditional generators such as those found in steam turbines, the dependence of the electrical frequency on rotational speed is not of significant concern as the flow the steam can be controlled to maintain a constant speed of rotation. However, for newer power generation approaches such as wind turbines, the speed of the rotor cannot be tightly controlled. As a result, the electrical frequency being produced is constantly changing. Therefore, the power produced by wind turbines is typically rectified (i.e., converted into a DC signal). The DC signal is then converted into a 60 Hz signal at the appropriate phase to provide power to the rest of the grid.

12.7   Problems

Problem 12.1: What is the fastest speed a three-phase induction motor can rotate if it is connected to a power source that operates at 100 Hz?

Problem 12.2: What is the synchronous speed for a three-phase induction motor with 4 poles being powered by a 50 Hz signal?

Problem 12.3: A three-phase induction motor is being driven at a frequency of 80 Hz, and the motor speed is 1000 rpm. How many poles does the motor probably have and what is the slip?

Problem 12.4: A three-phase induction motor with 4 poles is being driven at 45 Hz and is running in its normal operating range. When connected to a certain load, the motor rotates at a speed of 1070 rpm and outputs 800 W.

(a) What is the slip and developed torque?
(b) What is the new slip and motor speed if the developed torque is reduced to 4 Nm?

Problem 12.5: A three-phase induction motor with 6 poles is running in its normal operating range. The motor is operating at a slip of 0.1 while rotating at a speed of 1900 rpm, what is the electrical frequency used to power the motor?

Problem 12.6: A 76 Hz three-Phase induction motor is running at 1400 rpm when under full load conditions.

(a) Determine the most likely slip at full load.
(b) Determine the most likely speed if the load torque drops in half.

Problem 12.7: A three-phase induction motor with 4 poles is being driven at 90 Hz and is running in its normal operating range. When connected to a certain load, the motor rotates at a speed of 2500 rpm and outputs 1000 W.

(a) How many poles does the motor probably have?
(b) What is the slip?
(c) What is the torque supplied assuming negligible rotational losses?
(d) Now a new load is connected that requires 1500 W of power. What is the new speed of the motor?

Problem 12.8: A three-phase induction motor with 8 poles is being driven at 60 Hz and is running in its normal operating range. The motor is providing an output power, P_{out}, of 1.5 kW and has rotational losses of 100 W when rotating at a speed of 810 rpm.

(a) What is the slip?
(b) What is the developed torque, T_{dev}?
(c) If a new load is connected that requires 25 Nm of developed torque. What is the new slip and speed of the motor in rpm?

Problem 12.9: A three-phase induction motor is being driven at 60 Hz and is running in its normal operating range.

(a) When connected to a load of $T_{dev} = 20$ Nm, the motor rotates at a speed of 1000 rpm. What is the slip?
(b) What is the new motor speed in rpm if the motor is now connected to a load with $T_{dev} = 15$ Nm?

Problem 12.10: The speed-torque curve for a three-Phase induction motor is shown below along with speed-torque characteristics for the load connected to the motor. The motor has negligible rotational losses.

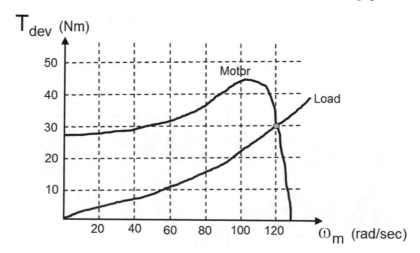

If the motor is being driven by a 50 Hz voltage source, find the following:

(a) The number of poles in the motor
(b) The slip when operating at steady-state with the load shown.
(c) The output power and the conductive/copper losses in the rotor bars.

Problem 12.11: For this problem, you have an 8 pole, three-Phase 60 Hz AC induction motor connected to a power supply in the wye-wye configuration as shown below where the source voltage $\tilde{V}_{an} = 120\ V_{rms}$ and $Z_{line} = 1.0 + j1.0\ \Omega$.

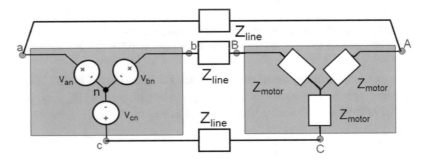

In this diagram, Z_{motor} is the impedance seen looking into a single-phase of the motor under normal operation as shown below.

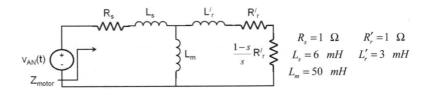

What is the current flow $i_{aA}(t)$ when the motor is connected to a load such that the slip is 0.1?

Problem 12.12: For this problem, you have a 4 pole, three-Phase 60 Hz AC induction motor connected to a power supply as shown below where the source voltage $\tilde{V}_{ab} = 120\ V_{rms}$ and $Z_{line} = 0.1 + j0.1\ \Omega$.

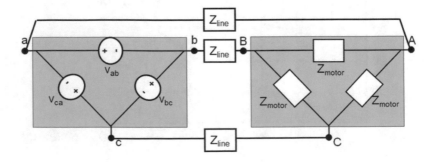

In this diagram, Z_{motor} is the impedance seen looking into a single-phase of the motor as shown below.

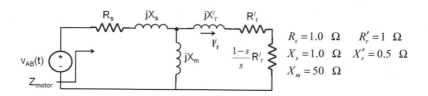

A 3-phase induction motor is being driven at 60 Hz and is running at a speed of 1400 rpm.

(a) What is the impedance of each phase of the motor, Z_{motor}?
(b) What is the RMS amplitude and phase of the current flowing in the top transmission line, $\tilde{I}_{aA}$?
(c) What is the output power assuming 100 W of rotational losses?
(d) What is the developed torque?

Problem 12.13: Find the starting torque and starting line current, $\tilde{I}_{aA}$, for the motor/circuit described in Problem 12.12.

Problem 12.14: A 3-phase induction motor with 6 poles is connected to a voltage source with $v_s(t) = 120\cos(80t)$ V_{rms}. The circuit has negligible line impedance. The motor has $R_s = 1\,\Omega$, $R'_r = 0.5\,\Omega$, $X_s = 0.8\,\Omega$, $X'_r = 1.5\,\Omega$, and $X_m = 50\,\Omega$. In addition, the motor is spinning at a speed of 250 rpm. What is the output torque of the motor assuming negligible rotational losses using the simplified Thevenin based circuit model for the motor?

Problem 12.15: For the circuit/motor described in Problem 12.14, what is the starting torque for the motor?

Problem 12.16: For the circuit/motor described in Problem 12.14, what is the maximum torque the motor can supply and the motor speed corresponding to this torque?

Problem 12.17: A 3-phase induction motor with 4 poles is connected to a voltage source with $v_s(t) = 80\cos(50t)$ V_{rms}. The circuit has negligible line impedance. The motor has $R_s = 2\,\Omega$, $R'_r = 0.6\,\Omega$, $X_s = 5\,\Omega$, $X'_r = 1\,\Omega$, and $X_m = 5\,\Omega$. Plot the speed torque curve for the motor using the complete circuit model as well as the simplified Thevenin based model.

Problem 12.18: A 3-phase induction motor with 6 poles is connected to a voltage source with an amplitude of 180 V_{rms} and a frequency of 100 Hz. The circuit has negligible line impedance and you can assume the simplified Thevenin based circuit model is valid. The motor has $R_s = 2\,\Omega$, $R'_r = 0.4\,\Omega$, $X_s = 3\,\Omega$, $X'_r = 1\,\Omega$, and $X_m = 100\,\Omega$. In addition, the motor is providing a T_{dev} of 5 Nm. What is the slip and speed of the motor in rpm?

Problem 12.19: For the circuit/motor described in Problem 18, what is the slip and speed of the motor in rpm if the motor is providing a T_{dev} of 20 Nm while operating close to its normal operating range?

Circuit Measurements

13

One of the most critical aspects of any engineering design is the careful measurement of the final device to ensure it is operating according to the designed specifications. Often there are multiple testing stages of various prototypes as the design is developed from concept to reality. Measurements on the prototypes are critical as they will guide changes to the design if specifications are not met. However, conducting proper measurements is not trivial as all measurement systems have inherent limitations. In this chapter, we will discuss the fundamentals in making measurements of electrical circuits and some of the practical limitations the measurement devices place on the accuracy of the measurements.

13.1 Digital Multi-meter Basics

The most basic measurement device for electrical circuits is the digital multi-meter. Modern multi-meters are used to measure resistances, capacitances, DC voltages/currents, and rms voltages/currents with some multi-meters having additional measurement capabilities. The fundamentals of each of these measurements are discussed below.

13.1.1 Resistance Measurement

Measuring the resistance is often used to confirm resistance values, check for short circuits, and confirm that appropriate electric contact has been established. It is often wise to confirm that a reasonably high resistance exists from the power terminals to ground before applying power to a newly built circuit. When set to measure resistance, the two probes of the multi-meter must be placed across the unknown resistance. The meter then measures the resistance by providing a constant known current and then recording the resulting voltage across the load via an analog-to-digital converter (ADC) as shown in Fig. 13.1. ADC's translate voltages into bits in a computer or microprocessor.

© Springer Nature Switzerland AG 2020
T. A. Bigelow, *Electric Circuits, Systems, and Motors*,
https://doi.org/10.1007/978-3-030-31355-5_13

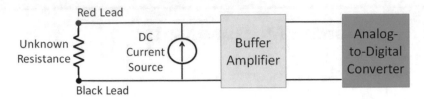

Fig. 13.1 Approximate circuit for multi-meter set to measure resistance

13.1.2 Voltage Measurement

Measuring the voltage between two nodes can be used for everything from confirming that the voltage between two nodes is at the expected values to check that a battery is not dead. To measure the voltage between two nodes, the two terminals of the meter are placed on each node. The positive voltage lead, normally the red lead, should be placed on the node that is anticipated to be at the higher potential as shown in Fig. 13.2. The voltage reported by the meter would then be the voltage difference between the two nodes.

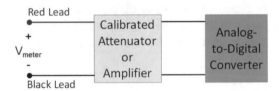

Fig. 13.2 Approximate circuit for multi-meter set to measure voltage

13.1.3 Current Measurement

While it is far more common to measure resistance or voltage, there are times when a measurement of current is desired. The current measurement is more challenging as one of the branches of the original circuit must be broken so that the meter can be inserted in the appropriate branch. The current to be measured must flow through both the meter and the branch. In addition, the positive probe must often be moved to a new connection on the meter prior to making current measurements. The current is converted to a voltage before being read by the ADC in the multi-meter as shown in Fig. 13.3.

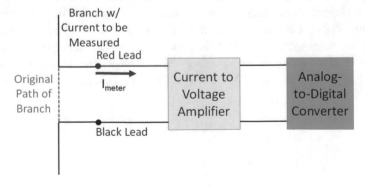

Fig. 13.3 Approximate circuit for multi-meter set to measure current

13.1.4 Capacitance Measurement

Measuring the capacitance is very useful as many capacitors may not be labeled with their capacitance value and capacitors do not have color codes like resistors. To measure capacitance, a constant current is applied to the capacitor just like the resistance measurement. The ADC in the multi-meter then records the voltage across the capacitor as a function of time. Knowing the current applied and the rate of change in the voltage allows the capacitance to be calculated.

$$C = \frac{I_{\text{meter}}}{(\partial V_{\text{meter}}/\partial t)} \tag{13.1}$$

13.2 Oscilloscope Basics

While it is possible to measure rms voltages using a digital multi-meter, it is far more common to use an oscilloscope when measuring voltages that vary with time. Oscilloscopes usually have multiple channels with each channel having a signal connection and a ground connection. When using an oscilloscope, the ground connection is connected to the circuit ground and the signal connection is attached to the node with the voltage waveform of interest. At times, a probe with a set attenuation is used between the oscilloscope and the node of interest. These probes reduce the loading on the circuit to enable more accurate measurements. The probes will also reduce the voltage to a range that will not damage the oscilloscope should high voltages be present. The presence or the absence of the probe should be noted and the settings of the oscilloscope should be manually adjusted by the operator prior to taking any measurements.

13.2.1 Triggering

The most important setting for the oscilloscope when measuring voltage waveforms is the triggering. Triggering defines time equal to zero for the oscilloscope. It is usually accomplished by detecting when a signal (i.e., trigger signal) exceeds a predetermined level (i.e., trigger level). The most common mistake is to attempt to use the desired measured signal as the trigger signal. This will introduce jitter onto the measured waveform degrading the measurement. A separate trigger signal from the function generator used to power the circuit, or a similar source should always be used for the trigger signal. Using the waveform you wish to measure as a trigger signal should only be done in extremely unusual circumstances when no other trigger signal is available.

13.2.2 Waveform Scaling

Once proper scaling has been established, the next step is to adjust the volts per division (vertical scale) and time per division (horizontal scale) to bring the desired features of the waveform into focus. If the divisions are too large, the waveform will be small and hard to visualize. Conversely, if the divisions are too small, then the waveform will exceed the view screen.

13.2.3 Knob-Ology

In addition to the basic oscilloscope settings, most scopes have numerous other features that can enhance measurement productivity. For example, the waveforms can be averaged over multiple trigger cycles to reduce the influence of noise. Signals from two different channels can be added or subtracted to compare the voltage signals over time. The Fourier Transform can be implemented to provide the frequency spectrum of the measured signal. Also, many scopes have built-in filters or anomaly detection schemes to capture and analyze unique signaling events. It is advisable to learn the basics of these features to take full advantage of the system's capability while conducting measurements.

13.3 Loading Errors

Whenever making measurements with any system, there is always the risk that the measurement itself will alter the state of the variables we are attempting to determine. As a result, we cannot trust our measurements unless we know the impact of the measurement system itself is minimal. In circuit measurements, the errors introduced are known as loading errors as they correspond to changes in current flow introduced by the meter or oscilloscope. The loading error for a circuit is quantified as

$$\% \text{ Loading Error} = 100 \cdot \left(\frac{\text{True} - \text{Measured}}{\text{True}} \right) \qquad (13.2)$$

13.3.1 Loading Errors of Voltage and Current Meters

Figure 13.4 shows the equivalent circuit for the real voltage and current meter. In this figure, the meter, M, corresponds to an ideal meter that would not alter the measurement. However, the circuit is impacted by the resistance of the meter, R_{meter}. For an ideal voltmeter, R_{meter} would be infinite and no current would be flowing into the meter. Likewise, for an ideal current meter R_{meter} would be 0 Ω and there would be no voltage drop across the meter when connected to the circuit allowing the meter to act just like the wire that was connected before the meter was inserted. For real meters, however,

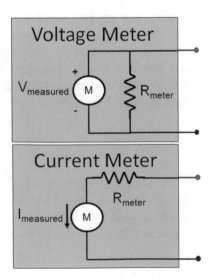

Fig. 13.4 Equivalent circuits for voltage meter (or voltmeter) and current meter (or ammeter) illustrating the internal resistance of the meter

R_{meter} is typically on the order of 10 MΩ for the voltmeter and 0.02–2 Ω for the current meter with the lower resistances for settings designed for higher current measurements.

Example 13.1 Find the loading error for the voltage between nodes a and b, V_{ab}, for the circuit shown in Fig. 13.5 assuming the measurement is made by a voltmeter with an internal resistance of 10 MΩ with the probes connected to node a and node b.

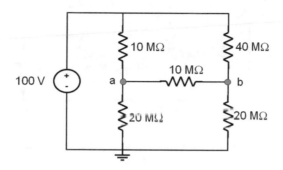

Fig. 13.5 Circuit for finding loading errors in Example 13.1

Solution: The first step is to find the true value before the meter has been connected. This can be done by solving the node-voltage equations for nodes a and b.

$$\frac{V_a - 100 \text{ V}}{10 \text{ M}\Omega} + \frac{V_a}{20 \text{ M}\Omega} + \frac{V_a - V_b}{10 \text{ M}\Omega} = 0$$
$$\Rightarrow 2V_a + V_a + 2V_a - 2V_b = 200 \text{ V}$$
$$\Rightarrow 5V_a - 2V_b = 200 \text{ V}$$

$$\frac{V_b - 100 \text{ V}}{40 \text{ M}\Omega} + \frac{V_b}{20 \text{ M}\Omega} + \frac{V_b - V_a}{10 \text{ M}\Omega} = 0$$
$$\Rightarrow V_b + 2V_b + 4V_b - 4V_a = 100\text{V}$$
$$\Rightarrow -4V_a + 7V_b = 100\text{V}$$

$$7 \cdot (5V_a - 2V_b = 200\text{V}) \Rightarrow 35V_a - 14V_b = 1400 \text{ V}$$
$$2 \cdot (-4V_a + 7V_b = 100\text{V}) \Rightarrow -8V_a + 14V_b = 200 \text{ V}$$
$$\Rightarrow 27V_a = 1600 \text{ V} \Rightarrow V_a = 59.2593\text{V}$$
$$V_b = \frac{100 + 4V_a}{7} = 48.1481\text{V}$$
$$(V_{ab})_{\text{True}} = V_a - V_b = 11.1111\text{V}$$

Now, we need to determine the impact of adding the meter to the measurement. The meter has an internal resistance of 10 MΩ, so a 10 MΩ resistor will need to be added between nodes a and b to represent the meter as shown below.

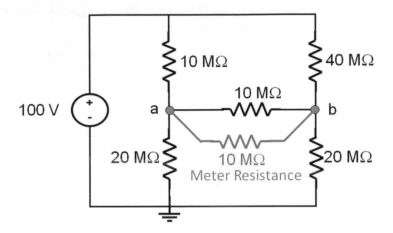

We can then solve the new node-voltage equations to get the voltage that the meter would actually measure were it to be connected between nodes a and b.

$$\frac{V_a - 100\text{V}}{10\text{ M}\Omega} + \frac{V_a}{20\text{ M}\Omega} + \frac{V_a - V_b}{10\text{ M}\Omega} + \frac{V_a - V_b}{10\text{ M}\Omega} = 0$$

$$\Rightarrow 2V_a + V_a + 4V_a - 4V_b = 200\text{V}$$

$$\Rightarrow 7V_a - 4V_b = 200\text{V}$$

$$\frac{V_b - 100\text{V}}{40\text{M}\Omega} + \frac{V_b}{20\text{M}\Omega} + \frac{V_b - V_a}{10\text{M}\Omega} + \frac{V_b - V_a}{10\text{M}\Omega} = 0$$

$$\Rightarrow V_b + 2V_b + 8V_b - 8V_a = 100\text{V}$$

$$\Rightarrow -8V_a + 11V_b = 100\text{V}$$

$$11 \cdot (7V_a - 4V_b = 200\text{V}) \Rightarrow 77V_a - 44V_b = 2200\text{V}$$

$$4 \cdot (-8V_a + 11V_b = 100\text{V}) \Rightarrow -32V_a + 44V_b = 400\text{V}$$

$$\Rightarrow 45V_a = 2600\text{V} \Rightarrow V_a = 57.7778\text{V}$$

$$V_b = \frac{100 + 8V_a}{11} = 51.1111\text{V}$$

$$(V_{ab})_{\text{measured}} = V_a - V_b = 6.6667\text{V}$$

The loading error due to the meter is thus given by

$$\% \text{ Loading Error} = 100 \cdot \left(\frac{\text{True} - \text{Measured}}{\text{True}}\right) = 100 \cdot \left(\frac{(V_{ab})_{\text{True}} - (V_{ab})_{\text{measured}}}{(V_{ab})_{\text{True}}}\right) = 40\%$$

The importance of the meter resistance and subsequent loading error can also be determined from the relative size of the meter resistance to the Thevenin resistance at the terminals where the voltmeter is connected. Since the Thevenin equivalent for a circuit can always be found for any two nodes in a circuit, we can find the Thevenin equivalent for the nodes where we are planning to connect our meter for the measurement as shown in Fig. 13.6. The true voltage in the absence of the meter would then be given by the Thevenin voltage, V_{Th}, while the voltage measured by the meter, V_{Measured}, would be given by a simple voltage divider as

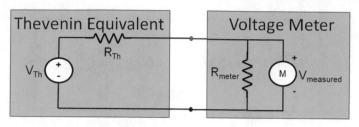

Fig. 13.6 Voltmeter connected to Thevenin equivalent circuit

$$V_{\text{measured}} = V_{Th} \cdot \left(\frac{R_{\text{meter}}}{R_{\text{meter}} + R_{Th}} \right) \tag{13.3}$$

Hence, the loading error for the voltmeter would be given by

$$\% \text{ Loading Error} = 100 \cdot \left(\frac{V_{Th} - V_{\text{measured}}}{V_{Th}} \right)$$

$$= 100 \cdot \left(\frac{V_{Th} - V_{Th} \cdot \left(\frac{R_{\text{meter}}}{R_{\text{meter}} + R_{Th}} \right)}{V_{Th}} \right) = 100 \cdot \left(1 - \left(\frac{R_{\text{meter}}}{R_{\text{meter}} + R_{Th}} \right) \right) \tag{13.4}$$

Example 13.2 Repeat Example 13.1 by first finding the Thevenin equivalent resistance between nodes a and b.

Solution: The Thevenin equivalent can be found by replacing the voltage source with a short circuit (deactivate the independent source) and applying a test voltage to the terminals as was discussed in Chap. 3. Redrawing the circuit for this purpose gives

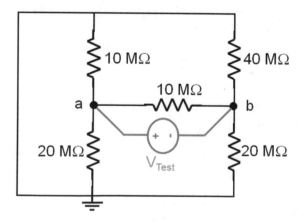

With the voltage source replaced by a short circuit, several of the resistors are now in parallel as they share both nodes.

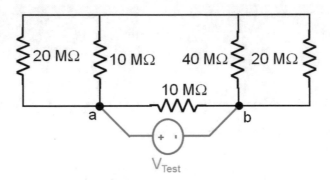

Replacing these resistors with their equivalent yields two more resistors in series as shown below.

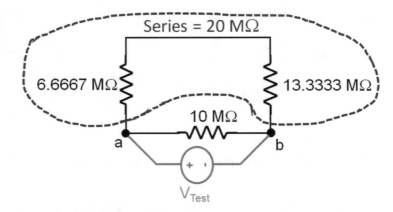

Hence, the test current flowing due to the test voltage is given by

$$I_{\text{test}} = \frac{V_{\text{test}}}{20\,\text{M}\Omega} + \frac{V_{\text{test}}}{10\,\text{M}\Omega} \Rightarrow R_{Th} = \frac{V_{\text{test}}}{I_{\text{test}}} = 6.6667\,\text{M}\Omega$$

$$\% \text{ Loading Error} = 100 \cdot \left(1 - \left(\frac{R_{\text{meter}}}{R_{\text{meter}} + R_{Th}}\right)\right) = 40\%$$

Example 13.3 If the internal resistance of a voltmeter is $2.5\times$ bigger than the output impedance of a circuit, what is the loading error?

Solution: The output impedance is another term for the Thevenin impedance. Therefore, the loading error would be given by

$$\% \text{ Loading Error} = 100 \cdot \left(1 - \left(\frac{R_{\text{meter}}}{R_{\text{meter}} + R_{Th}}\right)\right)$$

$$= 100 \cdot \left(1 - \left(\frac{2.5 R_{Th}}{2.5 R_{Th} + R_{Th}}\right)\right) = 100 \cdot \left(1 - \left(\frac{2.5}{3.5}\right)\right) = 28.5714\%$$

The Thevenin resistance can also be used to calculate the loading error introduced by the current meter. In this case, we would find the Norton equivalent circuit corresponding to where we want to insert the current meter as shown in Fig. 13.7. For the ideal case, the measured current would be the same as the Norton current as the meter should not redirect any of the current flow. However, since the meter has a finite resistance, the measured current will be slightly impacted. Specifically, $I_{measured}$ will be given by the current divider equation as

$$I_{measured} = \frac{1}{R_{meter}} \left(\frac{I_N}{\frac{1}{R_{meter}} + \frac{1}{R_{Th}}} \right) = \left(\frac{I_N}{1 + \frac{R_{meter}}{R_{Th}}} \right) \tag{13.5}$$

Thus, the loading error would be given by

$$\% \text{ Loading Error} = 100 \cdot \left(\frac{I_N - I_{measured}}{I_N} \right)$$

$$= 100 \cdot \left(\frac{I_N - \frac{I_N}{1 + \frac{R_{meter}}{R_{Th}}}}{I_N} \right) = 100 \cdot \left(1 - \left(\frac{R_{Th}}{R_{meter} + R_{Th}} \right) \right) \tag{13.6}$$

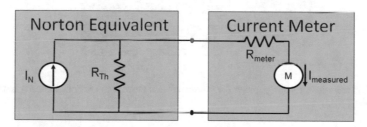

Fig. 13.7 Current meter connected to Norton equivalent circuit

Example 13.4 If the internal resistance of an ammeter is 1/3 the output impedance of a circuit, what is the loading error?

Solution: The output impedance is another term for the Thevenin impedance. Therefore, the loading error would be given by

$$\% \text{ Loading Error} = 100 \cdot \left(1 - \left(\frac{R_{Th}}{R_{meter} + R_{Th}} \right) \right) = 100 \cdot \left(1 - \left(\frac{R_{Th}}{\frac{R_{Th}}{3} + R_{Th}} \right) \right)$$

$$= 100 \cdot \left(1 - \left(\frac{1}{\frac{4}{3}} \right) \right) = 25\%$$

Lab Hint: In order to avoid having the voltage or current meter load the circuit corrupting the measurement, it is usually advisable to have the voltmeter resistance be at least 100 times greater than the output impedance of the circuit while having the ammeter resistance be at least 100 times smaller than the circuits output impedance. In most circumstances, the exact output impedance does not need to be calculated provided it can be estimated to within an order of magnitude prior to beginning circuit measurements.

13.3.2   Loading Errors of Oscilloscopes

The equivalent circuit for oscilloscope's terminals is shown in Fig. 13.8. The primary difference between the voltmeter and the oscilloscope is that one of the probe connections on the oscilloscope is always grounded whereas the voltmeter had two "floating" probes that could be connected to any node. This is very important to remember as any part of the circuit that is connected to the oscilloscope ground is usually automatically connected to the circuit's ground. This can cause short circuits, high current flows, and generally unpredictable results. Therefore, when using an oscilloscope, the ground for the oscilloscope should always be connected to the ground of the circuit. The remaining connection of the oscilloscope probe can then be connected to the other nodes in the circuit to measure the voltage waveforms at these nodes. The resistance, R_{scope}, and capacitance, C_{scope}, for the oscilloscope are typically on the order of 1 MΩ and 10 pF, respectively. However, these values will be altered if an attenuating probe is used in the measurements.

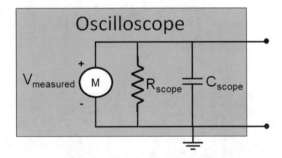

Fig. 13.8 Equivalent circuits for voltage meter (or voltmeter) and current meter (or ammeter) illustrating the internal resistance of the meter

Example 13.5 Find the loading error when an oscilloscope is used to measure the voltage between nodes a and b, $v_{ab}(t)$, for the circuit shown in Fig. 13.9 assuming the grounding node for the scope has been placed on node b and the internal resistance of the scope is 1 MΩ. Also, assume that the capacitance of the scope is small enough to be neglected.

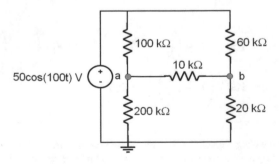

Fig. 13.9 Circuit for finding loading errors in Example 13.5

Solution: First, let's find the true voltage between nodes a and b before the scope has been connected using the node-voltage method as was done in previous examples.

$$\frac{\tilde{V}_a - 50\text{V}}{100k\Omega} + \frac{\tilde{V}_a}{200k\Omega} + \frac{\tilde{V}_a - \tilde{V}_b}{10k\Omega} = 0$$

$$\Rightarrow 2\tilde{V}_a + \tilde{V}_a + 20\tilde{V}_a - 20\tilde{V}_b = 100\text{V} \Rightarrow 23\tilde{V}_a - 20\tilde{V}_b = 100\text{V}$$

$$\frac{\tilde{V}_b - 50\text{V}}{60k\Omega} + \frac{\tilde{V}_b}{20k\Omega} + \frac{\tilde{V}_b - \tilde{V}_a}{10k\Omega} = 0$$

$$\Rightarrow \tilde{V}_b + 3\tilde{V}_b + 6\tilde{V}_b - 6\tilde{V}_a = 50\text{V} \Rightarrow -6\tilde{V}_a + 10\tilde{V}_b = 50\text{V}$$

$$\Rightarrow -12\tilde{V}_a + 20\tilde{V}_b = 100\text{V}$$

$$\left.\begin{array}{l} 23\tilde{V}_a - 20\tilde{V}_b - 100\text{V} \\ -12V_a + 20\tilde{V}_b = 100\text{V} \end{array}\right\} \Rightarrow 11\dot{V}_a = 200\text{V} \rightarrow v_a(t) = 18.1818\cos(100t)\,\text{V}$$

$$\tilde{V}_b = \frac{50 + 6\tilde{V}_a}{10} \Rightarrow v_b(t) = 15.9091\cos(100t)\,\text{V}$$

$$\Rightarrow (v_{ab}(t))_{True} = v_a(t) - v_b(t) = 2.2727\cos(100t)\,\text{V}$$

Now, let's find the voltage that will be measured by the scope. Since the ground of the scope has been connected to node b, the scope shorts node b to ground in addition to placing a resistance between nodes a and b. Therefore, the circuit with the scope connected is given by

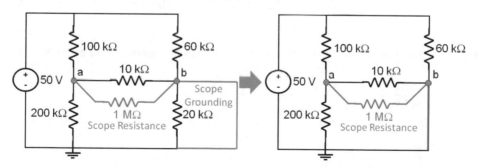

Hence, $\tilde{V}_b = 0\text{V}$ and $\tilde{V}_a$ can be found from

$$\frac{\tilde{V}_a - 50\text{V}}{100\,k\Omega} + \frac{\tilde{V}_a}{200\,k\Omega} + \frac{\tilde{V}_a - 0\text{V}}{10\,k\Omega} + \frac{\tilde{V}_a - 0\text{V}}{1000\,k\Omega} = 0$$

$$\Rightarrow 10\tilde{V}_a + 5\tilde{V}_a + 100\tilde{V}_a + \tilde{V}_a = 500\text{V} \Rightarrow 116\tilde{V}_a = 500\text{V}$$

$$v_a(t) = (v_{ab}(t))_{measured} = 4.3103\cos(100t)\,\text{V}$$

Thus, the loading error is given by

$$\% \text{ Loading Error} = 100 \cdot \left(\frac{\text{True} - \text{Measured}}{\text{True}}\right) = 100 \cdot \left(\frac{(v_{ab})_{True} - (v_{ab})_{measured}}{(v_{ab})_{True}}\right) = -89.6552\%$$

Clearly, the unintentional grounding of a node by the oscilloscope can have a significant impact on the measurement. If the voltage across a resistor not connected to the ground must be measured, then the correct approach would be to use two probes, one on each side of the resistor, connected to two different channels of the oscilloscope. The signals could then be subtracted by the scope itself to get the voltage across the resistor.

Example 13.6 Repeat Example 13.5 assuming two oscilloscope probes are used, each with a resistance of 1 MΩ, and that the probes are placed on nodes a and b with their grounds connected to the ground of the circuit.

Solution: The circuit with the scope probes connected is given by

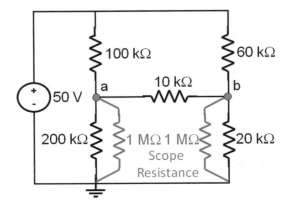

The measured voltages at nodes a and b can be found from the node-voltage equations as

$$\frac{\tilde{V}_a - 50\text{V}}{100\,k\Omega} + \frac{\tilde{V}_a}{200\,k\Omega} + \frac{\tilde{V}_a}{1000\,k\Omega} + \frac{\tilde{V}_a - \tilde{V}_b}{10\,k\Omega} = 0$$
$$\Rightarrow 10\tilde{V}_a + 5\tilde{V}_a + \tilde{V}_a + 100\tilde{V}_a - 100\tilde{V}_b = 500\text{V} \Rightarrow 116\tilde{V}_a - 100\tilde{V}_b = 500\text{V}$$

$$\frac{\tilde{V}_b - 50\text{V}}{60\,k\Omega} + \frac{\tilde{V}_b}{20\,k\Omega} + \frac{\tilde{V}_b}{1000\,k\Omega} + \frac{\tilde{V}_b - \tilde{V}_a}{10\,k\Omega} = 0$$
$$\Rightarrow 50\tilde{V}_b + 150\tilde{V}_b + 3\tilde{V}_b + 300\tilde{V}_b - 300\tilde{V}_a = 2500\text{V} \Rightarrow -300\tilde{V}_a + 503\tilde{V}_b = 2500\text{V}$$

$$\left.\begin{array}{l} 116\tilde{V}_a - 100\tilde{V}_b = 500 \text{ V} \\ -300\tilde{V}_a + 503\tilde{V}_b = 2500 \text{ V} \end{array}\right\} \Rightarrow \begin{bmatrix} 116 & -100 \\ -300 & 503 \end{bmatrix} \begin{bmatrix} \tilde{V}_a \\ \tilde{V}_b \end{bmatrix} = \begin{bmatrix} 500 \text{ V} \\ 2500 \text{ V} \end{bmatrix}$$

$$\Rightarrow \begin{bmatrix} \tilde{V}_a \\ \tilde{V}_b \end{bmatrix} = \begin{bmatrix} 17.6908 \text{ V} \\ 15.5214 \text{ V} \end{bmatrix} \Rightarrow (v_{ab}(t))_{\text{measured}} = 2.1695 \cos(100t)\text{V}$$

Hence, the loading error is given by

$$\% \text{ Loading Error} = 100 \cdot \left(\frac{(v_{ab})_{\text{True}} - (v_{ab})_{\text{measured}}}{(v_{ab})_{\text{True}}}\right) = 100 \cdot \left(\frac{2.2727 - 2.1695}{2.2727}\right) = 4.5435\%$$

13.4   Problems

Problem 13.1: If the internal resistance of a voltmeter is $100\times$ bigger than the output impedance of a circuit, what is the loading error?

Problem 13.2: If the internal resistance of an ammeter is $10\times$ smaller than the output impedance of a circuit, what is the loading error?

Problem 13.3: If the internal resistance of a voltmeter is $5\times$ bigger than the output impedance of a circuit, what is the loading error?

Problem 13.4: If the internal resistance of an ammeter is $7\times$ smaller than the output impedance of a circuit, what is the loading error?

Problem 13.5: If the internal resistance of a voltmeter is $6\times$ bigger than the output impedance of a circuit, what is the loading error?

Problem 13.6: If the internal resistance of an ammeter is $4\times$ smaller than the output impedance of a circuit, what is the loading error?

Problem 13.7: For the circuit shown below, determine the loading error when measured by a voltmeter with an internal resistance of 12 MΩ and the voltmeter probes are connected to node (a) and node (b).

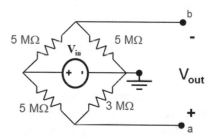

Problem 13.8: For the circuit shown below, what is the loading error if you use an ammeter to measure the current through R_3? The meter has a resistance of 3 Ω while $R_1 = 12$ Ω, $R_2 = 9$ Ω, and $R_3 = 5$ Ω.

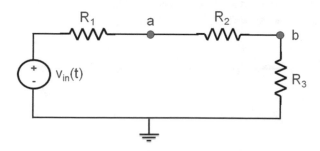

Problem 13.9: For the circuit shown below, what is the loading error if you use an oscilloscope with an internal resistance of 1 MΩ and the oscilloscope probe is connected with the probe clip-on (a) and the probe ground on (b)?

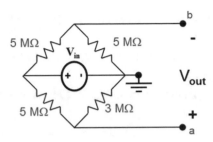

Problem 13.10: For the circuit shown below, determine the following if $V_{in} = 8$ V:

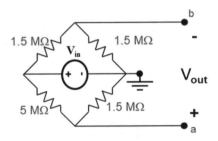

a. Find the true output voltage, V_{out}.
b. Find the measured voltage, V_{out}, when measured by an oscilloscope with an internal resistance of 1 MΩ and the oscilloscope probe is connected with the probe clip-on (a) and the probe ground on (b).
c. Find the measured voltage, V_{out}, when measured by a voltmeter with an internal resistance of 10 MΩ and the voltmeter probes connected to node (a) and node (b).

Problem 13.11: For the circuit shown below, what is the loading error if you use an oscilloscope to measure the voltage across R_2 with the positive terminal of the scope placed at node a and the negative terminal/ground of the scope placed at node b. The scope has a resistance of 0.5 MΩ while $R_1 = 2$ MΩ, $R_2 = 6$ MΩ, and $R_3 = 4$ MΩ.

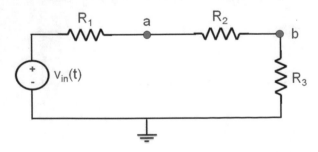

Problem 13.12: For the following circuit, determine the voltage measured by an oscilloscope with an internal resistance of 1 MΩ and a negligible capacitance when $v_{in}(t) = 100\cos(100t)$ V and

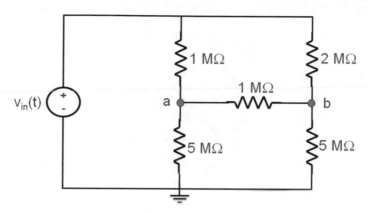

a. The oscilloscope probe is connected with the probe clip-on (a) and the probe ground on (b).
b. The oscilloscope used two probes with the Channel 1 probe clip-on (a) and the Channel 2 probe clip-on (b). The ground for both clips is on the circuit ground. The oscilloscope then subtracts Channel 2 from Channel 1.
c. What should have been the measured voltage?

Problem 13.13: For the circuit shown below, determine the loading error when the voltage $V_{ab} = V_a - V_b$ is measured by an oscilloscope with an internal resistance of 1 MΩ and a negligible capacitance when $v_{in}(t) = 10\cos(100t)$ V and the oscilloscope probe is connected with the probe clip-on (a) and the probe ground on (b).

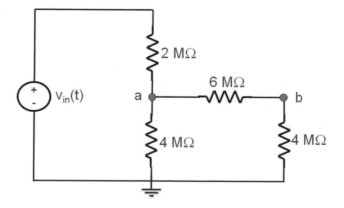

Problem 13.14: For the circuit shown below, determine the smallest possible loading error, when determining the voltage V_{out} if all you have to make the measurement is an oscilloscope with an internal resistance of 10 MΩ where $R_1 = 2$ MΩ, $R_2 = 3$ MΩ, $R_3 = 2$ MΩ, $R_4 = 5$ MΩ, and $V_{in} = 10$ V.

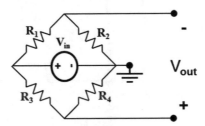

Index

© Springer Nature Switzerland AG 2020
T. A. Bigelow, *Electric Circuits, Systems, and Motors*,
https://doi.org/10.1007/978-3-030-31355-5

Printed in the United States
by Baker & Taylor Publisher Services